D1200552

PRECAMBRIAN OF THE NORTHERN HEMISPHERE

AND GENERAL FEATURES OF EARLY GEOLOGICAL EVOLUTION

Developments in Palaeontology and Stratigraphy, 3

PRECAMBRIAN OF THE NORTHERN HEMISPHERE

AND GENERAL FEATURES OF EARLY GEOLOGICAL EVOLUTION

by

L.J. Salop

All-Union Geological Research Institute (V.S.E.G.E.I.), Leningrad, U.S.S.R.

English translation edited by *G.M. Young*

ELSEVIER SCIENTIFIC PUBLISHING COMPANY

AMSTERDAM - OXFORD - NEW YORK, 1977

ELSEVIER SCIENTIFIC PUBLISHING COMPANY
335 Jan van Galenstraat
P.O. Box 211, Amsterdam, The Netherlands

Distributors for the United States and Canada:

ELSEVIER NORTH-HOLLAND INC.
52, Vanderbilt Avenue
New York, N.Y. 10017

ORIGINAL TITLE:
OBSHCHAYA STRATIGRAFICHESKAYA SHKALA DOKEMBRIYA

ISBN: 0-444-41142-9 (series)
ISBN: 0-444-41510-6 (vol. 3)

Printed in The Netherlands

PREFACE TO THE ENGLISH EDITION

I am pleased that Elsevier has agreed to publish the English version of my book, "Precambrian of the Northern Hemisphere and General Features of Early Geological Evolution". This will undoubtedly enlarge the circle of readers of the book and will permit more active and fruitful discussion of the problems with which it attempts to deal. The text for the English translation was revised and some new data were included. However, I should like to address some additional remarks to the readers:

(1) Precambrian geology is a rapidly developing field in every country. The abundance of new material makes constant revision of ideas necessary. Existing schemes for subdivision of the Precambrian do not meet the requirements of the new data. New stratigraphic terminology is badly needed. In this book I propose some new names for subdivisions of the Precambrian. I am well aware of the fact that new terms initially pose some problems and may even cause some disagreement among geologists. I should like to emphasize, however, that the essence of the proposed subdivision lies not in the names used for the units, but in the natural basis for the subdivision. I hope that the reader will focus his attention on this very important aspect of the problem.

(2) The major conclusions in the book are so-called "empirical generalizations". Some are based on theory but in other cases this remains to be done. I have tried to avoid speculation, so that the conclusions are based on factual evidence derived from the Precambrian rocks of the continents of the Northern Hemisphere. In particular I carefully avoided discussion of the problem of the "new global tectonics" as applied to the Precambrian period of geological evolution. In my opinion, this problem cannot be resolved on the basis of material from the Northern Hemisphere continents alone. Many of the facts presently available admit different interpretations. I hope to devote a separate book to this problem, including material on the Precambrian of all continents.

(3) The continuation of this work is a comparative analysis of Precambrian formations of the continents of the Southern Hemisphere. This work is almost completed and material on the African continent is ready for printing. I think it necessary to state that the major regularities established for the Northern Hemisphere continents are valid for the Southern ones and are thus of global extent. This mainly concerns subdivision of the Precambrian into large natural units (eras/erathems, sub-eras/sub-erathems) and definition of their age boundaries.

In conclusion, I extend my thanks to Dr. G.M. Young for his work in editing the English translation of the book.

Prof. L.J. Salop,
All-Union Geological Research Institute,
Leningrad

PREFACE

It is well known that the Precambrian occupies a period several times longer than the later Phanerozoic. Thorough studies of Precambrian formations, and especially their stratigraphy, are needed to elucidate the general laws of geological evolution. At present there is no internationally accepted stratigraphic scale for subdivision of the Precambrian. Some older formations of different ages are assigned to the same subdivisions (e.g. "Archean" or "Proterozoic"), and coeval strata are commonly described under different names. Geochronological limits of Precambrian "eras" are also variable so that cartography and correlation of Precambrian formations are difficult.

The problems of interregional correlation of Precambrian formations are also of practical value; some exogenic types of deposits (sedimentary and sedimentary—volcanogenic—metamorphic deposits) for example of iron, manganese, uranium, copper, gold and high-alumina raw materials, are associated with older strata of specific composition and age, and even many of endogenic (mostly of metasomatic type) deposits (phlogopite, lazurite, mountain crystal, etc.) are also confined to particular Precambrian formations.

A unified system for subdivision of the Precambrian is badly needed. There are many new geological and geochronological data. These data are considered sufficient to permit delineation of a general stratigraphic subdivision of older formations. A generalized subdivision of the Precambrian is proposed in this book, on the basis of comparative analysis of Precambrian formations of the East European, Siberian and North American platforms and associated fold belts. The less well-known Precambrian formations of peninsular India, and those of Phanerozoic fold belts are also discussed.

This work does not include detailed discussion of the stratigraphy of all regions of the Northern Hemisphere. It is assumed that the reader is familiar with the more important Precambrian sections or can use the included references for their study. We shall consider only the main features of the older strata and will give details only in the case of some new unpublished material, or in the case of some very important problematic data. The treatment of data on each region varies according to the amount of work done in the areas and according to the writer's knowledge of the material. The writer has had little opportunity for first-hand observation of areas outside the U.S.S.R. and has, in many cases, had to rely on published data, and review it using his own experience and knowledge.

Some problems discussed in this book have already been dealt with in earlier publications, but it was felt necessary to review or revise them as new data became available. Precambrian geology is a rapidly developing field so that future work may necessitate some change in the ideas expressed in this book. The writer, however, hopes that the main conclusions will remain valid.

The work on the Upper Precambrian of the East European platform was done in cooperation with K.E. Jacobson, who has studied the subdivision and correlation of the Precambrian deposits of the Russian plate, and their correlation with the Urals sections. K.E. Jacobson wrote some pages of the 7th, 8th and 9th chapters. I thank K.E. Jacobson and also my other colleagues and co-workers: Yu.R. Bekker, Yu.B. Bogdanov, V.K. Golovenok, A.Z. Konikov, K.N. Konyushkov, N.S. Krylov, V.Z. Negrutsa, S.N. Suslova, L.V. Travin and E.A. Shalek for help in this work and for fruitful discussion of problems of Precambrian regional geology.

In this work the rock and mineral ages, according to isotopic analyses, are based on the constants accepted in the U.S.S.R. ($\lambda_K = 0.557 \cdot 10^{-10} y^{-1}$ and $\lambda_{Rb} = 1.39 \cdot 10^{-11} y^{-1}$). The original age data from some original foreign papers are given in brackets. In many countries (with the exception of China and some others) another constant of K-decay is used ($\lambda_K = 0.585 \cdot 10^{-10} y^{-1}$) and in some papers another constant of Rb-decay ($\lambda_{Rb} = 1.47 \cdot 10^{-11} y^{-1}$) is used.

Prof. L.J. Salop,
All-Union Geological Research Institute,
Leningrad

CONTENTS

PART I

INTRODUCTION

CHAPTER 1

METHODS OF CORRELATION OF PRECAMBRIAN ROCKS

The methods of correlation of Precambrian rocks are dealt with in many works on Precambrian stratigraphy, so that detailed treatment is not warranted here. However, many aspects of these different methods are not well understood and some brief comments are necessary. Three main methods are used in correlation of Precambrian strata: isotopic, paleontological and geological.

Isotopic Methods

The isotopic method of age determination, based on radioactive decay, was initially accepted by some and rejected by others. The first group considered the physical principles to be the absolute truth, and believed that it was possible to solve all problems of age determination and rock correlation by the use of isotopic analyses. They often considered the measured age to be the age of the rock. The other group, in the face of many discrepancies between radiometric and geological data, tended to reject this technique in stratigraphic studies, or accepted it with great caution. At present differences of opinion are fewer because of the development of more sophisticated techniques, and better understanding of both geochemical and geological interpretation of isotopic analyses. However, it is still not certain that in all cases we have a true geological evaluation of radiometric data.

Different isotopic methods and their modifications are used for age determinations of the older formations. These are: potassium—argon (K—Ar), lead isotopes (U—Th—Pb), lead isochron (Pb—Pb), model lead, α-lead and rubidium—strontium (Rb—Sr). Comparative analysis of all these methods is beyond the scope of this work.

All of these methods have their advantages and disadvantages. Initially it was thought that reliable isotopic age determinations would make possible definitive correlation of the older supracrustal formations and that the main task was getting the necessary dates. However, it is only rarely that we can determine the age of sedimentation or even of diagenesis of sedimentary rocks. Such dating appears possible only for sedimentary strata of platforms and miogeosynclinal types, incorporating some potassium-bearing syn- and epigenetic minerals. Even in these cases we cannot be sure that we have the true age and not a "rejuvenated" age related to later processes. It is well known that glauconite, commonly used for sedimentary-rock dating, loses radiogenic argon at relatively low temperatures (about 150°C or even lower). In addition, the potassium content of glauconite varies because of exchange reactions. Under very low-grade metamorphism, glauconite becomes unstable, decomposes, and is replaced by other minerals. Many studies have shown that even fresh, unaltered, glauconite cannot be useful for age dating because of argon diffusion. There are cases where glauconite from unmetamorphosed strata has yielded ages older than glauconite from underlying rocks. For example, the glauconite age of Eocene sandstones from California, taken from the surface, was measured to be 43 m.y. Those from drill cores at depths of 2—4 km in the same deposits, decreased in a regular way from 33 m.y. to 18 m.y. (Evernden et al., 1960). A similar situation was found in glauconite dating of Upper Precambrian rocks of Kazakhstan (Karatau). In this case the glauconite of the upper part of the section, near the contact with fossiliferous Cambrian deposits, yielded an age of 420 m.y. Glauconite from the lower part of the succession yielded lower figures — as low as 350 m.y. (analyses made by L.I. Borovikov and G.A. Murina, personal communication, 1972). It is possible that in these, and similar cases, "rejuvenation" of the age is due to stronger and longer heating of the glauconite-bearing rocks at greater depths, but other explanations are not excluded.

There are many other cases where glauconite from fossiliferous deposits has yielded evidently "rejuvenated" age values. For instance, Cambrian brachiopod- and trilobite-bearing deposits of Alberta (Canada) yielded 413-300 m.y. (several determinations from different horizons). According to the accepted geochronological scale these values correspond to the Early Devonian—Permian (Stevens, 1965). Argon losses in glauconite are evidenced by a wide scatter of ages measured from different samples of the same deposits. Glauconites of the Kar'yernaya Group of the Yenisei mountain range gave ages ranging from 747 to 815 m.y. (five determinations); the Pogoryuy Group of the same area gave ages in the 750—1,630 m.y. range (ten determinations). In general the older the rocks the wider the scatter of glauconite ages obtained from them (Fig.1).

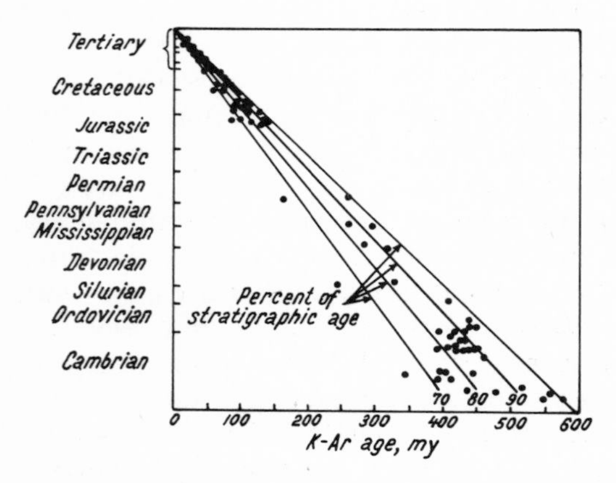

Fig.1. A comparison between K—Ar age determinations on glauconites and stratigraphic classification. After Obradovich (see Thompson and Hower, 1973).

The fact that some glauconite ages increase down-section indicates that the age data are not true. The thermal processes responsible for argon loss could have influenced the whole section, causing general "rejuvenation" of the glauconite ages. In this case, the difference in age of the older and younger deposits would be preserved. The above should be kept in mind when interpreting the dating of sedimentary strata using glauconite. In most cases the age data are "rejuvenated", and may indicate only the upper age limit of the strata. In particular, caution is required in the interpretation of results from older strata (Precambrian) and from altered or faulted rocks. Truer ages are to be expected for horizontal unaltered deposits in the upper parts of platform-cover sections (irrespective of age). Age determinations of geosynclinal strata using glauconite are not reliable. At present all the boundaries of the Riphean subdivisions were determined using glauconite from rocks of the miogeosynclinal complex in the Southern Urals — the Riphean stratotype. The correlation of the platform and miogeosynclinal strata using only age determinations made on glauconite is highly questionable.

"Rejuvenated" K—Ar ages are characteristic not only of glauconites from Precambrian rocks. Rubinshtein (1967), who first proposed the use of glauconite for dating sedimentary rock, came to the conclusion "that in most cases the data for the Paleozoic (and certainly for the pre-Paleozoic!) and Mesozoic glauconites are too young". He even suggested addition of about 15% to the available dates based on glauconite from Mesozoic rocks, but this is not technically justifiable. For Precambrian rocks a rather close approximation to the true age is obtained when using glauconite from the uppermost part of platform sequences.

Many problems arise in interpreting K—Ar age data from hydromicaceous minerals, shales, phyllites and some other similar materials. In addition to "rejuvenation" due to possible argon loss it is commonly impossible to determine whether the data give the age of diagenesis (epigenesis), or the age of primary or superimposed metamorphism. In many cases ages are misinterpreted because of the presence of detrital micas, feldspars and some other minerals.

It is possible to determine the age of unaltered carbonate rocks by the Pb-isochron method, but in the case of metamorphic rocks the interpretation of such data can lead to erroneous results.

Volcanogenic rocks, especially those of acid composition, are more easily dated. Rb—Sr isochron and Pb-isochron (by whole-rock analysis), and Pb-isotopic analysis (on zircon) usually give the age of unaltered or slightly metamorphosed volcanites. K—Ar analysis may give the true age of unaltered volcanics. However, volcanic rocks that appear unaltered under the microscope, may also have suffered argon loss. A good example is provided by the slightly altered acid volcanics (porphyry) of the Akitkan Formation in the Baikal area; K—Ar dating gave an age of 100—550 m.y., whereas Rb—Sr isochron analysis gave their true age of about 1,700 m.y.

Isotopic dating of Precambrian rocks can generally be done mainly on metamorphic and plutonic minerals. These ages indicate the time of metamorphism, metasomatism and intrusions; they reflect events subsequent to the deposition of sedimentary or volcanogenic rocks. In many cases the ages do not necessarily indicate plutonic events that directly followed sedimentation or volcanism, but may reflect later thermal events.

Migration of radiogenic elements and their decay products may take place during metasomatism, deformation, or under the action of various exogenic factors (weathering, solution, exchange reactions, etc.).

All of the above is mainly concerned with K—Ar analysis. Geological and experimental investigations have shown that argon loss from minerals and rocks may take place under low-temperature conditions if they are imposed for a long period. Under these conditions there is no evidence of recrystallization or metamorphic alteration of the rock. Such processes may not be detectable by simple petrographic methods. Thus, this method discloses only the age of rocks and minerals formed on the earth's surface or at relatively shallow depth, and which were not subjected to even slight but prolonged heating or deformation. In most cases K—Ar dating only provides the minimum age of metamorphic and intrusive rocks. This can differ greatly from their true age. Factors affecting the apparent K—Ar age cease to act on uplift of crustal blocks above a critical level which coincides approximately with the 300°C geoisotherm. When uplift of the earth's crust completes the tectono-plutonic cycle, the K—Ar analysis reflects (in a general way) the time of folding, metamorphism and plutonism (Salop, 1963). Relict dating may in some cases indicate the age of metamorphic and plutonic rocks because of local conditions favourable for argon retention in the mineral lattice.

All of these complications in regard to K—Ar analysis are partly applicable to other methods. The difference lies mainly in the fact that argon migrates more easily than some other daughter radioactive elements. K—Ar and Rb—Sr analyses on the same minerals (micas) revealed that argon and strontium migration from minerals in nature may take place simultaneously, but at different rates. In most cases the migration of strontium is slower, but sometimes argon is preferentially retained. The radius of strontium migration is known to be relatively small, and in the case of isochemical metamorphism redistribution of strontium in minerals may take place without a marked change of its content in a relatively small volume of the rock. This is the basis for application of the whole-rock Rb—Sr isochron technique. This method gives the age of the rock and/or the time of early and later metamorphism.

There are great limitations in direct application of Pb-isotope (U—Th—Pb) analysis in solving geochronological problems, for thermal and other processes that continue after mineral formation may cause migration of parent elements, and especially their decay products. Analysis of Pb-isotopes provides evidence of such migration by comparison of different isotopic ratios. Such analyses make possible the use of isochrons so that even discordant figures may give a mineral age by means of various isotopic ratios. In some cases alteration of a mineral or rock may be so profound that it is impossible to determine even the approximate age.

Hart et al. (1968) clearly showed that later thermal events strongly influence the migration of radioactive decay products, and cause "rejuvenation" of isotopic ages. These studies were conducted in the area of the contact aureole of the small Cenozoic Aldor pluton intruded into Precambrian gneisses dated by Pb-isotopic analysis at 1,650 m.y., in an area far removed from the zone of influence of the younger intrusion. In the area of the contact aureole the "ages" of different minerals from gneisses by K—Ar, Rb—Sr and Pb-isotope (according to various ratios) analyses decrease regularly as the Aldor stock is approached. Near the contact the values are similar to those of the intrusion. Only the zircon ages ($^{207}Pb/^{206}Pb$ ratio) remain relatively unaffected. Near the contact the zircon "age" decreases to 1,200 m.y., but at a distance of 100 ft. from the contact the age becomes "normal" (1,650 m.y.). Thermal influence on K—Ar and Rb—Sr dates and on zircon ($^{206}Pb/^{238}U$ ratio) was noted at a distance of about 3,200 ft. from the contact, where available data suggest that the temperature did not exceed 200°C. Thus, even a slight or short-lived increase in temperature near a small intrusive stock may cause a marked "rejuvenation" of isotopic dating. With strong and prolonged elevation of temperatures such as may occur during orogenic cycles, isotopic "rejuvenation" will be more intense and universal.

"Rejuvenation" of Rb—Sr mineral ages was described above. Studies by Pidgeon and Compston (1965) proved that strong contact metamorphism may cause complete homogenization of strontium isotopes in the wall rocks, so that even whole-rock Rb—Sr isochron analysis may not reveal the age of the earlier events.

Thus, direct use of the figures derived by isotopic analyses is commonly difficult or even impossible. We need abundant and varied geochronological data, with a solid basis of geological and geochemical facts. Single determinations, especially by K—Ar analysis, are of little value in dating high-grade polymetamorphic rocks; at best they provide a minimum age limit. Numerous dates from different rocks and minerals are necessary before we can begin to determine "relict" values. K—Ar dating methods normally provide the youngest ages. This is because the possibility of an "older" age is quite rare, as argon will be trapped in a mineral only when the outside argon pressure is higher than that within. In nature this is exceptional (Gerling et al., 1965a).

In cases where the age apparently was increased a check can be made by determining the "age" of cogenetic minerals with different temperature of argon loss. Anomalously old K—Ar rock ages are common in ultrabasic (and partly basic) magmatic rocks due to specific conditions of emplacement and crystallization of the magma (Salop, 1970a). In deeper parts of the crust radiogenic argon, which is expelled from the crystal lattice of potassium-bearing minerals under the effect of high pressure and temperature, may accumulate in open crustal fractures under low-pressure conditions. If magma is intruded into such cavities the crystallizing minerals will trap much of the argon present there (Salop, 1970a). Since such deep fractures commonly control the introduction of ultrabasic magma, the K—Ar dating which yields anomalously old ages is usually found in ultramafic and associated mafic rocks.

The most valuable techniques are the Pb-isotope, Rb—Sr isochron and Pb-isochron methods, but even these may sometimes merely reveal the time of the last high-grade metamorphism when isotopic equilibration took place or when homogenization of isotope distribution took place in cogenetic formations. In such cases it is difficult to determine the true age of the rock. However, the available data may permit an estimation of the time and nature of the main events. In some cases similar or identical isotopic ages are obtained from older basement rocks and an unconformably overlying metamorphic complex. In this case the isotopic age of the older rocks is "rejuvenated" by the later (superimposed) metamorphism.

In many parts of the world the oldest gneiss—granulite complexes and unconformably overlying low-grade volcanic complexes yield identical, or very similar (about 2,600—2,800 m.y.) dates. Granites which were emplaced in the older complex, and younger granites intruding the greenstone strata, give the same age. In rare cases both the older granites and older supracrustal rocks yield an age of about 3,500 m.y. These dates may be regarded as "relict" ages reflecting the period of early metamorphism and plutonism. The more common "age" of the older rocks (2,600—2,800 m.y.) reflects the time of superimposed thermal processes. In the Limpopo belt in South Africa, rocks of the gneiss—granulite complex have a Rb—Sr isochron age of 2,000 m.y. They are transected by the "Great Dike", which, in an area to the north, free from the 2,000 m.y. orogeny, has provided an age of 2,500 m.y. (Rb—Sr isochron). Thus the gneisses and granulites dated in the Limpopo belt are highly "rejuvenated" (Van Breemen et al., 1966), for in adjacent areas these rocks have yielded ages of about 3,500 m.y. Values of about 2,700 m.y. (Van Breemen and Dodson, 1972) were recently reported for gneiss of the Limpopo belt (Rb—Sr analysis).

Geological control on interpretation of radiometric data is essential. It is only with such control that radiometric data will provide a basis for global division and correlation of Precambrian rocks.

Paleontological Methods

Phytolites (stromatolites and microphytolites) occurring predominantly in carbonate-rich Upper Precambrian deposits have recently been used for subdivision and correlation of older strata. This relatively new technique is very popular among Soviet geologists. Unfortunately, this method in some cases has been applied without sufficiently careful analysis. Some empirical regularities in the vertical distribution of stromatolites and microphytolites are not adequately documented. There are many discrepancies between ages based on phytolites and ages based on geological and isotopic criteria. Some examples will be described in dealing with regional stratigraphy.

It is probable that many apparent contradictions are explained by the lack of detailed taxonomy, by incorrect determinations or by misinterpretation of geological and isotopic ages. The diagnostic features of many stromatolites and microphytolites are not fully studied or described. This may lead to subjective identifications of some fossils.

The Soviet workers originally contended that vertical zonation of stromatolites might be established by using high-rank taxonomic units such as "group" or "type". It was suggested that the oldest (Lower Riphean) phytolite complex was characterized by stromatolites of the *Kussiella* group in particular, the Middle Riphean complex by the *Baicalia* group, the Upper Riphean by the *Inzeria* and *Gymnosolen* groups, and the Vendian complex by the *Patomia* and *Linella* groups. Further investigation, however, evidenced a rather wider vertical range of stromatolite distribution, with some forms occurring at different stratigraphic levels. Thus, the established sequence is expressed by predominance of certain groups at definite levels, and in a general change from one phytolite complex to another with time. The complexes themselves are characterized by a number of stromatolite forms differing in "species" composition.

Phytolite taxonomy has nothing in common with true biological taxonomy, except for the use of Latin names. The major taxonomic units of stromatolites as established by Soviet workers are based on macroscopic differences in external shape. The significance of these features for classification is commonly debatable. Lower-rank taxonomic units ("species") are erected largely on the basis of microstructural features. However, the microstructures are not necessarily primary but may be masked by recrystallization processes, or even produced by them. It is important to note that even small (1 m diameter) stromatolite bioherms may include forms belonging to different stromatolite groups.

We do not understand the mechanisms of stromatolite and microphytolite evolution. These fossils are not plant remains in the literal sense, but are rather carbonate (sometimes originally siliceous) structures — rocks which formed in response to life activity of different species of blue-green algae and bacteria. The role of algae in the formation of stromatolite structures is not clear: do they actively secrete calcite, or does the carbonate precipitate from seawater on the algal mucus to form the so-called "mat"? The occurrence of originally siliceous stromatolites (for instance in the Precambrian Gunflint Formation, which is a chemogenic ferruginous—siliceous unit in Canada; Barghoorn and Tyler, 1965) tends to favour the second opinion. The change in shape of the structures is presumed to be related to evolution of plant remains and gradual increase in their concentration (biomass) with time. It remains possible that shape variations in algal structures depend largely on environmental conditions during sedimentation (hydrodynamic regime, salinity, climate, etc.). Logan (1961) demonstrated that the distribution of different forms of modern stromatolites in Shark Bay on the west coast of Australia is closely related to the facial environment, in particular to tidal currents.

Changes in stromatolite and microphytolite shape would thus be related to a general evolution of sedimentary rocks under the influence of geochemical, climatic and tectonic factors. Oncolites are relatively rare in Lower and, to a lesser degree, Middle Precambrian rocks. This may be due to predominance of rather quiet hydrodynamic conditions in the basins of that time. These quiet conditions might be the result of a lack of relief on the earth's surface at that time and possibly also of a poorly expressed climatic zonation (Salop, 1964). Variation in number and thickness of dark (organic-rich) and light (carbonaceous) laminae reflects a gradual increase in quantity of organic matter with time. Thus, the conophytons of younger deposits contain more dark laminae. This ratio is one of the bases of classification of the conophyton group, proposed by Komar et al. (1965). Phytolites of the oldest fossil-bearing Precambrian deposits are particularly poorly known.

Subdivision and correlation of Upper Precambrian rocks based on phytolites is a valuable and promising technique, but some uncertainties exist.

The spherically shaped microfossils called acritarchs are of minor stratigraphic value; once they were considered to be spores. All workers consider them to be organic in origin but their true nature is not known. Many believe them to be remains of planktonic blue-green algae. Acritarchs and other plant microfossils are common in Upper Precambrian formations, but they are poorly studied. Their use is restricted to younger Precambrian rocks. More work on these fossils is required.

Recently Precambrian rocks (including those up to 3,300 m.y. old) of many areas have yielded microscopic and submicroscopic (1—40 μm) spheroidal, complicated and irregular fossil forms. They represent fossilized cells and filaments of blue-green and other algae and bacteria. We know already that the oldest algae propagated by germination, and the algae occurring in the Late Precambrian possessed a sexual means of reproduction. The first simple-celled group belongs to the prokaryota; the second one, possessing cells with inner nuclei, to the eukaryota (Cloud, 1968). Doubt exists as to the organic nature of some of these structures, but in most cases their biogenic nature is certain or probable.

There are few animal fossil remains in the Precambrian. They are known from the youngest Precambrian units of several continents. These are mostly casts of soft-bodied animals, such as medusoids, rangeids and some of vermicular and problematic affinities. The only skeletal forms in Upper Precambrian rocks are forms with tubular skeletons, possibly pogonophores (*Paleolina*, *Sabellidites*) and vermicular forms or hyolites (*Anabarites trisulcatus* Miss.). The latter organisms occur only in beds close to the base of the Cambrian. These fossils are of limited use because of their rarity and restriction to the uppermost Precambrian. Some other structures occur in the Precambrian but their biogenic nature is not definitely established. The stratigraphic value of these problematic structures is, in most cases, questionable. Many structures, formerly described as Precambrian animals or plants, are now considered to be inorganic (Cloud, 1968; Hofmann, 1971).

Geological Methods

Geological methods provide the main basis for subdivision of older Precambrian strata. The fundamental aims are the subdivision of stratigraphic units, establishment of their boundaries, recognition of gaps in sedimentation and angular unconformities and also the relative proportions of stratified and plutonic rocks. The latter are commonly used as geological markers separating supracrustal rock complexes. In many cases only plutonic rocks are suitable for isotopic dating, so that they also serve as geochronological markers denoting definite stages in geological history.

Plutonic bodies may define various geological phenomena depending on when they formed in relation to the tectonic cycle. There are three main types of plutons, as follows: (1) intrusions coeval with sedimentation and volcanism — many hypabyssal and subvolcanic bodies are of this kind, as also are some ultrabasic and basic sills intruded into sedimentary and volcanic rocks; (2) intrusives of subplatform type formed close to the time of deposition of sedimentary and volcanic rocks or during gaps in sedimentation — some granites (rapakivi) and differentiated basic and alkaline rocks are of this kind; and (3) intrusive and ultrametamorphic (including anatectic) bodies formed later than the volcano-sedimentary complexes during orogeny in geosynclinal areas; both synkinematic and late-kinematic granitoids are of this kind. Isotopic dating of different types of plutonic rocks obviously reflects the time of different events.

The most important techniques for subdivision and correlation of Precambrian strata are the classical stratigraphic analysis and the lithoparagenetical ("formational") analysis. The basis for these techniques is the belief in the existence of irreversible changes in the tectonic evolution of the earth. Concomitant changes in the chemical composition and thermodynamic condition of the outer geospheres of the earth also contributed to the evolution of sedimentation processes with time.

It is well established that many Precambrian complexes in different areas (even on different continents) situated in similar positions in the stratigraphic column, and with similar ages according to isotopic analyses, show remarkable lithological similarities. This concept will be elaborated below. Some specific rock types such as jaspilitic iron formations, massive and oolitic, hematitic, and sideritic stratiform ores, gold- and uranium-rich conglomerates, tillites, and some others, characteristically occur at specific stratigraphic levels in the Precambrian. The proposed evolution involves not only the rock types, but also their paragenesis. The relative importance of some rock types changes with time. For example, the volume of volcanic rocks and dolomites decreases and the volume of coarse clastic rocks and limestones increases through time (Salop, 1964; 1970a).

There are some changes in the composition of plutonic rock, but they are not so obvious. Anorthosites are evidently confined to the Lower and Middle Precambrian; rapakivi granites and associated differentiated intrusions of gabbroid rocks to certain boundaries in the middel part of the Precambrian. There is a gradual decrease in synorogenic granitization and an increase in the volume of late-orogenic allochthonous granites. Granulite-facies metamorphism (in its regional meaning) is characteristic only of the Archean.

Some tectonic structures such as gneiss fold ovals are distinctive features in the Archean. In the future new and subtle features of the evolution of the stratigraphic record will no doubt be found. At present there have been attempts to subdivide iron formations into several different types. Some of them are definitely restricted to deposits of a particular age. Perhaps similar analysis will be possible with some other rock types. Simple stratigraphic (lithological) correlation is more important in Precambrian stratigraphy than it is in dealing with younger successions. This is probably due to the fact that, at least in the Early Precambrian, there was little tectonic differentiation or climatic zonation. Similar deposits were thus formed in different parts of the world. This method of correlation of the older strata must be used together with other geological techniques. Attention must be paid to geochemical and mineralogical data, facies changes, relationships with over- and underlying deposits, the nature of contacts (gaps, angular unconformities), etc. The correlation of specific formational types in some cases permits establishment of rock ages even when isotopic determinations and organic remains are absent. In some cases, this technique permits correction and refinement of results obtained by other methods. Thus, some empirically observed regularities concerning the vertical distribution of specific rock types (as proved by isotopic dating), may be important in objective subdivision and correlation of Precambrian sequences.

The main disadvantage of this essentially lithological method of correlation is its low degree of sensitivity. The evolution of sedimentation and of different rock types is an extremely slow process so that changes in the rock types can be used only for correlation of the largest subdivisions of the Precambrian.

No single method (among those discussed above) should be regarded as uniquely applicable; rather, a combination of all of them, when critically analysed, may produce a satisfactory and objective basis for subdivision and correlation of Precambrian rocks.

CHAPTER 2

GENERAL SUBDIVISIONS OF THE PRECAMBRIAN (ERA/ERATHEM AND SUB-ERA/SUB-ERATHEM)

Review of the Problem

For a long time metamorphic rocks were mainly studied from a petrographic point of view. Even at the beginning of this century, in some papers, and in many geological maps, Precambrian rocks were not subdivided according to their age, but were designated "metamorphic schist" or "crystalline schist" (gneiss) depending on the degree of recrystallization present. However, it was realized that it was necessary to fit such rocks into a stratigraphic scheme. Local stratigraphic subdivisions were applied to the Precambrian rocks of some areas.

The British geologist, Sedgwick, who established the Cambrian system in 1836, proposed in 1838 that all the older strata should be called Protozoic (Sedgwick, 1838). This was the first term proposed for the Precambrian. His compatriot, Murchison, in 1845 suggested that all the strata older than the Silurian, should be called the Azoic (Murchison et al., 1845). Yet another well-known British geologist, J. Dawson, in 1868 proposed that all the strata older than Cambrian should be known as the Eozoic (J. Dawson, 1868). This term was later used for the upper part of the Precambrian.

North American scientists in the seventies of the previous century began a detailed study of the Precambrian so that most of the general names for older strata were proposed by them. In 1872 Dana assigned all the Precambrian rocks to the astral eon and suggested that they be considered as Archean or Archeozoic (Dana, 1872). Emmons proposed in 1887 to use the term "Proterozoic" for Upper Precambrian rocks between the Archean and the Cambrian (Emmons, 1888). Walcott (1889) gave the name "Algonkian" to the same rocks. This latter name is usually used as a synonym for Proterozoic, and was commonly applied in the U.S.A. where the Precambrian was subdivided into Archean and Algonkian. Thus, a twofold subdivision of the Precambrian was developed and widely accepted.

Van Hise (1908), a prominent American worker in the Precambrian of the Great Lakes area (Canada and U.S.A.), suggested some criteria for subdividing these two units. He considered the Archean to be mostly magmatic rocks (granites, etc.), and the Algonkian (Proterozoic) to consist mainly of metamorphosed sedimentary rocks underlying the Cambrian. The Archean rocks were described as being more strongly folded, and highly metamorphosed, and separated from the overlying Algonkian by an unconformity. Further investigations in the Canadian shield showed that these criteria were not satisfactory and of limited practical value. It was early established that in the

Precambrian rocks of the Canadian shield there are, in fact, several angular unconformities. Also the older rocks assigned to the Archean include metamorphosed sedimentary rocks. At least some North American workers started to use the terms Archean and Algonkian (Proterozoic) for the types of deposits rather than for the age of the rocks.

In 1955 the American Commission on Stratigraphic Nomenclature rejected the terms "Archean" and "Proterozoic" (Algonkian), and accepted only one chronostratigraphic subdivision — "Precambrian". In different areas the Precambrian could be subdivided into three parts comprising Lower, Middle and Upper. Lawson earlier (1913—1916) proposed subdivision of the Precambrian of the Canadian shield into three parts and suggested the following names (from the base upwards): Ontarian, Huronian and Algonkian (Lawson, 1913). In 1935 Canadian geologists held another opinion. The Canadian National Committee on Stratigraphic Nomenclature recommended subdivision into Archean, which embraced all pre-Huronian rocks, and Proterozoic, including Huronian and some younger Precambrian strata (Alcock, 1934). Later the Geological Survey of Canada subdivided the Precambrian into Archean and Proterozoic. The latter was subdivided into three parts: the Lower, Middle and Upper Proterozoic. Thus, a fourfold subdivision of the Precambrian came into being. At the same time the Geological Surveys of Michigan and Minnesota, neighbouring states in the U.S.A., maintained a threefold subdivision of the Precambrian into Lower, Middle and Upper Precambrian.

The above review of the evolution of the subdivision and terminology used for the Precambrian is presented to explain the history of many now widely used general stratigraphic terms. The presently used stratigraphic subdivisions of the Precambrian will now be discussed. Neither in the Soviet Union, nor in other countries, are there universally accepted classifications of the Precambrian. In 1963 the Interdepartmental Stratigraphic Committee of the U.S.S.R. decided to retain two major subdivisions: Archean and Proterozoic (Resolution, 1965), but geochronological boundaries were not suggested because the opinions of the Stratigraphic Committee members on the problem were not unanimous. At the same time the Stratigraphic Committee pointed out that the Proterozoic may be subdivided into Lower, Middle and Upper units. The Commission on Absolute Age Determinations of Geological Formations at the Academy of Sciences of the U.S.S.R., in 1964 proposed subdivision of the Precambrian into four groups: Archean, Lower Proterozoic, Middle Proterozoic and Upper Proterozoic. Their upper age boundaries (corresponding to the tectonic—plutonic cycles) were respectively 2,600, 1,900, 1,600 and 570 m.y. (Afanas'ev et al., 1964). This scheme is commonly used at present. There is also another scheme proposed by the workers of the Geological Institute of the Academy of Sciences of the U.S.S.R. — B.M. Keller, G.A. Kazakov, M.A. Semikhatov and others (Keller et al., 1960). According to this scheme, in addition to Archean, Lower and Middle Proterozoic there is the Riphean which corresponds to the Upper Proterozoic. The

Riphean is subdivided on the basis of stromatolites and microphytolites into three parts: the Lower, Middle and Upper Riphean which were formed during the following periods: 1,550—1,350, 1,350—1,000, and 1,000—650 m.y. The Vendian (or Vendian complex) with the upper age limit taken at 550 m.y. is situated between Riphean and Cambrian. Recently Keller and Semikhatov (1968) proposed that the Vendian (including the Volyn' and Valday Groups) be considered a fourth subdivision of the Riphean (terminal Riphean). The name "Riphean" was introduced for the first time by Shatsky in 1945 for low-grade Precambrian rocks of the Urals and the Volga—Urals area of the platform adjacent to it (Shatsky, 1945). These rocks lie unconformably on older high-grade basement strata. The thick Upper Precambrian sequence (including several unconformities), in the Bashkir anticlinorium on the western slopes of the Urals, was taken as the Riphean stratotype. Shatsky earlier believed that Riphean strata formed during a rather short period (about 70—100 m.y.) but isotopic dating later showed that they represent a longer time period. Thus, it was proposed to have a separate Riphean Era (Erathem) equal to the other larger subdivisions of the Precambrian.

At the moment it appears that even the latest geochronological limits for the Riphean (550—1,550 m.y.), accepted by B.M. Keller and his coworkers, are too restricted for the Urals stratotype (Salop, 1970b; Salop and Murina, 1970). The data available for the Lower Riphean (Burzyan Group) suggest an age of more than 1,570—1,600 m.y., and its lower boundary is possibly as old as 2,600 m.y. This range for the Riphean suggests that it is too great to be kept as a single unit in a breakdown of the Precambrian. This is also suggested by the complex structure exhibited by the Riphean strata, the presence of a number of significant recently established disconformities, and by some unique features of Lower Riphean structure (Jacobson, 1968).

In addition to the above widely used schemes there are several other subdivisions of the Precambrian proposed by Soviet geologists. Ukrainian geologists mainly use the scheme proposed by Semenenko (1965). In this proposal the Precambrian is subdivided into five megacycles which are, in ascending sequence, as follows: Precambrian I, Precambrian II, Precambrian III, etc. Their lower boundaries are respectively 3,500, 2,600, 1,900, 1,600, and 1,100 m.y. This scheme was mainly based on isotopic data and reflects, in a general way, true major boundaries. They coincide with the Precambrian era boundaries adopted by other workers, and by the Commission on Absolute Age Determinations in 1964. But these megacycles do not have a precise geological meaning, and in many cases this poses problems.

Another scheme for subdivision of the Precambrian was proposed by Salop (1964). In this proposal all the Precambrian rocks of the world were subdivided into five large natural complexes corresponding in rank to the Phanerozoic erathems. Initially (Salop, 1964) the proposed erathems were: Archean, Lower Proterozoic, Middle Proterozoic, Upper Proterozoic and Epiproterozoic, and the tectono-plutonic cycles provided age boundaries as follows: 2,600—2,800, 1,800—2,100, 1,400—1,500, and 1,000—1,100 m.y. The upper boundary of the Epiproterozoic was considered to be a cycle falling into the 600—650 m.y. period. The concept of a universally applicable subdivision was the subject of many subsequent publications (Salop, 1968a, 1969, 1970b, 1972a). It was proposed that the Precambrian be subdivided into two eons: the Archean and the Proterozoic, which should include four eras which were given new names: Paleoprotozoic, Mesoprotozoic, Neoprotozoic and Epiprotozoic. The geochronological boundaries of the eras were changed on the basis of new isotopic data. The upper limit of the Archean is now taken at about 3,500 m.y., the upper limit of the Paleoprotozoic at 2,600—2,800 m.y., that of the Mesoprotozoic at 1,000—1,100 m.y., and that of the Epiprotozoic at 650—670 m.y.; the Eocambrian—Cambrian boundary is placed at 570 m.y. The Neoprotozoic is subdivided into three sub-eras (sub-erathems) (Salop, 1970b, 1972a). This subdivision is the subject of this work, and is described in detail below.

There have been other proposals for a universally applicable stratigraphic subdivision of the Precambrian of the U.S.S.R. However, most of them are not well founded and are rarely used by Soviet geologists.

Geologists studying the Precambrian of Africa and Australia have generally used the names of larger local units for subdivision and correlation. In some general publications on geochronology and tectonics they have used common designations using figures or letters. Choubert and Faure-Muret (1967) subdivided the Precambrian on the tectonic map of Africa into five eras with the following symbols (in ascending order): D, C_2, C_1, B, and A. The respective age boundaries are: 2,500—2,650 (Zagoride folding), 1,850—2,100 (post-Birrimian or Eburnean folding), 1,650—1,800 (Mayombe folding), and 900—1,100 m.y. (Kibaran folding). The Precambrian A—Eocambrian (Infracambrian) boundary is considered by these authors to be at about 620 m.y. (Katangan folding), and the Eocambrian—Early Cambrian boundary is at 550—600 m.y. (no folding in this case). There is a striking similarity between many of these proposed African orogenic boundaries and those accepted in the U.S.S.R. In this scheme and in some others, the era symbols are in the inverse order to that used for megacycles in Semenenkov's scheme, where the figure symbols increase from older to younger units.

The workers in India commonly use local names for the older sequences (Dharwar, Cuddapah, Delhi, Vindhyan "systems", etc.), but for the largest Precambrian subdivisions they use Archean or Proterozoic or subdivide the Precambrian into Lower, Middle and Upper Precambrian. Geochronological boundaries are not well established, and different authors use different values. The Precambrian stratigraphic subdivision proposed by Stockwell (1961, 1964, 1968) is the best of its kind outside the U.S.S.R. It is done with special reference to the Precambrian of Canada and is widely used there. In this classification the Precambrian of Canada is subdivided into two eons (super-eras), the Archean and Proterozoic. The Archean includes the Archean Era, and the

Proterozoic Eon is subdivided into three eras which were given new names (in ascending sequence): Aphebian, Helikian, and Hadrynian. The Helikian in turn is subdivided into two sub-eras: Paleohelikian and Neohelikian. The names are of Greek origin and refer to different stages of human maturity. The era boundaries are orogenic cycles of the first order: Kenoran (between Archean and Aphebian). Hudsonian (between Aphebian and Helikian), and Grenville (between Helikian and Hadrynian). The Helikian sub-eras are separated by the Elsonian orogeny which is of relatively minor importance. To establish the age boundaries of the Precambrian eras in the Canadian shield Stockwell took the average values (minus standard deviation) of the ages found by compiling numerous K—Ar mineral datings (mainly on micas) from granites and metamorphic rocks formed during the orogenic cycles. Thus, the Kenoran orogeny is considered to have taken place at about 2,500 (2,390) m.y., although the isotopic determinations present a scatter from 2,280 (2,200) to 2,800 (2,700) m.y. (Fig. 2). The age of the Hudsonian orogeny is

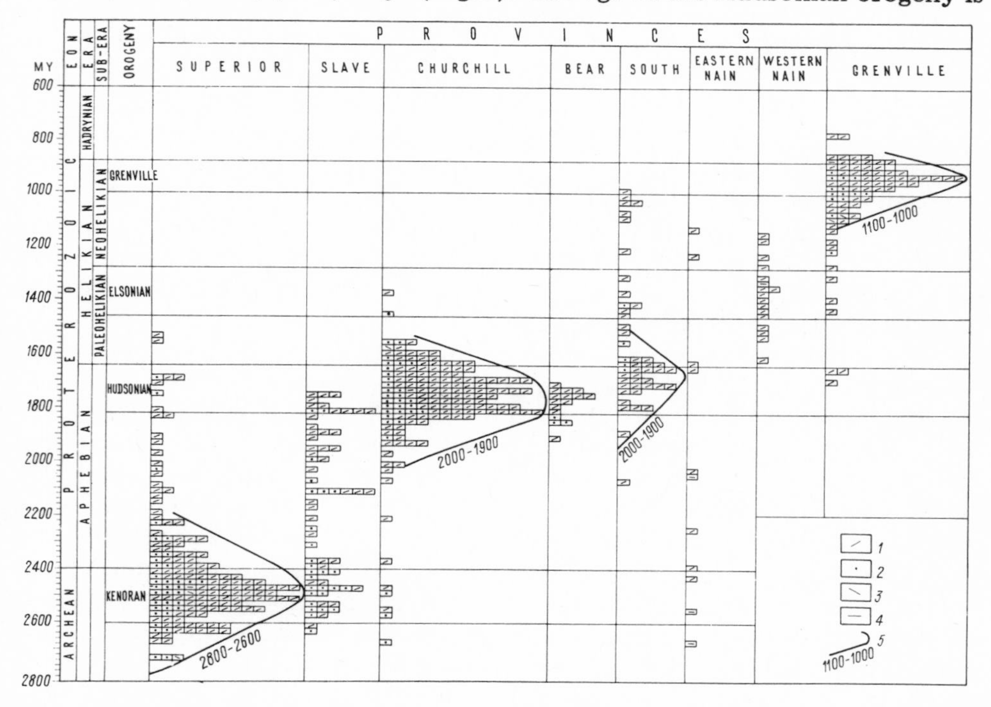

Fig. 2. Histogram of K—Ar dates on minerals from Precambrian granitoids and metamorphic rocks of the Canadian shield, showing the time of orogenic (tectono-plutonic) cycles. This figure is after Stockwell (1964), with some additions by Salop (curves showing the general shape of the histograms).
1 = one determination on biotite; *2* = the same on muscovite; *3* = the same on other micaceous minerals; *4* = the same on amphibole; *5* = the ages of orogenic episodes accepted here ("relict" values) in millions of years.

1,700 (1,640) m.y. — range: 1,600 (1,550) to 1,990 (1,920) m.y. — , the age of the Elsonian orogeny is 1,340 (1,280) m.y. — range: 1,300 (1,240) to 1,450 (1,400) m.y. — , and the age of the Grenville orogeny is 840 (800) m.y. — range: 840 (800) to 1,180 (1,130) m.y. The Proterozoic—Cambrian boundary is conventionally taken at 630 (600) m.y.

Stockwell's scale has many advantages and some disadvantages including the following: (1) it does not include the important oldest part of the geological record; (2) relationships between intrusions assigned to the Elsonian orogeny and Helikian strata are uncertain in many cases; poor evidence for considering the Grenville orogeny as the boundary between the Helikian and Hadrynian because such deposits are rare in the area affected by the Grenville folding; (3) in the scale there is no place for the youngest Precambrian deposits underlying the Cambrian ("transitional beds"); (4) the names of the eras are poor in that they do not reflect the age sequence, and they are of quite different character to other (Phanerozoic) eras that must be included in a universal scale; and (5) the dating of age boundaries of the Precambrian eras is probably erroneous.

The first, second and fifth points will be dealt with below. The third and fourth do not need further explanation.

Stockwell used only K—Ar dating and the average values of numerous and widely varied data as a basis for determination of age boundaries of eras. The "histogram method" of evaluation of the isotopic age of tectono-plutonic cycles is basically wrong. As stated above, K—Ar analysis of abyssal rocks usually gives "rejuvenated" values which do not correspond to the true age of the rocks but rather to the period when they cooled below 300°C and diffusion of argon stopped. Due to low heat conductivity of wall rocks the cooling of plutonic rocks may continue for a long period (millions of years). Also deep-seated plutonic—metamorphic complexes may take a long time to rise above the critical geoisotherm (300°C) — this may be long after their formation by successive orogenic processes. Thus, in order to pin down the time of various orogenic cycles it is not the average values but rather the oldest values (so called "relict" dates) that are important. Taking this into account the Kenoran orogeny took place at 2,600—2,800 m.y., the Hudsonian at 1,900—2,000 m.y., and the Grenville at 1,000—1,100 m.y. ago (Fig. 2). The Elsonian orogeny, as stated by Stockwell, is not reliably established either on geological or geochronological grounds. The dates that correspond to it on Stockwell's histograms are probably "rejuvenated" dates from tectono-plutonic cycles.

Many new age determinations from rocks of the Canadian (or Canadian—Greenland) shield suggest that these boundaries are more realistic when more refined Rb-isotopic and Rb—Sr isochron analyses are used. Recently Stockwell (1973) revised his geochronological scheme for subdivision of the Canadian Precambrian. In this new scheme, based on Pb-isotope and Rb—Sr analyses, the upper limits of the Archean, Alphebian, Paleohelikian and Neohelikian, or the dating of the Kenoran, Hudsonian, Elsonian and Grenville orogenies, correspond to 2,560 (or 2,690), 1,800 (or 1,850), 1,400, and 1,000—1,070 m.y., respectively. These values are similar to those suggested here as global orogenic cycles (the values in parentheses are based on a Rb—Sr decay constant of $\lambda_{Rb} = 1.39 \cdot 10^{-11}\ y^{-1}$).

Principles used in Subdivision of the Precambrian

The basis for subdivision of the Precambrian must be natural stages in the earth's evolution. These must be periods in geological history which are characterized by common tectonic, geochemical and physical environments to produce specific types of rock associations, tectonic structures, forms and types of magmatism, and to some extent control organic evolution. This principle is recognized by many workers although different people have different approaches to the problem. Some workers (Borovikov and Spizharsky, 1965; Rankama, 1970) believe that the most objective approach is a subdivision of the Precambrian according to the age of the rocks as expressed in astronomical time units (years or megayears), and evaluated by radiometric analyses regardless of geological events. Thus, there are two separate scales, a stratigraphic one (geohistorical), and a time scale. It is easier to date events in the field of socio-historical science where subdivision is based on changes in social structures and the calendar system of chronology is totally independent. The very act of establishment of this subdivision suggests the existence of definite natural stages with clear boundaries. Thus, this "calendar principle" is often impossible or problematic because of a lack of radiometric data or because of the complexity of their interpretation.

Some workers attach the greatest importance in global Precambrian stratigraphy to biostratigraphic data. Paleontological principles cannot serve as the basis for general subdivision of the Precambrian because of the scarcity of fossils and present uncertainty as to their stratigraphic value, particularly for those occurring in the lower part of the Precambrian. Organic complexes that are presently known (mainly stromatolites and microphytolites) may be used to characterize parts of the larger natural subdivisions or for adjacent units separated by a long time or by orogenic cycles. However, these fossils appear to be useful for description of the Upper Precambrian erathems, and for their subdivision into smaller units. This principle of description of natural stages in geological evolution (and the subdivisions corresponding to them) is equally applicable to the Phanerozoic. The Phanerozoic subdivisions were established by different authors at different times, so that the principles used were different in different cases. The major Phanerozoic subdivisions (systems and periods), are based mainly on organic evolution. Worldwide systemic correlation in the Phanerozoic was also based to a large extent on organic evolution, but also took account of several aspects of inorganic evolution of the planet. The principle of "natural subdivisions" in the Phanerozoic is not strictly adhered to because biological evolution is only partly associated with such environmental changes. However, in many works this principle is used. Biostratigraphic principles obviously cannot serve as the sole basis for subdivision of the Precambrian. It is only by consideration of a combination of geological, isotopic and paleontological data that geologists will arrive at a universally applicable scheme for subdivision of the world's Precambrian rocks. If the Precambrian could be subdivided on the principle of "natural stages" then this scheme would have some advantages over that used for the Phanerozoic. It is apparent that subdivision of the Precambrian cannot be made in as much detail as has been possible in the Phanerozoic, for the evolution of the earth as a whole was a much slower process than the evolution of living matter. Precambrian subdivision must accordingly be on a much larger scale. The proposed basic unit corresponds to the rank of era or sub-era (erathem or sub-erathem) in the Phanerozoic, but it represents a much larger period of time.

The most important global phenomena must be used in subdivision of the largest units, and phenomena of lesser importance should be used for subdivision of these units. The problem of definition of boundaries of these global units is also rather complex. The presence of unconformities, indicating tectonic movements and intrusive magmatism is the major factor used in all local Precambrian subdivisions. It might be thought that this tectonic aspect, for purposes of interregional correlation and subdivision of ancient strata should be applied with great caution. However, radiometric dating of older plutonic and metamorphic rocks formed during orogenic cycles shows that orogenic processes, in the Precambrian stages of the earth's evolution, were separated by long intervals, and were to a large extent, simultaneous on all continents. The geochronological data also show that such tectonic cycles became much more frequent and complicated in the younger part of the earth's history so that the same tectonic principle cannot be used to subdivide and correlate Phanerozoic deposits. Precambrian orogenic episodes appear to serve as boundaries separating larger natural stages in the geological evolution of the earth. This tectonic principle is the most fundamental used in subdivision of older stratigraphic units and is the basis for establishment of the boundaries of larger-scale subdivisions in many modern stratigraphic and geochronological classifications of the Precambrian.

It is usually assumed that the end of an orogenic cycle is the beginning of a new, long evolutionary stage with concomitant sedimentation and volcanism. However, this does not mean that the boundaries of the larger units (erathems) in the Precambrian must everywhere be represented by angular unconformities or stratigraphic breaks. Orogenic activity probably never caused the entire surface of the earth to be dried up and in some submerged regions sedimentation must have continued. In such regions the stratigraphic record would have been continuous, but even there tectonic events would have had some effect (change in type of sedimentation, local erosion, etc.). The major Precambrian tectono-plutonic cycles are interpreted as having taken place more or less simultaneously throughout the world. However, in different areas they may have a particular style and intensity. In areas of continuous sedimentation the delimitation of Precambrian erathem boundaries is difficult, but in many cases correlation is possible with adjacent areas where unconformities are present.

In most cases unconformities separating the larger Precambrian subdivisions may be present over very extensive areas (in some cases of continental dimensions). This is probably because of the long duration of the Precambrian diastrophic cycles. When all phases of such long-lived orogenies are considered together, then they make up an extremely widespread (almost global) unconformity. Such tectonic events were probably locally expressed, but may have occurred simultaneously in different areas (Fig.3). In many

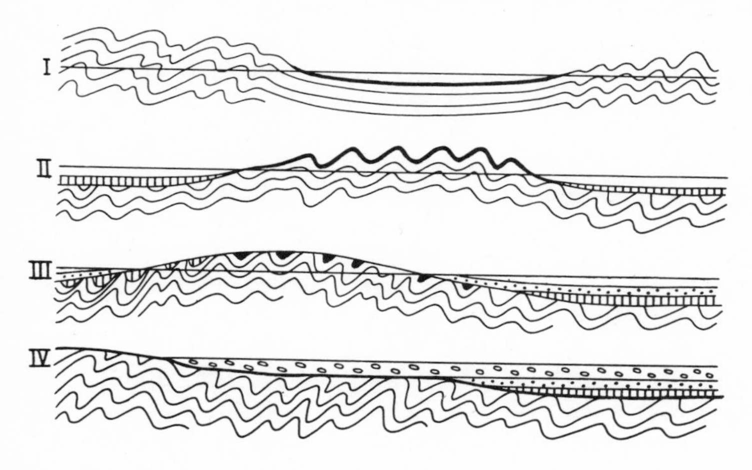

Fig. 3. Diagram to show the mode of formation of a single regional angular unconformity as the result of several unconformities originated during a long polyphase orogenic cycle with migration of zones of folding.
The numerals *I* to *IV* show different stages in the formation of a regional angular unconformity.

places a single unconformity (in some cases two) is present at the boundary of the larger Precambrian subdivisions. This is a unique aspect of the Precambrian, and is a major difference from Phanerozoic successions where several unconformities are present, corresponding to several phases of a simple orogenic cycle. Possibly, this difference is explained by loss of the upper parts of older supracrustal complexes due to erosion, and by the longer duration of the gaps, but this is probably not a major factor because there are well-preserved continuous Precambrian sections where there are no angular unconformities.

Isotopic dating of plutonic and metamorphic rocks commonly yields a wide scatter of about 100—200 m.y. Only a minor part of this scatter is due to analytical errors. The major part of the scatter is due to the great length of time involved in the tectono-plutonic cycles and extremely slow cooling which may have taken millions, or scores of millions of years.

Because of their long duration, the orogenic cycles might be considered unsuitable for subdivision of the Precambrian. However, the relatively quiet stages of Precambrian evolution involved even longer periods, of the order of 350—900 m.y. Also, in a number of areas tectono-plutonic processes that took place in shorter periods of time (by isotopic dating), still lie within the proposed boundaries of the global orogenic cycles. On the whole the upper age limit of the orogenic cycles is taken as the boundary of the large natural units (eras) of the Precambrian. In some areas the termination of diastrophism and beginning of a new sedimentary cycle appeared to have taken place somewhat earlier than the upper limit of the worldwide orogenic episode. The deposits of different areas, although belonging to a definite general stratigraphic unit may be of slightly different age in different places and may represent only part of the time period involved.

Proposed Global Scheme for Subdivision of the Precambrian: Brief Outline

The main purpose of this work is to establish natural stages in the earth's evolution during the Precambrian. On this basis, major units of a global stratigraphic scale are defined. A brief outline of the main aspects of the proposed scheme is presented here.

The Precambrian may be subdivided into five very large, natural units, corresponding in rank to the Phanerozoic eras, but of much longer duration. Global orogenic (tectono-plutonic) cycles are used as boundaries. This conclusion is based on analysis of material from ancient complexes of different areas of the world (in this book only the Northern Hemisphere is considered). The length of time involved in formation of the supracrustal and plutonic cycles is of the order of 350—900 m.y. These time periods are long (in some cases longer than the duration of all the Phanerozoic eras together), so that it could be argued that they should be assigned a rank higher than era. There appears to be a general tendency, both in the Phanerozoic and the Precambrian, for the times of sedimentation to be longer in older successions. The longer time periods involved in the Precambrian eras are simply a natural result of geological evolution of the planet, so that it is not considered necessary to raise the rank of these units.

The Precambrian erathems/eras are here combined into two super-erathems/super-eras (eonothems/eons): the Archean (or Cryptozoic) and the Protozoic. The rocks of these eons were formed under quite different physical and chemical environments representing different stages in the geological evolution of the planet. These are the permobile or "pan-geosynclinal" stage, when the whole crust of the earth was tectonically mobile, and the platform—geosynclinal stage, which continues up to the present. In the Archean (Cryptozoic) Eon only one Archean era is recognized. However, there are some reasons to believe that even older stratigraphic units may be discovered. The upper boundary of the Archean is marked by the Saamian tectono-plutonic cycle dated at about 3,500 (3,500—3,700?) m.y. This event brought about a fundamental change in the geological evolution of the entire earth.

The Protozoic Super-era is subdivided into four eras: Paleoprotozoic, Mesoprotozoic, Neoprotozoic and Epiprotozoic. The end of the Paleoprotozoic Era is marked by the Kenoran orogeny (2,600—2,800 m.y.), the Mesoprotozoic Era by the Karelian orogeny (1,900—2,000 m.y.), the Neoprotozoic Era by the Grenville orogeny (1,000—1,100 m.y.), and the Epiprotozoic Era by the Katangan orogeny (650—680 m.y.).

All these global orogenic cycles were accompanied by intense plutonic activity resulting in intrusions of basic and ultrabasic magmas and formation of large granitoid massifs that were emplaced at various stages of the tectonic cycle.

The Neoprotozoic Era is divided into three sub-eras (Early, Middle and Late) by two global cycles of platform or subplatform magmatism, locally associated with weak tectonic movements.

The first cycle, marking the upper boundary of the Early sub-era, occurred in the 1,600—1,750 m.y. period. It is characterized by abundant intrusions of gabbro norites, anorthosites, granophyric granites, rapakivi granites and alkalic rocks. This cycle is named the Vyborgian magmatic cycle (after the Vyborg rapakivi pluton on the northern coast of the Gulf of Finland). The second cycle took place at the end of the Middle Neoprotozoic Sub-era in the period between 1,300 and 1,400 m.y. ago. It is characterized by abundant small intrusions of diabase and gabbro diabases, mainly as dikes, stocks and sills, in some cases associated co-magmatic basic lavas. There are also a few intrusions of alkalic rocks, syenites and granodiorites. This magmatic cycle is called the Prikamian cycle (for the numerous diabase dikes in the Pri-Kama area). This complex may also be called Sanerutian (after the Sanerutian intrusive complex in Southern Greenland), or still better, the Kibaran, for in southern and equatorial Africa there was intensive tectogenesis during this interval (and not during the 900—1,100 m.y. interval as suggested by Choubert). These magmatic or poorly expressed orogenic cycles are of minor importance compared to those listed above. It is for this reason that they are used for subdivision of global stratigraphic units of sub-erathem rank. Deposits of the younger (Eocambrian) complex are separated by a significant stratigraphic break or angular unconformity from the Epiprotozoic. They lie either conformably, or with a small stratigraphic break, beneath fossiliferous Lower Cambrian units. Thus, the Eocambrian—Cambrian boundary is not defined on the basis of tectonics, but rather on the presence of a skeletal fauna in the lowermost Cambrian. On isotopic data this boundary is placed at about 570 m.y. so that the Eocambrian complex formed in about 80 m.y. and is similar in length to many Paleozoic periods (e.g., the Cambrian). The problem of stratigraphic rank of the Eocambrian, and its assignation to the Protozoic or Phanerozoic (Paleozoic) are debatable questions.

Thus, in the Precambrian there are eight proposed global stratigraphic subdivisions of different rank (Archean, Paleoprotozoic, Mesoprotozoic, Lower, Middle and Upper Neoprotozoic, Epiprotozoic and Eocambrian). Each unit possesses its own specific features which will be summarized following a regional analysis. Distinctive assemblages of fossils are characteristic of many of the units (Table 1).

Comparison of the proposed subdivision scheme with some of the other Precambrian schemes mentioned above is shown in Fig.4. The comparison is approximate because, in some cases, different authors have assigned rocks of different age to the same units.

In any developing science there is a need for some regulation in discarding old terminology and introducing new terms. The situation becomes critical in such rapidly developing branches of science as Precambrian stratigraphy. However, we cannot be satisfied with Precambrian stratigraphic terminology developed at a time when little was known about the rocks in question. Recent information necessitates change. Many of the old names can no longer be used in their original meaning. They must be changed or adapted to meet the new concepts. Thus, names such as "Lower, Middle and Upper Proterozoic" are no longer used. These terms are rejected because they do not give much idea of the great length of time involved in each. The length of time represented by each of these subdivisions is greater than each of the Phanerozoic eras and some are even greater than all of them. Also, because it is necessary to subdivide the Late Proterozoic into two independent eras — the Late Proterozoic proper (sensu stricto) and the Epiprotozoic — the term could have a double meaning. Finally, the names of the eras should be short (one word) and based on one and the same principle, as is the case for Phanerozoic eras. The terms suggested in this report are considered preferable to the terms proposed for the Proterozoic by Stockwell (Aphebian, Helikian and Hadrynian), because they give a clearer picture of the stratigraphic succession involved.

The name Riphean is also considered to be unsuitable for a Precambrian era (Late Proterozoic). It is based on a different principle from that on which the other Precambrian and Phanerozoic eras were erected. Also, more importantly, the Riphean is too all-embracing (in the Urals stratotype) for it includes subdivisions attributable to the Eocambrian and to the three Proterozoic eras, including the Mesoprotozoic. Use of terms such as "Early Proterozoic" and likewise "Lower Proterozoic" could also lead to difficulties in writing (for example, "in the earliest Early Proterozoic" as opposed to "in the Early Paleoprotozoic" or "in the lowermost Lower Proterozoic" instead of "in the lowermost Paleoprotozoic").

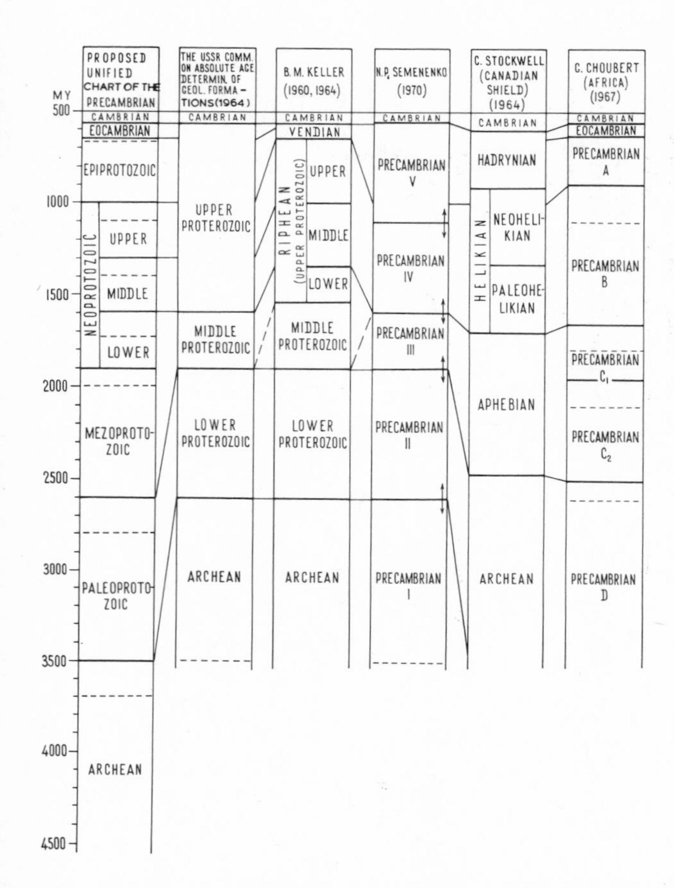

Fig. 4. Comparison of different schemes for subdivision of the Precambrian. Hatched lines in the columns show the lower age limits of the orogenic cycles. The age boundaries on Stockwell's scale (based on the K—Ar method) are shown according to the decay constant for potassium, accepted in the U.S.S.R. ($\lambda_K = 0.557 \cdot 10^{-10}\ y^{-1}$).

For the oldest super-era or eon (Salop, 1969, 1970a) the term Archean (Archean Eon) is used. This term is also widely used by both European and Soviet geologists for the strata included under this name in the proposed scheme. It might, however, be appropriately named the Cryptozoic (hidden life). Undoubtedly, living matter existed in the oldest strata, though we do not know in what form, for organic remains are either absent, or have not yet been discovered. The term "Cryptozoic" was also proposed for the whole Precambrian (corresponding to Phanerozoic for the post-Cambrian), but this usage of the term is incorrect because in the Protozoic there are organic remains of definite systematic rank. If the term "Cryptozoic" is accepted for the oldest super-era, then the term "Archean" will refer to its one era. Another possibility for a name for the Archean—Cryptozoic super-era is the

TABLE 1

Scheme for global subdivision of the Precambrian

Super-erathem (eonothem)	Erathem	Sub-erathem and complex	The age limits of the units and their duration in m.y.	Orogenic cycles, and their age limits in m.y.*	Paleontological characteristics**
			570		
		Eocambrian (complex)	(80)		IV ("Vendian") phytolite complex, algae *Epiphyton*, and *Renalcis*, casts of non-skeletal animals ("Ediacara fauna"), pogonophoras (*Sabellidites*); in the upper part — *Anabarites trisulcatus* Miss. hyolite(?)
			650	650	
	Epiprotozoic		(350)	Katangan 680	IV ("Vendian") phytolite complex, casts of medusoids
			1,000	1,000	
Protozoic		Upper Neoprotozoic	(300)	Grenville 1,100	III ("Upper Riphean") phytolite complex
			1,300	1,300	
	Neoprotozoic	Middle Neoprotozoic	(300)	Prikamian (Kibaran) 1,400	II ("Middle Riphean") phytolite complex
			1,600	1,600	
		Lower Neoprotozoic	(300)	Vyborgian 1,750	I ("Lower Riphean") and (or) II ("Middle Riphean") phytolite complexes

		1,900	1,900	
	Mesopro-tozoic	(700)	**Karelian**	I ("Lower Riphean") phytolite complex
			2,000	
Protozoic		2,600	2,600	
	Paleopro-tozoic	(900)	**Kenoran**	Rare, poorly studied stromatolites, and microphytolites (oncolites)
			2,800	
		3,500	3,500	
Cryptozoic (Archean)	Archean	(> 900?)	**Saamian**	Definable organic remains are not known
			3,700	

*Major orogenic cycles are lettered in bold print.
**In the Paleoprotozoic and Mesoprotozoic strata microscopic remains of blue-green algae (prokaryota) are reported; the eukaryota probably developed in the Middle Neoprotozoic.

term Agnostozoic (agnostus = unknown) proposed by Irving (1887). It indicates the unknown nature of the older bios.

The second Precambrian super-era is called the Protozoic (Protozoic Super-era or Eon), the first general name for the Precambrian (Sedgwick, 1838). This term is used in preference to the commonly used term "Proterozoic", because it is shorter and thus more convenient in composing names for the eras. Also, by using this name some confusion may be avoided due to the fact that in some countries (Canada, India, etc.), rocks belonging to the lower part of the Protozoic Super-era (as defined here) are included in the Archean. The Greek word "protos" means first, and in complex words it reveals the simplest features — Protozoa, for example. Thus the evolution of the earth is subdivided, according to the life present, into three eons: Cryptozoic (hidden life forms), Protozoic (first life forms) and Phanerozoic (clearly developed life forms). The names of the eras in the Protozoic Super-era (Eon) are based on the same principle as that used for Phanerozoic era names: Paleoprotozoic (ancient life forms), Mesoprotozoic (the middle life forms) and Neoprotozoic (new life forms). In the term "Epiprotozoic" the prefix "epi" means over, close or near. It signifies that the era has the highest position within the super-era.

The term "Eocambrian" (the dawn of the Cambrian) is widely used in different countries for the youngest Precambrian strata underlying the Cambrian, though it was originally proposed by Scandinavian geologists for deposits which we now attribute to the Epiprotozoic. We prefer this term to the widely accepted (in the U.S.S.R.) term "Vendian" because the latter term is now used for deposits of very great vertical range, including, in addition to the Eocambrian proper, Epiprotozoic and even Neoprotozoic rocks. The term "Vendian" was proposed by Sokolov (1952a) for strata corresponding approximately to the Eocambrian as defined here.

The Stratotypes

This chapter is concluded with a brief statement on the problem of stratotypes of the proposed global Precambrian subdivisions. For Phanerozoic deposits this problem is not so urgent because the correlated units have lower stratigraphic rank (system and lower), and correlation is based mainly on paleontological criteria. In correlation of older successions more emphasis is placed on lithological comparisons so that standard sections or stratotypes of the units are essential. The ideal stratotype should have the most complete section, well-defined upper and lower boundaries, it should be fossiliferous (for the Upper Precambrian), and well defined by isotopic dating. Global stratotypes should be chosen for each era and sub-era of the Precambrian. Different types of strata (platform, miogeosynclinal and eugeosynclinal) occupy large areas, so that it is necessary to choose local stratotypes for each such area. This is necessary because formational correlation is possible only within the same megafacies. Local stratotypes for the Precambrian are proposed below. It is important that these local formational stratotypes should themselves be correlatable.

PART II

THE GLOBAL STRATIGRAPHIC SCHEME: ITS BASIS, AND THE CHARACTERISTICS OF THE MAJOR PRECAMBRIAN SUBDIVISIONS

CHAPTER 3

MAJOR AREAS OF PRECAMBRIAN STRATA IN THE NORTHERN HEMISPHERE AND AVAILABLE STRATIGRAPHIC DATA

In the Northern Hemisphere there are four major areas of Precambrian rocks: the East European, Siberian, Indian and North American platforms, together with surrounding fold belts of various ages.

The East European craton is one of the largest in the world. Precambrian rocks make up its basement in the Baltic and Ukrainian shields and are also an important component of the lower part of the sedimentary cover of the larger part of the platform known as the Russian plate. Within the latter these deposits are known from drill holes of variable depth. Older strata are also present in the fold belts that fringe the platforms. They are most complete and widely distributed in the Hercynian fold belt of the Urals. They also occur, but less abundantly, in the Caledonian fold belt of Northern Europe, where it is commonly difficult to separate Precambrian strata from Lower Paleozoic metamorphic strata. Major exposures of Precambrian rocks in the Hercynian (Variscan) fold belt on the western margin of the platform are confined to the Bohemian Massif. In the Alpine belt of the Caucasus and in the Carpathians, Precambrian rocks are exposed only in a small area in the axial zones of some anticlinoria. Their stratigraphy is poorly known. Thus, for the purpose of subdividing the Precambrian the sections will be confined to the platform and the Urals fold belt.

For many years geologists from many countries have studied the Precambrian rocks of the East European platform. As a result of these studies, in areas of the Baltic and Ukrainian shields, and also of the Russian plate, Soviet and other geologists have put forward some important general concepts in relation to the subdivision and correlation of the Precambrian. At the end of the last century the Finnish geologist J. Sederholm proposed the first scientific scheme for subdivision of the Precambrian rocks of the Baltic shield, based on tectonic, or more precisely, tectono-plutonic criteria (tectono-plutonic cycles were used as boundaries for subdivision of the Precambrian). These criteria, together with others (discussed later) are widely used in modern Precambrian stratigraphy. It was proposed that the Lower Precambrian should be subdivided into the Riphean erathem and Vendian complexes (Shatsky, 1945; Sokolov, 1952a) as a result of studies of the older formations of the Russian plate, and of the adjacent Urals fold belt. The boundaries of these units changed with time, but their subdivision was a step in the right direction. The most important was the establishment of the Vendian complex as the uppermost Precambrian, transitional to the Paleozoic erathem. Studies of Upper Precambrian (Riphean) fossils in the Urals area revealed four successive stromatolite and microphytolite complexes (Krylov, 1966), so that the Riphean was subdivided into three parts and the Vendian complex was also considered as a separate unit (Keller and Semikhatov, 1963). This was the starting point for widespread application of paleontological techniques for interregional correlation of the Upper Precambrian. The Precambrian of the East European platform and of its surrounding fold belts is relatively well known, but some important stratigraphic problems of the Precambrian remain unsolved. There are many reasons for such a situation: poor exposure in many areas, complex geological structures, lack of detailed knowledge, general methods used in subdivision of the older strata, misinterpretation of isotopic age data, lack of communication between different scientific schools, communication barriers, etc. At the present time there is no universally accepted stratigraphic scheme for subdivision of the Precambrian, either for the whole platform and surrounding fold belts, or for some of its individual areas. The region under discussion includes the Ukrainian shield, Finland, the Scandinavian Peninsula and the Caucasus. Subdivision of the Precambrian rocks of the plate cover is far from satisfactory, although they are known from some hundreds or even thousands of drill holes.

The first preliminary attempts at correlation of the Precambrian rocks of the whole platform and adjacent areas were attempted by Salop (1958a, b), and by Solontsov (1960). These were not based on isotopic dating, since at that time such data were few. In 1961 the Interdepartmental Stratigraphic Committee of the U.S.S.R. adopted a correlation scheme for the Upper Precambrian of the U.S.S.R., and in 1962 it adopted a more detailed correlation scheme for similar rocks of the Russian (East European) platform. It was published in 1965 (Resolution, 1965); then, in 1968 correlation charts for all the Precambrian rocks of the U.S.S.R. (including the East European platform and the Urals) were compiled by Salop in collaboration with the V.S.E. G.E.I. geologists (1968b). In this work, as in all the previous ones, some grave mistakes were made, mainly because of the incomplete nature of the geological and geochronological data and because of misinterpretation of isotopic age determinations and paleontological data. Because of revision of geochronological boundaries of the Precambrian eras it is necessary to revise the subdivision and correlation of the older strata in Eastern Europe. Many chapters in the second part of this book deal with this problem. When compiling a new scheme of Precambrian correlations in the East European platform and surrounding fold belts (Table I*) many new data on the forty-six major sections of the older strata were used. Their locations are shown on Fig. 5.

The Siberian platform is the most important structural element in Asia and the largest cratonic block of the Asian continent. Its area is a little less than that of the East European platform and much greater than that of the Indian platform. The modern Siberian platform is bounded by surrounding fold belts of different ages; the Upper Paleozoic Taymyr, the Mesozoic Verkho-

---

*Tables enumerated by Roman numerals (*I—V*) are not incorporated in the text, but are enclosed as inset plates at the end of the book.

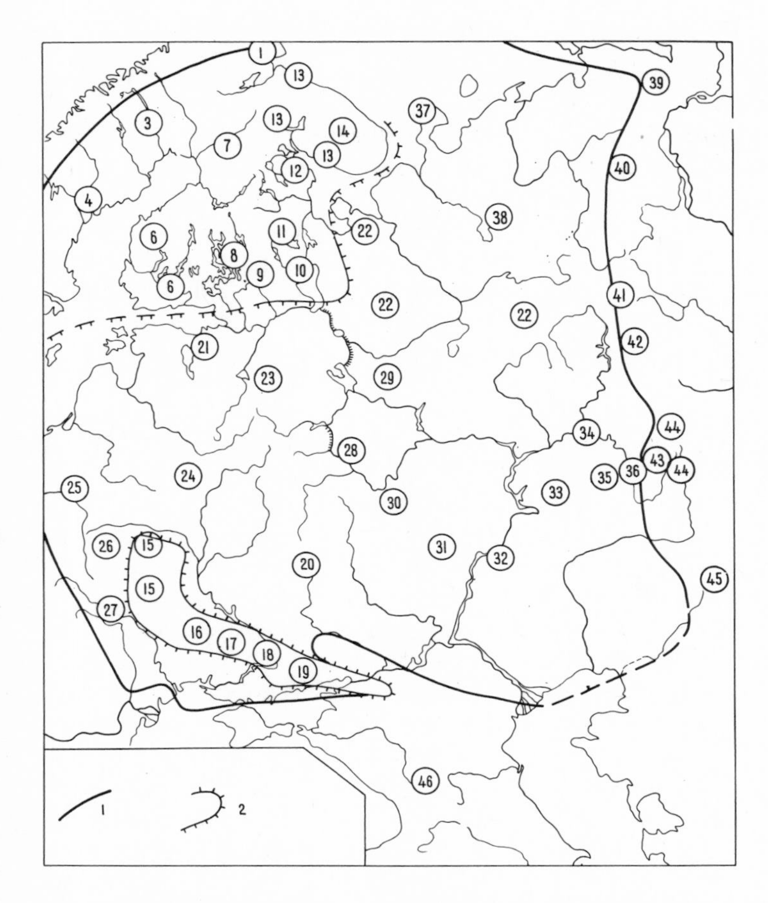

Fig. 5. Stratigraphic position of the principal sequences of older strata shown in the correlation chart for the Precambrian of the East European platform and surrounding fold belts (Table I).
1 = platform boundaries; *2* = shield boundaries; figures inside circles show the locations of the sections in the correlation chart. Sections 2 and 5 are situated beyond the western margin of the diagram.

yansk, the Protozoic Stanovoy fold belt, which was strongly activated in the Paleozoic and Mesozoic and the two Upper Protozoic to Lower Caledonian (Baikalian) Baikal and East Sayan fold belts. In the west the platform is overlain by sediments of the West Siberian plate.

The Precambrian basement of the Siberian platform is exposed only on its margins — to the north within the limits of the Anabar shield, and in a number of small uplifts (Udzhins, Kuoyka—Daldyn, Olenek and Kharaulakh), to the southeast within the limits of the Aldan shield and Yudoma—Maya uplift, to the south in the uplift of the Sayan region, and to the west in the

Yenisei anteclise and the Khantai—Rybninsk uplift. The Precambrian formations of the latter two structural areas form part of the Yenisei fold belt which was stabilized at the end of the Precambrian (just before the Cambrian Period started). It is for this reason that they are considered as part of the platform basement, and not as part of the surrounding fold belt.

Precambrian strata in the interior parts of the platform lie at great depth. They are exposed only in a few drill holes on the periphery of the platform, in marginal uplifts of the platform, mainly in the southern part of the Irkutsk amphitheatre and on the northern slopes of the Aldan shield (anteclise).

Older rocks are more extensive in the fold belts that surround the Siberian platform than in those that surround the East European platform. Precambrian rocks in all the fold belts of Eastern Siberia (with the exception of the Verkhoyansk belt) make up a large part of the area. They are well exposed and the sections are very complete, so that they are suitable for geological (and stratigraphic) studies.

Geological investigation of the Precambrian in Eastern Siberia (especially in southern areas) is now advanced to the point where it is as well known as such "classical" Precambrian areas as the Baltic or Canadian (Canadian—Greenland) shields. Many features of that region are very favourable to the study of the Precambrian, so that it has been possible to advance some new concepts of Precambrian geology using data obtained from Siberia. In this region supracrustal strata are widespread, and exposure is good. In a number of areas there is a clear transition from platform-type deposits to those of geosynclinal aspect. Also, fossiliferous Lower Cambrian rocks are widespread. Thus, in the Baikal mountain area Precambrian mio- and eugeosynclinal belts were first differentiated. Within these fold belts some zones and subzones with very particular rock types were recognized. Transitions between geosynclinal- and platform-type deposits were established, and some general unidirectional trends in the genesis of Precambrian sediments were recognized (Salop, 1958b, 1964—1967).

Later a similar classification was carried out for many other Precambrian areas of Siberia. In studies of the Precambrian of the Aldan shield a new peculiar type of fold system was recognized. These were called "gneiss fold ovals". They are quite characteristic of the older (Archean) complexes of many shields in different continents (Salop, 1971b). The organic origin and stratigraphic importance of microscopic remains which are here called microphytolites were proved for the first time in Upper Precambrian deposits of Siberia (Reitlinger, 1959; Zhuravleva, 1964), and the use of biostratigraphic techniques in Precambrian studies was based essentially on determinations of different complexes of stromatolites (Semikhatov, 1962; Korolyuk, 1963, 1966; Komar, 1964, 1966; Nuzhnov, 1967) in rocks from this region.

There are many available data, therefore, on which to base a scheme of subdivision and correlation of the older rocks of the Siberian platform and its surrounding fold belts. Previously, in the absence of these modern correlation techniques this problem was practically insoluble. However in the 1930's V.A. Obruchev (1935—1938), pioneer in the study of the Precambrian in Siberia, successfully correlated the older rocks of Southern Siberia on the basis of lithological comparison and degree of metamorphism. Later the same basis was used by S.V. Obruchev (1958, 1963), and Salop (1958a, 1963) who also took into account some isotopic age determinations. More recently these older stratigraphic correlations were also based on lithoparagenetical ("formational") analysis (Salop, 1964). Considerable advances in correlation of the older strata were made following the introduction of biostratigraphic (paleontological) methods.

The first working scheme for the correlation of the Precambrian in major areas in Siberia, based on modern correlation techniques, was adopted at the Interdepartmental Meeting on Correlation of the Precambrian and Cambrian Terrains of Middle Siberia, held at Novosibirsk in 1965. This scheme was useful in Precambrian studies of some areas. It was also a success in terms of technique, and it signified a new stage in investigation of the Precambrian. The proposed scheme, however, also had some errors and inconsistencies.

In 1966 Salop compiled correlation charts for Precambrian strata throughout the U.S.S.R., eliminating some of the previous errors. These charts, published in 1968 (Salop, 1968a), are still largely valid today. However, in the light of new data, some of the statements, especially on the age and position of some units in the correlation scheme, had to be revised.

Correlation of major sections of Precambrian supracrustal assemblages of the Siberian platform and surrounding fold belts, based on various correlation techniques including new geological and geochronological data is shown in Table II. Locations of the major sections shown in this chart are given in Fig.6.

The greater part of the second largest ancient craton of Asia, the Indian platform, is underlain mainly by Precambrian rocks. In the southeastern part of peninsular India the Precambrian is exposed in a vast territory comparable to the Baltic shield. Extensive exposures are present in the northwestern part, and in the northern part of the peninsula in Bundelkand, and in Kaimur. There is also a thick cover of Cenozoic rocks and Upper Cretaceous basalts in a wide zone of the Himalayas piedmont, and in a vast region forming the northwestern part of the Deccan Plateau. In this region some Paleozoic and Mesozoic sediments of platform type overlie the Precambrian rocks but these are insignificant by comparison with other platforms.

Metamorphic Precambrian rocks are exposed among younger rocks in the Himalayan fold belt bordering the Indian platform in the north. These rocks are poorly studied and hardly subdivided and there are few isotopic analyses. For these reasons these sequences are not presently useful for elucidation of general problems of Precambrian stratigraphy.

The Precambrian of the Indian platform is less well known than that of the other platforms of the Northern Hemisphere. Recently, however, some new and comprehensive material has become available, and some new features have been described that may be of great importance in the establishment of a universally applicable subdivision of the Precambrian.

Unfortunately, the results of many of these new studies are either not published or are known only from synopses. Thus the compilation of a correlation chart for the various areas of India is difficult at this time. This aspect will be dealt with in a later work. The Precambrian geology of India bears many similarities to that of the Southern Hemisphere.

The makeup of the Precambrian of peninsular India is identical in many ways to that established for other larger Precambrian areas. The trends observed elsewhere, and used in establishing a scheme for subdivision of the Precambrian stratigraphic scheme are true for all the platforms of the Northern Hemisphere.

Important exposures of Precambrian strata are present in the Central Asiatic fold belt area in a vast region between the Siberian and the Indian platforms, but unfortunately, stratigraphic studies have just begun in this area. The Precambrian rocks are well studied and differentiated from the widely developed Paleozoic rocks only in the western and northwestern margins of the Central Asiatic fold zone, in Central Asia, Kazakhstan, and the Altai—Sayan fold belts. The local nature of some of the Precambrian strata and the poor amount and quality of isotopic dates lead to different interpretation of their relationships and there are many different correlations.

The Precambrian is widely developed and well studied in some provinces of China, but unfortunately, new data are not available at this time, and the old ones are somewhat suspect, mainly because most of the isotopic dates (which are relatively scarce) were done by K—Ar analysis. The situation is a little better in the case of the Precambrian of North Korea.

Exceptionally valuable material for establishing a global Precambrian stratigraphic scheme is provided by Precambrian sections of the New World in the vast North American platform and surrounding orogenic belts which include virtually the whole of North America and Greenland. In addition to the large Canadian, or more precisely, the Canadian—Greenland shield, the North American plate includes the Arctic region, the Great Plains, the Eastern Rocky Mountains and the Midcontinent. In the west and southwest, the platform is framed by the huge Alpine (Mesozoic—Cenozoic) fold belt of the North American Cordillera, to the north by the Caledonian—Early Hercynian Innuitian fold belt, in the east by the Caledonian East Greenland fold belt and in the southeast by the Caledonian—Hercynian Appalachian fold belt.

The Precambrian is widely developed in all of these structural elements of the continent. They are most extensive in the area of the Canadian—Greenland shield. The part of the shield which occupies the northern part of the North American continent is probably one of the most extensive continuous areas of Precambrian rocks in the world. It is only in the Hudson Bay region that older strata are largely covered by water and Paleozoic platform deposits

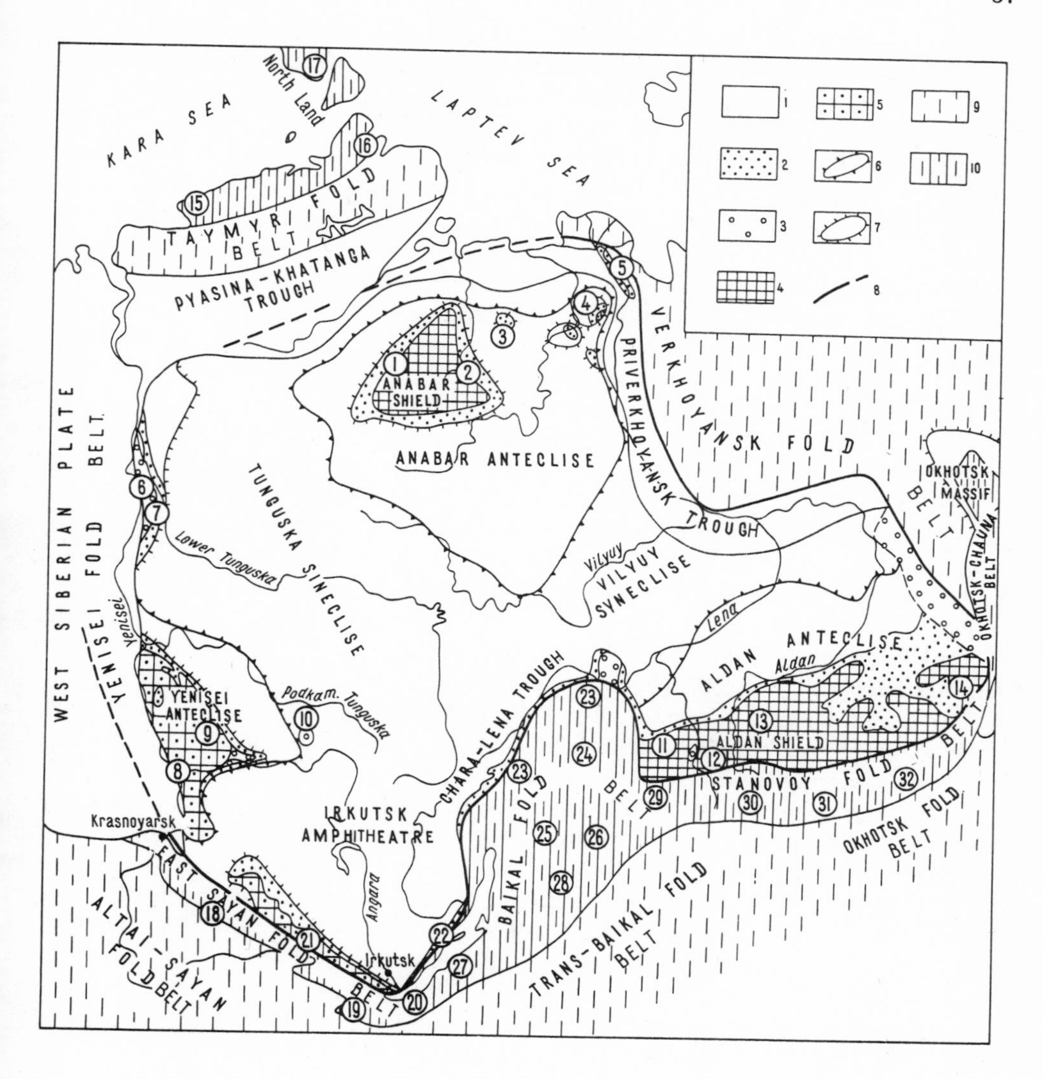

Fig.6. Location of the principal sequences of older strata shown in the correlation chart of the Precambrian of the Siberian platform and surrounding fold belts (Table II). *1* = Phanerozoic platform cover; *2* = Precambrian platform cover; *3* = Precambrian subplatform strata of aulacogen type; *4* = outcrops of Archean basement in the platforms (shields); *5* = outcrops of Protozoic folded basement in the platform; *6* = boundary of anteclise within platform; *7* = boundary of platform uplifts; *8* = platform boundary; *9* = folded areas surrounding the platform; *10* = Precambrian outcrops in fold belts surrounding the platform; figures in circles show the locations of the sections in the correlation chart.

(Hudson Bay syneclise). The eastern part of the shield is less well exposed. The Baffin and Labrador seas hide close structural relationships between the Precambrian basement of the North American continent and Greenland. Greenland is almost completely ice-covered except for a narrow coastal strip.

In the North American plate Precambrian rocks everywhere form the basement, and, to some extent, make up the lower part of the platform cover. Many good exposures are located on the southwestern margin of the platform, in the Eastern Rocky Mountains and Colorado Plateau, where the ancient platform was reactivated, faulted and uplifted during the Mesozoic—Cenozoic diastrophism, which strongly affected the adjacent geosynclinal system of the Cordillera. Minor exposures of older rocks are present in the Arbuckle and Uichita Mountains and also in the Ozark Plateau, and in Texas in the Midcontinent area. Throughout a vast area of the Great Plains the Precambrian is deeply buried and covered by thick Phanerozoic deposits.

In the Cordilleran fold belt Precambrian strata are widespread in the Rocky Mountains (Purcell, Absaroka, Great and Little Belt, Beartooth, Wasatch, etc.), in Southern Nevada and adjacent areas of California, in Central Arizona (Mazatzal Mountains) and in some other regions. Here they are exposed in the cores of large anticlinoria or form horst structures in younger folded Phanerozoic terrains. It is probable that some of the highly altered supracrustal and plutonic rocks present in a wide band along the Pacific rim in the Coast Ranges (British Columbia) and presently regarded as metamorphosed Paleozoic or even Mesozoic rocks, are in fact largely Precambrian. However, their age and structure are still poorly known.

In the Innuitian fold belt (Canadian Arctic Islands) Precambrian exposures are reported in a small region in the northeastern part of Ellesmere Island. In the Eastern Greenland fold belt these deposits are widespread. Finally, in the Appalachian fold belt they are exposed in the southeastern part of Newfoundland, in some localities of Nova Scotia and New Brunswick, in the Blue Ridge and in the Appalachian Valley and Ridge.

The most important regions for the establishment of a Precambrian stratigraphy are in the Canadian—Greenland shield. This is a classical region for the Precambrian. Many schools of North American, Canadian and Greenland (Danish) geology were established by the students of the Precambrian in these regions. Many of the names widely used in subdivision of the Precambrian were proposed by American and Canadian geologists.

The Precambrian strata in North America as well as in other areas of the world are not all equally well studied. Most detailed studies have been carried out in densely populated mining areas (Southern Canada and adjacent areas of the U.S.A., Appalachians, Colorado Plateau and some parts of the Rocky Mountains). Less-detailed studies have been done in Northern Canada (with the exception of the Great Slave Lake area and the Labrador trough), and still less is known about Arctic Canada. There is currently a great deal of interest in the field of Precambrian geology and this has encouraged geological mapping and other studies in northern and Arctic regions. It is hoped that a more even coverage will soon be available. Isotopic analyses have played a special role in Precambrian studies of North America. Considerable advances have been made in this field in Canada and the U.S.A. At the moment many thousands of isotopic dates have been obtained and in recent years the quality of these analyses has been very high. In many cases these data permit true interpretation of various older data (including K—Ar analyses).

Correlations by lithoparagenetical ("formational") and paleontological methods are insufficiently used in North America and Greenland. Although both American and Canadian geologists have studied stromatolites for a long time, these are hardly used for interregional correlation. The classification and methods used in their studies are to a large extent different from those proposed by the Soviet workers. Some Canadian attempts to use the Soviet methods of study and classification of these fossils have been made (Hofmann, 1969). These did not appear to produce good results in regard to stratigraphic correlation. However, some attempts at comparison of stromatolites from the Precambrian of Canada and the U.S.A. with those of the U.S.S.R. were made by some Soviet workers (I.N. Krylov, M.A. Semikhatov). The microphytolites of North America and Greenland have hardly been studied, but spherical and filamentous microscopic remains, which mostly represent blue-green algae, have been studied in some detail in the Precambrian of these regions. These fossils are of some help in interregional correlations of Precambrian strata.

Up to the moment there has been no attempt at detailed correlation of the Precambrian rocks over large areas of the continent, or throughout the continent as a whole. There are many publications on Precambrian rocks of North America and Greenland, and in many cases they are of a very high scientific calibre. At best there are some generalizations regarding possible correlations in different areas. Some correlation charts have been compiled for rather small regions which have common formations, or correlation charts showing several sections of one group have been produced. In compiling the new geological map of Canada (1969) the authors made some preliminary correlation of the Precambrian strata, but a correlation chart was not published, and the stratigraphic correlations shown on the map by colours and symbols are rough, and in many cases debatable.

The Precambrian correlation charts compiled for this book include most of the North American continent and Greenland (Table III). As a basis for these charts original works were consulted. The most important of these are given in the reference list. This chart comprises correlation of seventy-eight sections in different areas. These include the major, typical Precambrian sections of the continent and Greenland, but some other sections are included for purposes of comparison and for elucidating some problems of the Precambrian in different areas. Section locations are shown in Fig. 7.

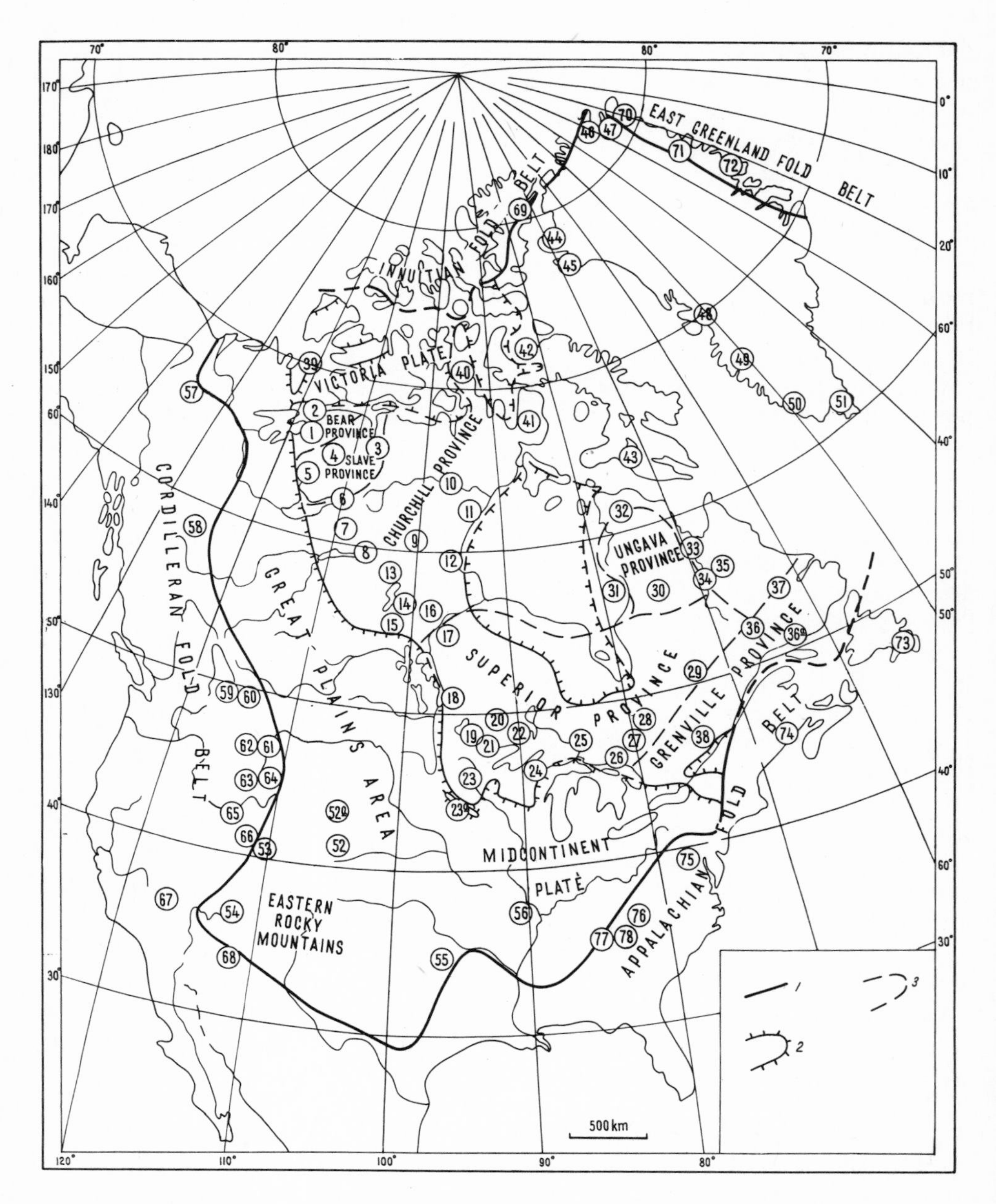

Fig. 7. Locations of the principal sequences of older strata shown in the correlation chart of the Precambrian of the North American platform and surrounding fold belts (Table III). *1* = platform boundaries; *2* = shield boundaries and boundaries of extensive outcrops of Precambrian rocks close to the shield; *3* = boundaries of some tectonic provinces within the shield; figures in circles show the locations of the sections in the correlation chart.

The correlation charts for three major Precambrian regions in the Northern Hemisphere will hopefully inspire some new radiometric age determinations and revision of accepted correlation of these ancient strata. This in turn will lead to the necessary revision of many characteristics of geological evolution, tectonics and exogenic metallogeny of the Precambrian of some areas. This book is dedicated to a certain task and all of these problems cannot be dealt with. The scope of this work does not even permit adequate detailed justification of the adopted classification. This would require a thorough analysis of every section and of many isotopic datings. It is, however, hoped that careful examination of the plates will, in most cases, show the reasons behind the main principles of correlation used in different terrains. In the following chapters the basic ideas shown in the correlation charts are expanded somewhat. The charts themselves are based on thorough analysis of geological and geochronological data.

CHAPTER 4

THE ARCHEAN

This era includes the oldest Precambrian strata, formed more than 3,500 m.y. ago before the Saamian orogeny. In many areas, however, these strata were subjected to polyphase metamorphism associated with various younger orogenic cycles in the Protozoic, or younger fold belts of the Phanerozoic. Thus, in many cases radiometric dating of Archean metamorphic rocks and granites yields an age younger than the true one. In such cases the Archean age of the rocks can be judged by the fact that they may be correlated with analogous strata in adjacent regions where the true age has been radiometrically determined. Usually this type of comparison is fairly reliable, because the Archean strata in different parts of the world, even in different continents, are very similar in composition, stratigraphic sequence, and metamorphic grade. In many cases we can trace the older rocks from an area with "relict" dates to one with "rejuvenated" dates.

In many regions where very old dates, corresponding to the time of original metamorphism or the time of formation of the rocks have been obtained, younger ages are also indicated by radiometric analyses of rocks which have undergone various stages of rock transformation. The "relict" ages are usually obtained by the Pb-isotope, Pb-isochron and Rb—Sr isochron methods. K—Ar dating usually gives "rejuvenated" ages. Sometimes K—Ar dating also reveals the time of early events, but in cases where successive metamorphic episodes were very intensive and were accompanied by loss and gain of elements, all the methods yield only the time of the latest events. Such "isotopic rejuvenation" can usually be detected because the Archean rocks are commonly unconformably overlain by younger metamorphic Precambrian strata, which yield the same age as the underlying rocks. In such cases the isotopic age obtained from the Archean rocks must represent the time of the latest events.

Regional Review and Principal Rock Sequences

Archean rocks are widely distributed in many areas of the Northern Hemisphere, but they are mostly exposed within the limits of the shields, and this material is of major importance in studying their stratigraphy. The main characteristics of the Archean strata of different areas are discussed below, together with discussions on their age. The conclusion of the chapter is a summary of the main characteristics of the era.

Europe

The oldest Precambrian strata are represented in Europe by various highly metamorphosed supracrustal complexes and plutonic rock types (mostly granites). These are mainly developed in the East European platform. They are exposed in large areas of the Baltic and Ukrainian shields, and are known from drill holes or by geophysical work in many localities of the Russian plate, in particular in the North and East European part of the U.S.S.R., where they are present as large continuous massifs in the platform basement. Archean strata are less extensive in the Phanerozoic fold belts, but are exposed in some median massifs or in the cores of some large anticlinoria and in some tectonic blocks. The stratigraphy of the Archean complexes is only well known in shield areas.

In spite of many detailed studies the Archean stratigraphy of these regions is less well known than in the Aldan shield of the Siberian platform, where the world stratotype for the Archean is exposed. This is largely due to the fact that in the Baltic and Ukrainian shields Archean strata were reworked by later tectonic movements and metamorphism, which made their structure still more complicated, and in some cases masked relationships with the overlying Precambrian complexes. These younger events also caused "rejuvenation" of the isotopic ages.

One of the most complete sections that may be accepted as a stratotype for the *Baltic shield* Archean and for the East European platform was described by Bondarenko and Dagelaysky (1968) from the central Kola Peninsula. The formations of that region belong to the Kola Group. There are three formations (upwards): Pinkel'yavr, Chudzyavr and Volshpakh.

The Pinkel'yavr Formation (between 600 and 2,000 m thick?) is characterized by interbanding of migmatized garnet- and sillimanite-bearing biotite gneisses, pyroxene granulites, two-pyroxene—hornblende schists and pyroxene amphibolites. It also contains some horizons of magnetite quartzites, magnetite-bearing amphibole and pyroxene schists. The Chudzyavr Formation (500 m thick) is mainly represented by amphibolites and basic schists with rare interbands of magnetite schist. Many rock types are characterized by high CaO content, and for this reason this formation is also called "the formation of rock rich in Ca". The Upper Volshpakh Formation (>2,000 m thick) is composed of alumina-rich garnet—sillimanite, garnet—biotite and some other garnet-bearing gneisses and schists, commonly interbedded.

In the present stratigraphic scheme there is an "ultrametamorphic complex" composed of various migmatites, granite gneisses, diorite gneisses, and charnockites in particular. This complex is placed at the base of the Archean section by Bondarenko and Dagelaysky, and is considered to be separated from the overlying rocks by an unconformity. The existence of this unconformity, however, is not proved, and it is probable that this ultrametamorphic complex represents highly granitized rocks (largely basic schists and gneisses) of the Pinkel'yavr Formation.

The basal stratigraphic position of the Kola Group in the Precambrian of the Kola Peninsula is indicated by the presence of basal conglomerates of the unconformably overlying Paleoprotozoic Tundra Group, metamorphosed 2,600—2,700 m.y. ago. Several K—Ar dates on biotites and one on amphibole from the Kola Group rocks and intruding granites, show a wide scatter from 1,700 to 3,590 m.y. Possibly even the oldest of these dates are related to the period of original metamorphism of the Archean rocks and the other figures are due to some later remobilization processes. The Pb-isochron (Pb—Pb) method yielded an age of 3,200 m.y. for the Kola Group gneisses (Maslenikov, 1968). Zircon from the older granite gneiss (Voronya River) gave 3,300 m.y. (Gerling and Lobach-Zhuchenko, 1967) on the isochron (concordia) of Ahrens—Wetherill. Zircon from granulite of the Kola Group gave an age of 2,740 m.y. on a concordia diagram (Bibikova et al., 1973). The Rb—Sr isochron method gave an age of about 2,700 m.y. for these gneisses. As was stated by the authors, this date "may possibly be regarded as the time of retrogressive metamorphism of the amphibolite facies" (Gorokhov and Gerling, 1971, p.68).

The Kola Group is reliably compared with the granulite complex of the Kola Lapland and Belomorskaya (White Sea) Group of Karelia and the southern Kola Peninsula (Salop, 1971a). Thus, the lower, highly granitized part of the Pinkel'yavr Formation (including the so-called "ultrametamorphic complex") corresponds to the Keret' Formation, which is composed of granitized gneisses. The upper part of the Pinkel'yavr Formation, together with the Chudzyavr Formation, is correlated, on the basis of composition, with the Khetalambino Formation (amphibolites and amphibole gneisses) of the Belomorskaya Group, and with basic rocks of the granulite complex ("basic granulites"). Finally, the Volshpakh Formation is compared with the Loukhi (Chupa) Formation which consists of alumina-rich garnet gneisses of the Belomorskaya Group, and with leucocratic garnet-bearing granulites of the granulite complex ("acid granulites"). Radiometric (K—Ar and Pb-isotope) analyses of the rocks of these gneiss complexes yield a wide scatter from 1,700—1,800 to 2,700 m.y. The age of 2,700 m.y. was obtained from zircon from the Belomorian gneiss (Bibikova et al., 1973). Possibly these ages reflect superimposed Paleoprotozoic and Mesoprotozoic metamorphism.

Only one amphibole date from metabasite cutting the Belomorskaya Group gave a result close to the time of the Epiarchean diastrophism. It gave an age of 3,300 m.y.

Granite gneisses (with a ghost stratigraphy of supracrustals) underlie Paleoprotozoic sedimentary and volcanic strata (the age of which is determined by both geological observations and radiometric dating) and are attributed to the Archean in Northern Karelia.

The granulite complex extends from the western part of the Kola Peninsula into Finnish Lapland (Lappi), the Belomorskaya Group and into more southerly areas of Northern Finland, where it is designated the "Heta and

Tunsa—Savukoski gneiss complex". These terrains have their counterparts in some other areas. In Southern Finland there is the Usimaa—Turku—Raumo gneiss—granulite complex. In the southwestern part of Sweden there are the so-called "pre-Gothian gneisses" which include widely developed pyroxene gneisses, schists, amphibolites and quartzites included in the gneisses. In Southern Norway these rocks include the Østfoll gneiss—granulite complex. The basis for attributing them to the Archean is given in a separate paper (Salop, 1971a). The Archean age of all these complexes is based on their low stratigraphic position in the section in relation to the unconformably overlying Paleoprotozoic strata. These rocks also bear a striking similarity to the gneiss complexes of the Kola Peninsula and Belomor'ye (White Sea area). Radiometric dating usually reveals highly "rejuvenated" values.

Highly metamorphosed supracrustal rocks which can be correlated on the bases of stratigraphic position, isotopic age and composition, with Archean strata of the Baltic shield, are widely developed in the Ukraine. In the southwestern part of the *Ukrainian shield* they are known as the "Bug Group" (Bobkov et al., 1970) or "Pobujian complex" (Laz'ko et al., 1970). The stratigraphy of the Bug Group has been studied by many workers who use different names for the various units within the group. The rock succession established by them is considered reliable (Polovinkina, 1960; Drevin, 1967; Laz'ko et al., 1970). The completeness of the section and the extent of the studies done on the Bug Group equal those on the Kola Group, and for these reasons it could serve as the East European stratotype. However, its relationships with the overlying rocks are not as well established as they are in the Kola Peninsula.

In the Lower Bug Group the basic schists or gneisses largely consist of hypersthene and plagioclase from which enderbites and charnockites later developed as a result of granitization (Pobujian Formation). These are overlain by pyroxene—plagioclase schists interlayered with biotite and sillimanite gneisses, amphibolites, and locally, quartzites (Dniester—Bug Formation). Conformably overlying these there are strata of amphibole and pyroxene plagiogneisses with layers of marble, calc-silicate rocks, garnet, sillimanite gneisses and quartzites, together with some lenses of magnetite-rich rocks (Teterev—Bug or Khoshchevataya—Zaval'evo Formation). Possibly the section is crowned with garnet—biotite plagiogneisses, but the relationships with the underlying rocks are not certain.

K—Ar isotopic analyses on micas and accessory minerals of the Bug Group gneiss, and granite cutting these rocks, usually give an age in the range of 1,900—2,200 m.y., but a few Pb-isotope analyses on accessory minerals from granites yield 2,300—2,600 m.y. Whole-rock Pb-isochron analysis of pyroxene plagiogneisses gave an age of 2,750—2,800 m.y., but the same figures are recorded for biotite gneisses of the Teterev Group which unconformably overlies the Bug Group, and belongs to the Paleoprotozoic (see below). These dates were obtained in the laboratory of V.S.E.G.E.I. (A.D. Iskanderova) on specimens from the collection of A.D. Dashkova. Thus, all the figures for the age of the Bug Group rocks are probably related to later metamorphic events. Recently A.D. Iskanderova (V.S.E.G.E.I. Laboratory) produced a Pb—Pb isochron analysis on marble from the Teterev—Bug Formation. It yielded 3,600 ±800 m.y. The model age of these rocks, determined by Hauterman's method, is 3,300 m.y. The same result (that is a Pb—Pb isochron age of 3,600 m.y.) was reported by Ukrainian geochronologists (Eliseeva et al., 1973).

In the central part of the Ukrainian shield the Ros'—Tikich Group corresponds to the Bug Group. In the former the following formations are recognized: the Volodarsk Formation (metabasite) is quite similar to the Pobujian and Dniester—Bug Formations; and the Upper Belotserkovsk Formation, composed of metabasitic calc-silicates, corresponds to the Teterev—Bug Formation. Some interlayers and lenses of magnetite silicate rocks are reported in metabasites of the Volodarsk Formation.

To the east, within the limits of the so-called Kirovograd block, the Ingul Group is assigned to the Archean. Several formations are assigned to this group. The Reyevskaya strata which consist mainly of amphibolitic rocks, and the Mayaksk and Zelenorechensk Formations, which are mainly composed of highly altered metabasites (basic schists, amphibolites and gneisses), occur as bands and ghost relics among ultrametamorphic granite gneisses. Magnetite silicate rocks and stratified magnetite-quartzite interbeds are less abundant than the metabasites. The composition of the Ingul Group corresponds to that of the Volodarsk Formation and to two lower formations of the Bug Group. K—Ar and Pb-isotope ages of minerals from the Ingul Group rocks are "rejuvenated", and yield a wide range of ages from 1,800 to 2,750 m.y. The oldest age is given by Pb-isotope analyses on zircons from biotite and hypersthene plagiogneisses. Belevtsev et al. (1971) suggested that these figures represent the age of parent rocks from which the zircon was derived. They suggested that it was deposited in the form of a sediment and that the sedimentary rocks were later altered into gneisses. This supposition is doubtful because detrital or authigenic zircons when dated in such highly metamorphosed rocks usually give the age of late intensive metamorphism. For example, spheroidal (detrital?) zircon from acid granulites in the Kola Peninsula yielded an age of about 1,900 m.y. (Tugarinov et al., 1968), corresponding to the time of the Karelian orogeny. These granulites are considered to be Archean because they are unconformably overlain by the Paleoprotozoic Tundra Group (older than 2,800 m.y.) and the Mesoprotozoic Pechenga Group (older than 1,900—2,000 m.y.) (Salop, 1971a). The hypersthene plagiogneisses were probably not formed from sedimentary rocks, but rather from basic or intermediate volcanics, so that the presence of detrital zircon in them is unlikely.

Many workers (Kalyaev, 1965; Dobrokhotov, 1967; Kalyaev and Komarov, 1969; Belevtsev et al., 1971) compare the Ingul Group (that is the Zelenorechensk and Mayaksk Formations) with the slightly metamorphosed iron-rich rocks of the Mesoprotozoic Krivoy Rog Group. Well-reasoned criticism of such views was presented by Grechishnikov (1971). The compared groups differ not only in their stratigraphic succession and degree of metamorphism, but also in containing different types of iron-rich rocks. In the Ingul Group these rocks are interlayered with metabasites (amphibolites and hypersthene plagiogneisses), and finely banded varieties of jaspilite are virtually absent, whereas in the Krivoy Rog Group the iron-rich rocks occur among paraschists (phyllites), metabasites are lacking and they are typically finely banded jaspilites.

In the Dnieper region the Orekhov—Pavlograd (Aul, Pridnieper) Group is considered to be Archean. The group comprises two very thick formations: the lower one, Novopavlograd Formation, is composed of amphobolites, amphibole or pyroxene—amphibole gneisses, crystalline schists with sheet-like bodies of metaultrabasites, interbeds of hypersthene—magnetite and amphibole—magnetic schists and coarsely banded magnetite quartzites. The upper one, the Orekhov Formation, is made up of different types of gneiss, largely biotite and biotite—amphibole gneisses, but with some sillimanite-bearing varieties with calc-silicate bands (Bobkov et al., 1970). Highly granitized rocks of the Orekhov—Pavlograd Group are sometimes known as the "Orekhov ultrametamorphic complex".

Gneissic rocks of the Dnieper region are similar to Archean strata of the central and western parts of the Ukrainian shield discussed above. The Novopavlograd Formation is comparable to the metabasites of the Ingul Group, and the lower parts of the Ros'—Tikich and Bug Groups (Table I). The Orekhov Formation may be correlated with the Belotserkovsk Formation of the Ros'—Tikich Group and the carbonate rocks of the Teterev—Bug Formation of the Bug Group. The Orekhov—Pavlograd Group is the only one in the Archean of the Ukraine for which there are relict K—Ar dates giving the time of early (Epiarchean) metamorphism. In addition to the "rejuvenated" ages (mainly in the range of 1,800—2,300 m.y.) for this group of rocks (and for the granites within it) these rocks have yielded a figure of about 3,500 m.y. This date was obtained by the K—Ar method on amphiboles from granitoids (gneiss diorite) in the vicinity of the town of Yamburg (Ivantishin and Orsa, 1965), and from amphibolites in the area of the Konsk and Belozersk magnetic anomalies (Ladieva, 1965). These amphibolites are usually assigned to the so-called Konsko—Verkhovtsevo (or Konsko—Belozersk) Group. However, strata of various ages may have been erroneously included in this group. These include Archean amphibolites and gneisses, Paleoprotozoic metavolcanics and Mesoprotozoic sedimentary rocks. It is now certain that in the Belozersk area the amphibolite dated at 3,500 m.y. is older than the Belozersk iron-ore-bearing group and the granites that cut them (data of the Dnieper Geology Trust geologists M.V. Mitkeev and E.M. Lapitsky).

The gneiss complex developed in the Azov region and in the easternmost part of the Ukrainian shield and known as the "Priazov Group" or "Priazov gneiss—migmatite complex" is closely associated with the Orekhov—Pavlograd Group of the Dnieper region. Its stratigraphy is well known (Esipchuk, 1968; Polunovsky, 1969; Usenko et al., 1971), and the various successions described and named by different authors correlate well.

The Priazov Group may be subdivided into three subgroups in ascending order as follows: the Lozovatka, Korsak—Shovkai and Karatysh. The lower one (Lozovatka), corresponding to the Lozovatka Formation in the scheme suggested by Usenko et al. (1971), is very thick (>4,500 m) and has a complex succession. It is composed largely of pyroxene and amphibole gneisses, various basic schists and products of granitization (charnockites, gneissic granites and migmatites). It is comparable to the lower part of the Bug Group (Pobujian and Dniester—Bug Formations) and its correlatives.

The overlying Korsak—Shovkai Subgroup (2,000—4,000 m), corresponding to the formation of the same name in the stratigraphic breakdown used by Usenko et al. (op.cit.), is subdivided into three formations: the Temryuk, Bogdanovsk and Dem'yanovsk (the formation names are after Polunovsky, 1969) which are composed of biotite, biotite—amphibole and pyroxene gneisses (migmatites) with marble, calc-silicate rocks, granitic gneiss, magnetite silica schists and rare magnetite-quartzite horizons. In the Temryuk Formation bands of corundum gneiss are reported, and some rocks of the Dem'yanovsk Formation are apatite-rich. This unit is correlated with the carbonate-bearing parts of the Bug Group. The Karatysh Subgroup is situated still higher in the section (in the succession proposed by Polunovsky this unit is called the Karatysh gneiss—migmatite complex). It is characterized by widespread development of migmatite and gneissic granite formed after garnet—biotite gneisses. It is similar to the upper garnet—biotite gneiss unit of the Bug Group.

There appears to be a metamorphosed conglomerate with granitoid and amphibolite pebbles among the gneisses forming the upper part of the Priazov Group. However, the exact nature of this unit, and the exact stratigraphic succession are not certain. Some geologists do not consider it to be a conglomerate, but a tectonic breccia, while others are of the opinion that it is a conglomerate, younger than the Priazov Group.

K—Ar age determinations on micas and amphiboles from gneisses of the Azov region gave a wide scatter of results from 1,900 to 2,860 m.y., and the dates indicating the age of the original metamorphism have not yet been obtained.

Archean rocks are known from drill holes in many areas in the basement of the *Russian plate*, but most of the holes are shallow so that there are insufficient data to work out the stratigraphy of the older strata. The Oboyan Group, consisting of gneiss, amphibolite and migmatite with iron-silicate ores (associated with amphibolite), in the area of the Kursk magnetic anomaly, is attributed by some to the Archean complex. The Oboyan Group is transgressively overlain by the Paleoprotozoic volcanic Mikhaylovsk Group (Polishchuk, 1970). Granitoids among the gneisses (of the Saltykov and Yakovlev complexes) yield K—Ar ages in the range of 2,000—2,800 m.y. It is certain that all these figures represent later thermal processes, because an age of 2,700 m.y. was obtained from amphibole from rocks of the overlying Mikhaylovsk Group, and 2,730 m.y. for pyrite from rocks of the still younger Kursk Group (Tugarinov and Voytkevich, 1970). In several other areas of the Russian plate (in Eastern Byelorussia the Neman complex, Eastern Poland, Volyn'—Podolia and in a vast region in the eastern part of the plate) gneiss complexes discovered in drill holes are probably Archean. Different types of gneisses are common, including hypersthene plagiogneisses, migmatites, granulites, amphibolites, basic schists, charnockites and gneissic granites. These are all common and characteristic rocks of the older Precambrian.

In the Phanerozoic fold belts of Europe, Archean strata are reliably recognized only in the Caledonides of Great Britain and Norway, and also in the Hercynides of Central Europe (Bohemian Massif), in the Urals and, possibly, to some extent, in the Alpine belt in the southern part of the continent.

In *Britain* older Precambrian rocks are exposed in the northwestern part of Scotland and in the Outer Hebrides in the so-called northwestern craton which was the foreland of the Caledonian orogeny, and also in the marginal parts of that fold belt. The Archean rocks there form part of the polymetamorphic Lewisian complex (Sutton and Watson, 1951; J.G.C. Anderson, 1965; Dearnley and Dunning, 1968). The Archean in this region is represented by various grey biotite, amphibole and pyroxene plagiogneisses, amphibolites, granulites and metahyperbasites, metamorphosed and granitized under granulite and amphibolite facies and strongly deformed during pre-Scourian diastrophism. The first orogenic deformation was accompanied by development of both concordant sheets and cross-cutting dikes which were folded during later Scourian diastrophism, and accompanied by metamorphism and formation of various palingenetic rocks and migmatites.

The age of the Scourian metamorphism is estimated to be 2,600—2,700 m.y. on the basis of dates obtained by several isotopic methods (including the Rb—Sr isochron method). It is for this reason that the pre-Scourian orogeny is considered to be more than 3,000 m.y. old (Dearnley and Dunning, 1968). Detailed metamorphic, tectonic and magmatic studies permitted establishment of two more Precambrian orogenic cycles superimposed on the older one in the gneiss—granulite complex in Scotland. These are the Inverian (1,900—2,200 m.y. ago) which was accompanied by folding, emplacement of potassic granites and formation of a second generation of pegmatites, and the Laxfordian (approximately 1,600—1,720 m.y. ago?). Folding was accompanied by granitic intrusion, generation of a third pegmatitic phase and amphibolite-facies metamorphism which caused retrograde alteration of the pre-Scourian gneisses.

Probably some highly metamorphosed rocks in the Northern Highlands of

Scotland are also Archean. These rocks are commonly regarded by British geologists as strongly altered correlatives of the Protozoic Moine Group rocks, but due to the complex structure of the area their relationships with either the Moine or the Lewisian complex are not clear. Some geologists believe that the Lewisian gneisses commonly form the cores of folds among the Moine schists.

In the *Caledonides of Norway* there are many areas composed of rocks metamorphosed under granulite and amphibolite facies. Their composition and genesis are very similar to those of Archean rocks in other parts of Europe, particularly in the Baltic shield. Their stratigraphic position, as in the Highlands, is obscured by complex folding, and for that reason they are often considered to be Lower Paleozoic. In some areas, however, the Precambrian age of these rocks has been established by radiometric methods, in spite of the fact that they were strongly reworked by Caledonian, or earlier, folding and by metamorphism. Thus, in the Lofoten Islands granulite-facies gneisses yielded Rb—Sr isochron ages of 2,800±85 m.y., and 2,495±210 m.y. The same rocks, under amphibolite facies, gave figures in the range of 1,705—1,840 m.y. (Griffin and Heier, 1969). Probably all of those figures are related to different stages of superimposed metamorphism (Kenoran and Karelian cycles of diastrophism). According to B. Windley (personal communication, 1973) new dating of gneiss from the Lofoten Islands by Heier, using Rb—Sr isochron analysis, has yielded an age of about 3,500 m.y.

In the *Bohemian Massif* in the Variscan zone of Central Europe, the Moldanubian complex is probably Archean. It is mainly comprised of sillimanite—biotite and cordierite—biotite gneisses, and migmatites. Amphibolites, granulites, quartzites and marble are less abundant (Zoubek, 1965; Jenĉek and Vajner, 1968). This complex is subdivided into two groups — the lower, Monotonous Group and the upper, Varied Group. The first (several thousand metres thick) is represented largely by gneisses with granulites in the upper part. The second (3,000—5,000 m thick) is composed of stratified gneisses interbanded with some other rock types. It is subdivided into four parts (in ascending sequence): (1) gneisses with quartzites; (2) gneisses with graphitic quartzites and granulites (erlans); (3) gneisses with marbles, calc-silicate rock and granulites (erlans); and (4) homogeneous gneisses (Jenĉek and Vajner, 1968). Formerly it was thought that the Varied Group unconformably overlays the Monotonous one, but recent work does not support this idea. Some workers consider the Bohemian granulites to have been derived from acid volcanics.

K—Ar and some Pb-isotope age determinations of the Moldanubian complex yielded "rejuvenated" values (350—800 m.y. and in a few cases about 1,700 m.y.). Originally these figures were interpreted as proof of the Paleozoic or Late Precambrian age of the rocks. At present Czech workers attribute the Moldanubian complex to the Lower Precambrian (Archean or Lower Proterozoic) in view of the presence of an unconformably overlying thick succession of slightly metamorphosed Upper Precambrian strata. Peculiar folds are present in the crystalline rocks of the Moldanubian complex and its correlatives in the Bohemian Massif. On geological maps of Czechoslovakia and adjacent countries it is apparent that the major fold structures (Fig. 8) are similar to the gneiss fold ovals (amoeboids) which are uniquely characteristic of the Archean (Salop, 1971b). The gneisses and granulites in the North Bohemian Massif, in the Granulite and Rudny Mountains of Saxony (German Democratic Republic), which are similar to those of Bohemia (including charnockites) are probably also Archean.

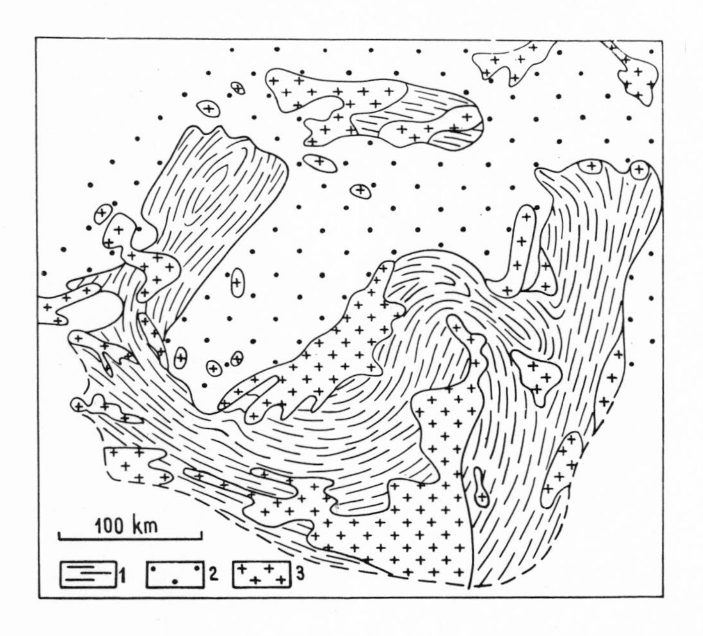

Fig. 8. Orientation of fold structures in the Precambrian crystalline complex of the Bohemian Massif, according to Škvor (1968).
1 = crystalline schist, gneiss and granite gneiss (Moldanubian complex and analogues) of the Archean; 2 = Protozoic and Lower Paleozoic low-grade to unaltered rocks; 3 = post-Archean granite.

The granulite-facies rocks of probably Archean age are exposed north of the pre-Mesozoic crystalline core of the *Pyrenees* and either form part of the platform basement bordering the Mesozoic fold belt on the north, or they may be median massifs (Zwart, 1968). Their age has not been determined either by geological or by radiometric methods.

In the *Hercynides of the Urals*, Archean strata may be present in the area of the Bashkir anticlinorium and in Mugodzhary. In the first area the lower part of the Taratash complex, designated by Smirnov (Abdulin and Smirnov,

1971) as the Arshin Group (complex), is tentatively attributed to the Archean. This group (>1,500 m thick) comprises various gneissic migmatites with predominant biotite, biotite—sillimanite, garnet—biotite and amphibolitic varieties, including sheet-like amphibolitic bodies. Biotite—amphibole—magnetite rocks are also present. Among the gneisses of the group are numerous concordant bodies of gneissic granite and metamorphosed gabbroids. The α-Pb (zircon) age of gneiss migmatites interfingered with the amphibolites is 3,200—3,320 m.y. (Krasnobaev, 1967) but recent data on zircons (Pb-isotope method) yielded lower values (Tugarinov et al., 1970). K—Ar dating revealed that metamorphism of Karelian age (1,900—2,000 m.y.) was superimposed on rocks of the Arshin Group. According to Smirnov, the ferruginous Tukmalin Group (complex), which is possibly Paleoprotozoic, unconformably overlies the Arshin Group.

In Mugodzhary the Kaindin complex is conventionally considered to be Archean. It is composed of various migmatized gneisses and some amphibolites, quartzites and marbles. It occupies the lowest stratigraphic position in the Precambrian section of this area.

Asia

Archean rocks are present in many parts of this vast continent. They occupy extensive areas on both the Siberian and Indian platforms, occur in a number of fold belts bordering the platforms, in the wide Central Asiatic fold belt and in the eastern part of the continent.

In Siberia the older cratonic cores of the Siberian platform are composed of Archean rocks (the Angara, Chara and Aldan cratons). Archean rocks that were strongly affected by later movements and thermal processes are also exposed in fold cores in some fold zones. In some cases they are quite extensive and are thought to represent the basement of deeply eroded older geosynclinal systems (e.g. in the Stanovoy Range fold belt).

Within the Siberian platform, Archean rocks are exposed in the Aldan and Anabar shields and in the Kansk block of the older Yenisei fold belt.

The *Aldan shield* is an exceptionally good area for studying the stratigraphy, lithology and tectonics of the Archean. These rocks are exposed over a vast region (one of the largest in the world) and in many areas they have not been strongly affected by post-Archean folding or metamorphism.

The gneiss complex of the Aldan shield, known as the "Aldan Group", is the most complete and best-studied Archean unit. It is therefore proposed as a global stratotype for the Archean Era. The Aldan Group is subdivided (Salop and Travin, 1971, 1974) in the following manner (from base to top):

Iyengra Subgroup

(1) Kurumkan Formation: quartzites, locally sillimanite-bearing, with some interbeds of sillimanite gneiss (>1,000 m).

(2) Ayanakh Formation: the lower part is pyroxene—amphibole and

amphibole schists and gneisses with quartzite interbeds; the upper part is finely interbedded garnet—biotite, biotite—sillimanite, cordierite—biotite and hypersthene-bearing gneisses with intercalations of two-pyroxene schist, amphibolite and quartzite (1,350 m).

(3) Suontit Formation: quartzite with sillimanite gneiss interbeds at the base and top, and interbanding amphibolite and quartzite in the thickest middel part (1,600—2,000 m).

Timpton Subgroup

(4) Nimgerkan Formation: varied composition, viz., amphibolites (often pyroxene-bearing), amphibole, hypersthene, two-pyroxene gneisses and schists with thin interbeds of quartzites, and garnet-bearing gneisses (1,200—1,300 m).

(5) Ungra Formation: amphibolites, pyroxene amphibolites, two-pyroxene schists, gneisses, enderbites and charnockites (1,900—2,500 m).

(6) Fedorov Mines Formation: basic schists and gneisses (similar to those of the Ungra Formation) with some horizons of limy (diopside-bearing) schists, dipside-bearing metasomatic rocks, marbles, calc-silicate rocks, and locally quartzites (2,300—3,100 m).

(7) Seym Formation: basic schists and gneisses, with, in the lower part, a thick (up to 600 m) unit of garnet—biotite gneiss (1,500—1,800 m).

Dzheltula Subgroup

(8) Kyurikan Formation: rhythmically interbanded garnet—biotite, garnet—pyroxenite, biotite, two-pyroxene, diopside and other types of gneiss with some layers of marble, calc-silicate rocks and graphite gneisses (1,700—2,100 m).

(9) Sutam Formation: monotonous schists, gneisses and leucocratic granulites; the gneisses in some cases contain graphite, sillimanite and cordierite (>2,000 m).

Total thickness of the group is 15,000—16,000 m.

Thick horizons of quartzite are present in the lower part of the group (Iyengra Subgroup). Basic (partly ultrabasic) schists and gneisses are predominant in the middle part of the group (Timpton Subgroup), and garnet-bearing gneiss is the dominant rock type in the upper part of the group (Dzheltula Subgroup). Carbonate units are typical of the upper half of the Timpton Subgroup, and in the lowermost part of the Dzheltula Subgroup. All the rocks of the group, with the exception of the quartzites and marbles, are strongly migmatized and locally granitized. The quartzites are commonly feldspathized (microclinized).

The basement of the Aldan Group is not known. Various rocks of this group, including cross-cutting gneiss granites, are unconformably overlain by less metamorphosed sedimentary and volcanic Paleoprotozoic (Subgan and Olondo), and Mesoprotozoic rocks (Udokan Group and correlatives).

K—Ar ages on amphiboles and Pb-isochron whole-rock ages of the Epiarchean orogeny (plutonism and metamorphism) of the Aldan shield yield about 3,500 m.y. This is "relict" dating (Manuylova, 1968; Rudnik and

Sobotovich, 1968; Salop and Travin, 1974) of the rocks of the Aldan Group and included granitoids. The same methods on both minerals and rocks of this group commonly yield lower values (as young as 1,700 m.y.). These dates have been interpreted as evidence of the younger age of the Aldan Group (Tugarinov et al., 1965b) or of some parts of it (Rudnik and Sobotovich, 1968). This idea is, however, contradicted by all the geological data, which suggest that all the units of the Aldan Group belong to a single stratigraphic complex (Salop and Travin, 1971), and indicate that it occupies the lowest position in the Precambrian succession of Siberia. The fact that Paleoprotozoic rocks unconformably overlying the Aldan Group are cut by granites and pegmatites dated at 2,600—2,800 m.y. old, indicates the great age of the gneiss complex of the Aldan shield. It is likely that the age of 3,500 m.y. indicates the time of original high-grade metamorphism and ultrametamorphism of the Aldan Group rocks, and the lower age values are due to "rejuvenation" phenomena associated with two later episodes of lower-grade metamorphism (2,600—2,800 and 1,900 m.y. ago), and also to some other later processes (Salop and Travin, 1974).

The stratigraphic subdivision of the Aldan Group is based on detailed studies in the central part of the Aldan shield (in the Central Aldan mining area). Recent work in the eastern part of the shield (by L.V. Travin and others) showed that the same units as those in the central part of the shield are present, although they were earlier given different names. These studies also showed that in this region the Archean units above the Sutam Formation are missing.

In the western part of the shield (Olekma and Chara River basins) close correlatives of the Aldan Group (particularly the middle part) are gneiss—granulite complexes which have the following local names: Kurulta Group, Olekma Group, Chara and Tora strata. Their stratigraphic successions are not yet established. Some authors (Frumkin, 1968; Mironyuk et al., 1971) have suggested that the Olekma and Kurulta Groups are younger than the Aldan Group.

Another extensive area of Archean rocks is present in the northern part of the Siberian platform in the *Anabar shield*. The stratigraphy of the older complex of this shield (known as the "Anabar Group") is poorly known because there has been little work on the complex tectonics of the region. On the basis of work by Rabkin (Rabkin, 1960; Rabkin and Lopatin, 1966) and his co-workers, the Anabar Group is subdivided into three conformable subgroups (or groups): the Daldyn, Verkhneanabar and Khapchan.

The Daldyn Subgroup (about 5,000—6,000 m) consists of the Bekelyakh and Kilegir Formations. It is composed of mesocratic and melanocratic two-pyroxene—hypersthene plagiogneisses, granulites with interbanded pyroxene—magnetite schists, and high-alumina schists. A peculiar feature of this subgroup is the presence of significant amounts of quartzites (up to 15%), especially in the Kilegir Formation. In this respect this unit is comparable to the lower (Iyengra) subgroup of the Aldan stratotype. The Verkhneanabar Subgroup (5,000—8,000 m) is composed of monotonous mesocratic pyroxene — largely hypersthene — plagiogneisses, pyroxene schists, amphibolites and charnockites and migmatites formed from them. In both composition and stratigraphic position it is very similar to the lower part of the Timpton Subgroup of the Aldan stratotype.

The Khapchan Subgroup (up to 6,000 m) is subdivided into two formations: a lower one, Khaptasynnakh, and an upper one, Billeekh—Tamakh. The first is made up of bands of biotite—garnet gneisses, diopside—scapolite rocks, calc-silicate rocks and marbles scattered through pyroxene plagiogneisses. The presence of limy rocks and the rhythmic nature of the interbanding are reminiscent of the upper part of the Timpton Subgroup and the lower part of the Dzheltula Subtroup of Aldan. The Billeekh—Tamakh Formation, which for the most part is composed of garnet and biotite—garnet (locally graphite-bearing) gneisses, can be compared with the Sutam Formation at the top of the Archean section in Aldan.

Only K—Ar ages from crystalline rocks of the Anabar shield are available. Dating of micas (biotite and muscovite) usually yields values in the range 1,850—2,000 m.y. from gneisses of the Anabar Group, from muscovite pegmatites and from two-mica granites (Protozoic). Amphiboles from the Verkhneanabar Subgroup yielded an age of 2,300—2,500 m.y. and pyroxenes from the Daldyn Subgroup have an age range of 2,530—2,980 m.y. These figures suggest two later periods of retrograde metamorphism during the Paleoprotozoic and Mesoprotozoic. These metamorphic episodes were accompanied by intrusion of various plutonic rocks.

The Archean age of the Anabar Group is mainly based on reliable comparison with the Aldan Group. This correlation is also supported by the presence, in the rocks of the Anabar Group, of peculiar tectonic structures (gneiss fold ovals) which are characteristic of the Archean complex of the Aldan shield. The upper boundary of the group is defined by the presence of unmetamorphosed platform deposits of the Lower Neoprotozoic (Mukun Group) which unconformably overlie the gneisses.

In the southwestern part of the Siberian platform (Yenisei Ridge) Archean rocks form the major part of the *Angara—Kansk block* — a median mass in the Protozoic Yenisei fold belt. This polymetamorphic complex is known as the Kansk Group. These rocks were first subjected to granulite-facies conditions and in places, to a later period of amphibolite-facies metamorphism. They subsequently underwent retrograde changes under epidote-amphibolite and greenschist-facies conditions. Locally this sequence of events is quite clear.

According to Parfenov (1963) the Kansk Group may be subdivided into three formations (in ascending sequence): the Kuzeyevo, Atamanovka and Kalantat. The Kuzeyevo Formation (2,000—3,000 m thick) is composed of pyroxene—plagioclase schists, and to a lesser extent, of garnet-bearing gneisses (migmatites) which locally contain corundum, cordierite and spinel. The Atamanovka Formation (up to 4,000 m thick) is made up of plagiogneisses (migmatites) that are interbanded with less abundant pyroxene schists. The Kalantat Formation (up to 2,000 m thick) is composed of biotite gneisses and amphibolites with horizons of marble, and calc-silicate rocks. This unit is in the zone of the most intensive retrograde metamorphism, which led some workers to think that this formation is merely an altered part of the Atamanovka Formation. This idea is considered to be erroneous because of the composition of the formation (in particular the presence of carbonate-rich rocks in it, and their absence from the Atamanovka Formation).

Sheet-like bodies of recrystallized anorthosites, pyroxenites and metanorites (?), and also numerous bodies of hypersthene granites (charnockites) are in some places associated with pyroxene schists and gneisses of the Kansk Group.

The Kansk Group has a faulted contact with younger Precambrian strata, including Paleoprotozoic sedimentary and volcanic rocks of the Yenisei Group. The rocks of the Kansk Group are much more intensively metamorphosed.

K—Ar ages from the Kansk Group rocks are always "rejuvenated", and indicate the time of later events. The whole-rock Rb—Sr age on granites and pegmatites in the Kansk gneisses is 2,500 m.y., but the upper age limit of these rocks is not certain (Volobuev et al., 1964). An age of 4,200±500 m.y. was obtained from monazite and zircon from charnockites of the Kansk Group (Volobuev et al., 1970). Because of a high common lead content, calculation of this value was done by Hauterman's method. The precision is very low. If the negative correction is applied, then the age obtained (3,700 m.y.) is close to that of the Archean metamorphism in Aldan and to that of many other Archean regions throughout the world.

Both in lithology and stratigraphic position, the Kuzeyevo and Atamanovka Formations of the Kansk Group may correspond to the lower part of the Timpton Subgroup of the Aldan stratotype. The Kalantat Formation probably corresponds to its upper part (Fedorov Mines Formation) which is characterized by the presence of carbonate-rich rocks.

In the marginal zone of the Siberian platform Archean rocks are present in all the fold belts, with the exception of that at Taymyr where recent work has shown that high-grade metamorphic Precambrian rocks, earlier considered Archean, are in fact Paleoprotozoic and Mesoprotozoic.

In the *East Sayan fold belt*, Archean strata are represented by the Sharyzhalgay and Slyudyanka Groups which are developed in the eastern area in the Sharyzhalgay marginal platform uplift and in the Garga block, and by the Biryusa Group in the Biryusa, Kansk and Arzybey blocks of the western area.

The Sharyzhalgay Group (several thousand metres thick) is subdivided into four formations. The three lower ones (Schumikha, Zhidoy and Zoga) are composed of pyroxene (hypersthene and two-pyroxene), amphibole—pyroxene, amphibole, and biotite (usually garnet-bearing) migmatitic gneisses, schists and amphibolites, in variable proportions. The upper formation

(Kitoy) is different in that it also includes laminae and lenses of marble and calc-silicate rocks. Magnetite silicate ores are associated with basic rocks of the Zoga Formation.

The Slyudyanka Group (several thousand metres thick) is composed of various gneisses (migmatites), basic pyroxene schists and amphibolites intercalated with thick units of marble and calc-silicate rocks. Interbands of diopside—quartz—carbonate—apatite bearing rocks occur locally in the carbonate rocks. The complexity of the tectonic structure has led to various interpretations of the stratigraphic sequence of the group. However, it appears to be divisible into three units which, in ascending sequence, are the Kultuk, Pereval and Kharagol Formations. The middle one is largely composed of carbonate rocks.

Due to the fact that many areas of the Sharyzhalgay and Slyudyanka Groups are somewhat separated from each other (they occur close to a large fault zone), correlations are debatable. It is agreed by most, however, that the Slyudyanka Formation is younger. Some (Buzikov et al., 1964; Dodin et al., 1968) consider this formation to lie above the Sharyzhalgay Formation, and to form part of a single Archean complex, but others (Elizar'ev, 1964; Shafeyev, 1970) have suggested a great stratigraphic break between them, and attribute the Slyudyanka Group to the Proterozoic.

Recent work by A.L. Dodin and V.K. Mankovsky, and A.Z. Konikov (unpublished) tends to support the first idea. These recent studies also support Korzhinsky's (1937) ideas concerning the low-grade metamorphic rocks in the central part of Khamar-Daban. These were included by A.A. Shafeyev in the Slyudyanka Group, but were considered by Korzhinsky to be younger Paleoprotozoic, and to be separated from the Archean complex by wide fault zones and schistose zones of retrograde metamorphism. In the outcrop area of the Sharyzhalgay Group, gneiss—carbonate schists similar to those of the Slyudyanka Group commonly conformably overlie gneisses.

Pb-isochron ages from gneisses of the Sharyzhalgay Group have given figures up to 3,000 m.y. (Sobotovich et al., 1965; Manuylova, 1968). Dating by the K—Ar method has yielded much lower values. Rocks of the Slyudyanka Group have mainly been dated by the K—Ar method. These data show a wide scatter, with a maximum age of 2,600 m.y. (micas from gneisses in the area of the Erma and Cheremshanka River basins). Much older ages were obtained by the K—Ar method on pyroxene from rocks of both groups, but due to a very low potassium content such dates are probably of very low precision. Probably all the age values for the Sharyzhalgay and Slyudyanka rocks are "rejuvenated" by later thermal processes. This is to be expected in areas where the Archean rocks are present in fold belts that have gone through several orogenic episodes.

Different stratigraphic subdivisions have been proposed for the Archean rocks of the western part of the East Sayan fold belt. These are generally known as the Biryusa Group. The stratigraphic subdivision proposed by

Dibrov (1964), is widely used, but recent work has shown that it cannot be applied throughout the whole region. Perhaps the breakdown proposed by Konikov (1962) is more widely applicable. According to this scheme the Biryusa Group is subdivisible into three formations which, in ascending sequence are: the Elgashet (>2500 m thick) composed of amphibole, pyroxene—amphibole, and biotite plagiogneisses (crystalline schist), and amphibolites; the Reshet (800 m thick) made up of marbles and calc-silicate rocks with local interbands of quartzitic schists and amphibolites; and the Golumbeyka (>1300 m thick) composed of garnet-bearing biotite and two-mica gneisses with subordinate interbands of quartzite and calc-silicate rocks. The two lower formations are compositionally similar to the Sluydyanka Group. The Golumbeyka Formation probably forms an addition to the Archean section of the East Sayan region.

There is no common opinion as to the age of the Biryusa Group. Some people consider it to be Archean (Buzikov et al., 1964), some Lower Proterozoic (Dibrov, 1964; Dodin et al., 1968). Radiometric dating does not provide a solution to the problem. K—Ar dates are widely scattered (from 500 to 2600 m.y.) and give a strong impression of superimposed events at different times. There is also good evidence of retrograde metamorphism. Close similarity between the Biryusa Group and the Slyudyanka Group, and the presence in the former of rocks with relict minerals of granulite facies, suggests that it may be Archean.

In the *Baikal fold belt* (Salop, 1964—1967) and in other fold belts bordering the platform, Archean rocks are mainly exposed in median massifs (see Fig. 23). In the Baikal block the older complex, known as the Pribaikal Group, is subdivided into two subgroups. The lower one, Talanchanskaya, is mainly composed of amphibolites and locally pyroxene—amphibole gneisses (migmatites), and the upper one, Svyatoy Nos, is characterized by abundant marble bands (calc-silicate rocks) with amphibole and pyroxene gneisses, and schists with some rare quartzites. The Svyatoy Nos Subgroup outcrops on the western shore of Lake Baikal. It is known there by the name "Priolkhon complex". The Pribaikal Group rocks have largely undergone amphibolite-facies metamorphism, but in some places there are rocks with mineral assemblages typical of the granulite facies. In zones of differential movements, retrograde metamorphic rocks with relict minerals of both facies are developed.

The Pribaikal Group is in fault contact with other Precambrian rocks. Its older age is indicated, however, by an abrupt change in metamorphic grade. The Archean age of the group is also indicated by the fact that Paleoprotozoic sedimentary and volcanic rocks close to the Pribaikal gneiss exposures have undergone greenschist-facies metamorphism, and include some conglomerates with gneiss clasts. The radiometric (K—Ar) ages yield various rejuvenated values (450—1900 m.y.).

Archean rocks of the South Muya and North Muya blocks, known as the

Vitim Group, also contain two subgroups. These are the Ileir and Tuldun, which are similar in composition to the corresponding units of the region around Baikal. However, one difference is the abundance of hypersthene plagiogneiss and schist in the Ileir rocks of the South Muya block. Geological, and partly radiometric data, suggest an older age for the Vitim Group. The basal metaconglomerates of the Paleoprotozoic volcano-sedimentary complex (Salop, 1964—1967) lie with angular unconformity on gneiss and marble of the Tuldun Subgroup and include gneissic granites in the southeastern margin of the North Muya block (Samokut River). K—Ar isotopic dating in most cases yields rejuvenated values, but some relict dates appear to be close to the time of original metamorphism of the Archean rocks. Thus, pyroxene amphibolites of the Tuldun Formation in the South Muya block gave an age of 3200 m.y. (Manuylova, 1968).

Archean supracrustal and plutonic rocks are widely distributed in the *Stanovoy fold belt*, on the southern margin of the Aldan shield. However, recently it has been suggested that the Archean is represented by small tectonic fragments among younger (Lower Proterozoic according to some workers, and Upper Archean according to others) metamorphic rocks called the Stanovoy Range Group. It is now known that major parts of the Stanovoy Range area are underlain by Archean rocks, including the Stanovoy Range Group (Olekma Group, Mogot and Lapri Formations, and the Dzhugdzhur gneiss complex).

In the Stanovoy Range area there are extensive areas of Paleoprotozoic metamorphic sedimentary and volcanic rocks. These areas are mostly fault-bounded tectonic wedges within the Archean. The Archean rocks of the Stanovoy Range are considered by most people to be correlative with those of the Aldan shield. The main difference is that in many regions they have undergone strong retrograde metamorphism so that granulite-facies assemblages are preserved only as relicts. These metamorphic effects are most strongly expressed in regions where Paleoprotozoic and Mesoprotozoic migmatites and granite gneisses are abundantly developed (the older Stanovoy Range and Kuandin granitoids).

In the Stanovoy Range area there are Archean rocks at amphibolite facies but this does not preclude correlation with the Aldan Group, for the same metamorphic grade is typical of the Aldan Group in a number of places in the Aldan shield area. In the Stanovoy Range area there are local occurrences of rocks of granulite facies. These regions have been interpreted as Archean blocks among younger rocks of amphibolite facies. However, rocks of both facies grade into each other along strike. The differences in metamorphic grade can largely be explained by strong retrograde processes.

The Archean rocks of the Stanovoy Range area differ from those of the Aldan shield in that they are intensely reworked by Paleo- and Mesoprotozoic events. However, in the region known as the "structural suture" at the contact with the Aldan shield, typical Archean gneiss fold ovals are present. The "suture" itself is actually a rather wide transitional zone (see Fig.12).

The Archean gneisses of the Stanovoy Range area are poorly known and there is no universally accepted stratigraphic subdivision for this region. There are many local subdivisions, but more work is needed. However, the Archean section of the Stanovoy Range, at least generally, resembles that of the Aldan Group (Table II). In the former there are units corresponding to the Iyengra Subgroup, except for its lowermost part (most of the Kurulta Zverevo Groups) and to the lower part of the Timpton Subgroup (Mogot and Lapri Formations, the Stanovoy Range complex in the central part of the area, Khudurkan Formation in the western part, etc.). In the western part of the Stanovoy Range there are correlatives of higher units of the Aldan Group. For instance the Alvanar and Sivakan Formations are probable correlatives of the upper part of the Timpton Subgroup and the Chilcha and Kudulikan Formations correspond to the Dzheltula Subgroup.

Isotopic age dating of rocks in the Archean Stanovoy Range (mostly by K—Ar analyses) gives an extremely wide scatter from 150 to 3,500 m.y. (Manuylova, 1968). This is due to the polycyclic nature of tectonic movements in this area and to strong superimposed Mesozoic tectonics and magmatism. It is probable that the oldest values (about 3500 m.y.) on amphiboles from gneisses and amphibolites, are relict ages and indicate the time of Archean diastrophism.

In the Mesozoic *Verkhoyansk* (or *Verkhoyansk—Kolyma*) *fold belt* Archean rocks form the large Okhotsk median massif. According to the data of Grinberg (1968) they are various pyroxene and two-pyroxene, amphibole, and biotite (cordierite- and garnet-bearing) gneisses and schists, together with amphibolites with rare quartzite interbeds. On the basis of proportions of different rock types present three formations are recognized within this 7,000 m thick gneiss complex. Neoprotozoic platform strata unconformably overlie the peneplaned surface of the gneisses. K—Ar age determinations fall into the 1,230—1,880 m.y. range. Whole-rock Pb-isochron analysis of a crystalline schist yielded a 3,700±500 m.y. age for the metamorphism (Sobotovich et al., 1973). This gneiss complex is generally similar in composition to the Timpton Subgroup with which it is commonly correlated.

When correlating Archean rocks of the fold belts surrounding the Siberian platform with the Archean stratotype there is a striking similarity in both lithology and stratigraphic sequence. In a median massif in the fold belts surrounding the platform, only correlatives of the lowest units of the Aldan Group (Iyengra Subgroup) appear to be missing. Schists and plagiogneisses of basic composition are widely developed. These correspond to the lowest part of the Timpton Subgroup. Overlying gneiss—carbonate units, which are also widely distributed, are comparable to rocks in the upper part of the Timpton Subgroup, and the lower part of the Dzheltula Subgroup of Aldan. In Eastern Sayan there are rocks analogous to garnet gneisses at the top of the Aldan Group (Golumbeyka Formation of the Biryusa Group).

Relatively small areas of Archean rocks are reported in the *Soviet Far East*.

They do not form part of the fold belts around the Siberian platform, but nevertheless, the Archean rocks of these regions are very similar to those of the Aldan shield. Gneiss—granulite complexes are probably the oldest rocks in this region. They occur in the Taygonos Peninsula on the north shore of the Okhotsk Sea and in the Khankai Massif in the southern part of the Ussuri region. Thermal processes strongly overprinted the rocks of these complexes at various times (including the Mesozoic), so that radiometric (K—Ar) dating always yields "rejuvenated" values (Firsov, 1965). The gneiss complex of the Taygonos Peninsula has been subdivided by Lipatov into two "groups" of uncertain relationship. Lipatov compared one of these groups, the Kossov, which is made up of melanocratic (amphibole and pyroxene) plagiogneisses, schists, granulites and calc-silicate rocks, with the Timpton Subgroup of the Aldan shield. The other, the Purgonoss, which consists of garnet—biotite, hypersthene—biotite and sillimanite-bearing plagiogneisses, was correlated with the Dzheltula Subgroup of the Aldan region.

The older rocks of the Khankai Massif (Iman Group) also comprise two formations. The lower one (Ruzhinsk) is peculiar in that it contains graphitic marble bands among its gneisses and schists (some of which are hypersthene-bearing). The overlying Matveyevsk Formation is mainly composed of garnet—biotite gneisses and subordinate amphibole and pyroxene gneisses and schist with interbanded hypersthene—magnetite rocks and marbles. The Ruzhinsk Formation is probably correlative with the Timpton Subgroup, and the Matveyevsk Formation with the lower part of the Dzheltula Subgroup of the Aldan stratotype. However, it is also possible that together, they are correlative with the Timpton Subgroup. Metamorphic, sedimentary and volcanic rocks of both Paleoprotozoic and Mesoprotozoic age unconformably overlie the Iman Group (Shatalov, 1968).

In the extensive *Central Asiatic fold system* older rocks are exposed in many median massifs situated in fold belts of various ages. Their stratigraphy is poorly known at present. They are represented by the Bekturgan Group of Ulutau, the Zerenda Group of the Kokchetav Massif in Kazakhstan, the Aktyuz and Kemin Formations in Northern Tien-Shan, the gneiss complex of the Takhtalyk Range, the Atbashi Formation of the Atbashi Range, the Karategin in Middle and Southern Tien-Shan, the Vakhan Group in Pamir, various gneisses forming the basement of the Tarim Massif and by many gneisses occurring as tectonic wedges in the mountain ranges of the Himalayas and Tibet (Bel'kova et al., 1969; Zaytsev and Filatova, 1971).

All of these units consist of metamorphic rocks of granulite and amphibolite facies. They are made up of various gneisses, schists, amphibolites, marbles and granulites which were intensely migmatized and granitized. In some cases their age cannot be determined geologically or geochronologically because of tectonic contacts with the surrounding rocks, and because of superimposed retrograde metamorphism which led to complete rejuvenation of isotopic ages. K—Ar dating usually gives the age of the last metamorphism or uplift.

In areas that underwent extensive recent uplift such as Pamir, Tibet and in the Himalayas, the K—Ar method commonly gives values of some scores of millions of years. In Paleozoic fold belts they are usually in the 250—600 m.y. range, but some ages as old as 1,800 m.y. have been obtained. A few determinations by the Rb—Sr isochron method on the same rocks in Pamir yielded values as old as 2,500 m.y., and by the Pb-isochron method on gneisses of the Karategin area, up to 2,900 m.y. However, in many areas geological relationships clearly show that the gneiss (gneiss—granulite) complexes are the oldest Precambrian units.

Extensive exposures of Archean rocks occur in many regions of *China*, mostly in the northern part, in Shantung, Liaoning, Sinkiang, Shansi and in a number of other places. All of these form parts of the oldest Precambrian complex in a small platform known as the North Chinese or Chinese—Korean platform. They are highly metamorphosed rocks which were originally metamorphosed under granulite- and amphibolite-facies conditions, and subsequently suffered strong retrograde metamorphism. The most characteristic is the Tai Shan complex in Shantung province. It is composed of biotite and amphibolite plagiogneisses, amphibolites and granulites, usually highly granitized and migmatized. The stratigraphic subdivision made by Ch'en yü-Ch'i et al. (1964) probably needs some modification in the light of recent work. The age of these rocks has been determined only by the K—Ar method on micas from migmatites, granites and granulites. Most of the results fall within the range of 1,800—2,300 m.y., but dates as old as 2,500 m.y. also exist. It is probable that these dates were "rejuvenated", because the gneiss complex suffered younger metamorphism about the time of the Karelian orogeny.

The gneiss—granulite complexes of Liaotung and Shantung Peninsulas, and the charnockite-bearing gneiss—granulite complex in the Kuruktag Mountains (Mount Karatekunul) and constituting the base of the Tarim platform, are probably comparable to the Tai Shan Group.

In most areas of China Archean rocks are unconformably overlain by platform and subplatform deposits of Neoprotozoic and Epiprotozoic age. However, in some places (Shansi Plateau, Wutaishan) they are unconformably overlain by sedimentary and volcanic rocks of Paleoprotozoic age (the Utai Group, etc.) (Regional stratigraphy of China, 1963).

In *North Korea* the Ranrim gneiss complex is considered to be Archean. Together with the Paleoprotozoic Machkholen complex, it forms the basement of the Korean platform. This complex is largely comprised of granite gneiss, migmatite, amphibolite and granulite. It is in fault contact with the Machkholen metavolcanics and is unconformably overlain by Neoprotozoic rocks (Sanvon Group, etc.).

Vast areas of Archean rocks are present in Southern Asia in the *Indian platform*. In both India and Ceylon sedimentary and volcanic rocks of the Dharwar "system", together with various gneisses, charnockites and gneissic granites which are given the common name "Peninsular gneisses" are considered to be Archean. The "Peninsular gneisses" are considered to be largely made up of plutonic rocks intruding supracrustal units of the Dharwar "system" or ultrametamorphic rocks formed during granitization of the Dharwar sedimentary—volcanogenic sequence. For this reason the "Peninsular gneisses" are shown in most stratigraphic schemes as the youngest Archean strata and are situated above the Dharwar "system" (Rama Rao, 1940, 1962; Krishnan, 1960a,b).

The Dharwar sequence appears to include alternation of rocks of different metamorphic grade both in vertical and horizontal senses. Even within small areas, highly altered rocks such as hypersthene and garnet—sillimanite gneisses occur together with rocks such as phyllites, greenschists and sandstones with well-preserved blastopsammitic texture, ripple marks and cross-bedding. In many areas slightly metamorphosed Dharwar strata occur as small bands in vast areas of migmatites and gneissic granites. It is possible that rocks of different age and different degree of metamorphism are assigned to the Dharwar "system". There are also published reports of Dharwar sequences that include metamorphosed conglomerates containing granite and gneiss pebbles. The published data on the Precambrian geology of India suggests that formations of different age are included within the Dharwar "system" (Salop, 1964, 1966).

In the Kolar goldfield, between Madras and Bangalore in the eastern part of the Shimoga—Dharwar schist belt, that is in the stratotype of the Dharwar "system", the basal part of the Dharwar greenstone succession contains metamorphosed conglomerates. These unconformably overlie gneissic granites (Champion granites) and include pebbles and boulders of the underlying gneissic granites, and various gneisses, migmatites and other rock types which make up the "Peninsular gneiss" complex (Fig. 9). Formerly, Indian geologists considered these rocks to be tectonic breccias. Various geological relationships in these complexes, together with new isotopic dates, suggest that only the "Peninsular gneiss" and associated strongly metamorphosed supracrustal rocks should be considered Archean. The Dharwar "system" is assigned to the Lower Proterozoic, or, as it is called here, the Paleoprotozoic (Salop, 1966). This view is supported by some Indian and Soviet geologists (Nautiyal, 1965; Radhakrishna et al., 1967; Moralev and Perfil'ev, 1972). In a new geological map of India the "Peninsular gneisses" are attributed to the Archean, and the Dharwar "system" is considered to be Lower Proterozoic.

The older Archean strata of the Indian platform constitute a complex of supracrustal and magmatic rocks metamorphosed under granulite- and partly amphibolite-facies conditions. They are commonly granitized and transformed into granitic gneisses and migmatites. This complex is widely developed and the name Hindustan complex is proposed for it.

The Hindustan complex is made up of plagiogneisses and granulites alternating with leptite-like gneisses, amphibolites, calc-silicate rocks, marbles, quartzites and various aluminous schists, including some with garnet and silli-

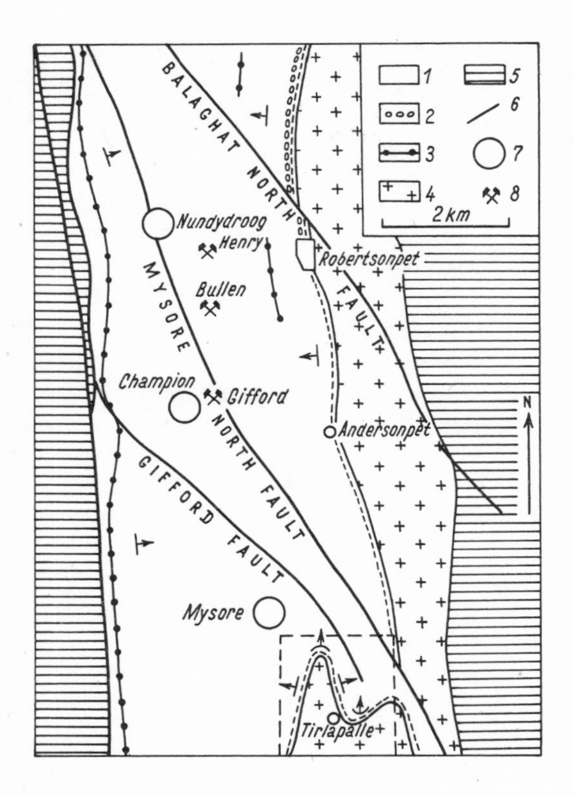

Fig. 9. Geological sketch map of the Kolar gold field according to Krishnaswamy et al. (1964), amended by Salop.
1 = Dharwar "system" of metamorphosed volcano-sedimentary rocks; *2* = Dharwar "system" basal conglomerate; *3* = jaspilite of the Dharwar "system"; *4* = Champion granite and gneissic granite, *5* = gneissic granite and migmatite ("peninsular gneisses"); *6* = faults; *7* = major mines; *8* = major open-pit mines.

manite (commonly graphitic) which, in India, are known as khondalites. These rocks are widely developed in East Ghats, in the state of Madras and in Ceylon. In many other areas of India such rocks occur in small areas among gneissic granites and migmatites derived from them. Charnockites and enderbites occur together with hypersthene gneisses and granulites. They are widely distributed in the southern and western parts of the peninsula (charnockites were first described from India). The charnockites and enderbites appear to have formed by granitization, and possibly in part by selective melting of foliated hypersthene gneisses.

The older supracrustal complex of India and Ceylon is similar in many respects to the ancient complexes of other regions of the world. There is a close association of pyroxene granulites with alumina-rich gneisses (khondalites) in many regions. In India, as in many other Archean regions basic crystalline schists are interbanded with magnetite—pyroxene crystalline schists, magnetite quartzites and irregular bodies of magnetite silicate ores. In the Hindustan complex such ores are rare and of small dimensions, in contrast to extensive jaspilitic banded iron ores of the upper part of the Dharwar succession. In correlating metamorphosed rocks, here assigned to the Hindustan complex with the Dharwar sequence, much emphasis was placed on the presence of iron ore in both. However, two similar (but not identical) epochs of iron-ore formation are known in many parts of the world — one in the Archean and one in the Paleoprotozoic.

The "Peninsular gneisses" are the most characteristic element of the Hindustan complex. They are widely developed in different parts of India, particularly in southern and eastern areas where they underlie about half of the region. On geological maps of India they are usually shown as "undivided granites with gneisses". They appear to consist mainly of light-grey gneissic granites with abundant dark inclusions of biotite, biotite—amphibole, garnet—sillimanite, and graphitic gneisses, amphibolites and granulites. Foliated migmatites are also typical.

Granites of various kinds, including gneissic varieties, are also included in the "Peninsular gneisses". They are given various names in different parts of India, such as the Champion gneiss, the Bundelkand gneiss, etc. The majority of the Hindustan granite gneisses are probably of pre-Dharwar (Archean) age. This is shown, not only by observations in the Kolar area, but also by geological relationships shown on maps of Southern India (Mysore state, Madhya-Pradesh, etc.). On these maps the Dharwar deposits are clearly situated in synclines separated by vast outcrops of the "Peninsular granite gneisses". The Dharwar rocks are commonly of greenschist facies, whereas the granite gneisses include large areas of granulites and khondalites. The possibility remains, however, that some post-Dharwar (Paleoprotozoic) granites are also included in the "Peninsular gneisses". There is abundant evidence of post-Dharwar mobilization of older granite gneisses to produce local intrusive contacts with the overlying Dharwar rocks.

The upper age limit of the Hindustan complex is determined by the unconformably overlying Dharwar succession and by the presence, in basal conglomerates of the Dharwar sequence, of clasts that were obviously derived from the complex itself.

Radiometric ages from rocks of the Hindustan complex give a wide scatter due to later metamorphic processes, especially rocks which have undergone Paleoprotozoic reactivation. K—Ar dates range from 600 to 2,500 m.y., but are commonly between 1,900 and 2,300 m.y. (Sen, 1970). The Rb—Sr method (Crawford, 1969) usually gives the age of the post-Dharwar metamorphism at about 2,600—2,800 m.y. An isochron, based on eight samples of gneiss and granite from Southern India (Mysore), yielded an age of 2,585 ±35 m.y. Rb—Sr isochron dating of charnockites in Nilgiri (Southwestern India) (four samples among several) gave an age of 2,615±80 m.y. but some other dates (assuming an original $^{87}Sr/^{86}Sr$ ratio of 0.700) were as great as 3,200 m.y. A Rb—Sr isochron on five samples of the older (by analytical data) gneisses of Southern India gave an age of 3,065±75 m.y. Gneisses underlying greenstones and iron-rich units (Dharwar analogues) in the Singbhum area have an age of up to 3,450 (3,320) m.y. based on K—Ar data (Sarkar et al., 1967). This value is probably a relict one because it is greater than that from rocks formed as a result of the post-Dharwar metamorphism and plutonism. Few data of this age have been obtained from the Archean of India. The intensity of Paleoprotozoic thermal processes on Archean rocks and the "rejuvenation" of isotope ages of the older rocks is shown by the fact that in the Kolar area the pre-Dharwar Champion granites, and also the Dharwar metamorphic rocks, and granites cutting them, have given the same or very similar ages (2,600—2,800 m.y.). The strong pervasive post-Dharwar metamorphism presents problems in attempts to investigate early thermal events in the history of the Indian Archean rocks. The true age of Archean metamorphism and plutonism in India can be indirectly inferred. The age of banded gneisses of the Bundelkand type is revealed by the age of the detrital zircon from metasandstones of the Mesoprotozoic Aravalli Group which unconformably overlies the gneisses. The Pb-isotopic age of the zircon is 3,500 ±500 m.y. (Tugarinov and Voytkevich, 1970).

North America

Canadian and American geologists consider the Keewatin-type volcano-sedimentary (largely volcanic) greenstone assemblages and the overlying (largely clastic) Timiskaming-type deposits to be Archean. These rocks are in many areas intruded by granites that are 2,600—2,800 m.y. old. It has already been suggested (Salop, 1970b) on the basis of data published by Canadian geologists, that within the *Canadian* (*Canadian—Greenland*) shield, much older crystalline rocks (various gneisses and granites) are unconformably overlain by the Keewatin-type assemblages. One important fact is the occurrence, in the Keewatin- and Timiskaming-type deposits, of thick conglomerates containing large granitic pebbles and boulders. Such conglomerates with various kinds of granitoid clasts are recorded among the Keewatin-type volcanics in the Kerr Parish area of Ontario (Prest, 1952). In Northwestern Ontario, in the Bee Lake area, they occur as interbeds in poorly sorted arkoses in Keewatin-type lavas (Goodwin and Schklanka, 1967). Near the Ontario—Manitoba border metamorphosed granitic conglomerates have been recognized in a number of localities. They are recorded, for example, in the eastern bank of Pierson Lake where granitoid clasts (up to 30 cm across) comprise 70% of the framework. The remainder of the clasts are quartzites (?) and vein quartz. Light-grey biotite granites predominate among the clasts, followed by grey granodiorite and brown granite. Considering their geological situation, and the composition of the clastic material, these rocks may possibly be assigned to the lower (basal) part of the Keewatin-type sequence. A lack of volcanic clasts is a characteristic feature of these conglomerates. In this area there are exposures of clastic rocks (quartzite and greywacke) close to the contact between Keewatin-type volcanics and large granite massifs (Derry, 1930).

Similar conglomerates are exposed in the Rice Lake area. They form part of the Rice Lake Group (of Keewatin type). In the area north of Lake of the Woods, on the northern shore of Little Crowduck Lake, close to the contact with Keewatin-type lavas and felsic gneisses, vertically dipping metamorphosed conglomerates are exposed. They comprise well-rounded clasts of granite and felsic gneisses (up to 60 cm across) (Greer, 1930). There are many such localities near the Ontario—Manitoba border. All of these occurrences and other similar conglomerates are assigned to the Keewatin-type complexes by Canadian geologists. They are not usually considered to form the lowermost units. However, the possibility remains that many of them may, in fact, be situated in the basal part of the Keewatin. Very old (see below) granites are characteristically widely distributed in areas close to these conglomerates. Donn et al. (1965) assigned the granitic conglomerates of the Rice Lake Group to the Manigotagan terrigenous unit, which underlies volcanics at the base of the group. Horwood (1935) described metaconglomerates at the base of the Heis Group in the area north of Lake Winnipeg. These conglomerates include granodioritic and tonalitic boulders up to 60 cm across. Granodiorites, similar in appearance and mineralogical composition to the clasts, underlie the Heis Group, and are widely distributed in the surrounding area, together with younger granitoids of different composition.

Ermanovics (1970b) reported conglomerate-like rocks with gneiss and granite-gneiss clasts in the area of the Lower Wanipigou River (east shore of Lake Winnipeg) at the contact between the Rice Lake volcanics and banded gneisses. These banded gneisses may be older than the Rice Lake Group. In addition to these banded gneisses some other granodioritic rocks that are widely developed in the area may also be Archean for they appear to be overlain unconformably by feldspathic quartzites interbedded with metaconglomerates. These quartzites are assigned by Stockwell (1938) to the San Antonio Formation. This unit also unconformably overlies volcanics of the Rice Lake Group.

Canadian geologists consider the granodiorites to intrude the volcanics, and consider the San Antonio Formation to be much younger. Radiometric dating of mineralization in quartz veins cutting both the San Antonio Formation and the Rice Lake Group, however, gave the same age: 2,720 m.y. (Ermanovics, 1970b). Cross-bedded feldspathic quartzites similar to those of the San Antonio Formation are also present among the Rice Lake volcanics. It is possible that many of the granodiorites on the eastern shore of Lake Winnipeg are older than the Rice Lake Group (and the San Antonio Formation). The reason for their being regarded as younger than the volcanics is that they were partly remobilized during the Kenoran (Paleoprotozoic) orogeny, and intruded the lower parts of the volcanic succession in some areas.

The Keewatin-type conglomerates with granitic clasts occur in many other areas of the shield. They are present in the volcanic Yellowknife Group in the Mackenzie District. In the same province in the Itchen Lake area Bostock (1967) reported conglomerates containing well-rounded clasts of massive and gneissose granites, granodiorites, diorites and vein quartz together with some volcanic clasts. In the Keewatin District granitoid boulders are present in metamorphosed Keewatin-type conglomerates near the northwestern coast of Hudson Bay (C. Bell, 1966). In Quebec (Labrador Peninsula) Keewatin-type conglomerates contain pebbles of red granite gneiss and granites (Eade, 1966a). Conglomerates of higher metamorphic grade (with granitic pebbles) are present on the west coast of Hudson Bay (south of Ferguson Lake). These are known as the Mackenzie Lake Group (R. Bell, 1971) and are considered to be younger than the Kaminak Keewatin Group (of Keewatin-type). These conglomerates could be interpreted as the basal strata of the Keewatin-type assemblage, unconformably overlying older gneisses and granite gneisses.

There are numerous exposures of conglomerates with granite and syenite pebbles in the terrigenous Pontiac Group which underlies and is in part, equivalent to, the Keewatin volcanics in the southern part of Quebec in the Noranda—Malartic area (MacLaren, 1952; Holubek, 1968). According to Holubek the conglomerates of this area which are situated close to the base of the succession contain larger granitic clasts. On the western shore of Kinojewis Lake these rocks occur directly on the granitic basement. A remarkable example of occurrence of basal Keewatin-type units on older granites was described by Jolliffe (1966) from the Steep Rock Lake area of Southern Ontario (in this region there is a large manganese—iron deposit). This occurrence will be described below in the chapter on the "Paleoprotozoic".

Baragar (1966) reported important data concerning the existence of a pre-Keewatin granite-gneiss complex in the Mackenzie District. In a region about 100 km east of the town of Yellowknife the lower part of the Yellowknife Group (a Keewatin analogue) is composed of a thick succession of greenstones (pillow lavas), conformably overlain by greywackes. The lavas have a sharp contact with underlying granite gneisses, and both are cut by a basic dike swarm. These dikes have a chemical composition similar to that of the Yellowknife lavas, to which they are probably genetically related. These dikes are not present in the greywackes.

In the southeastern part of the area, a massive granite pluton with associated pegmatites also intrudes the Yellowknife Group. These granites assimilate the basic dikes. The age of the "younger" granites is 2,540 m.y. by the Rb—Sr isochron method, and that of the older granite gneiss is about 2,600 m.y. Both granites give approximately the same age (W. Barager, personal communication, 1970). In periods between granite intrusions there was commonly a period of intense denudation and intrusion of diabase (subvolcanic) dikes. Thus it is possible that the older granite gneisses were "rejuvenated" by intrusion of younger granites. A zonal type of metamorphism was superimposed on the volcano-sedimentary assemblage in response to the second stage of granite formation. Baragar considers the older granite gneisses to have been a basement that was overlain by the volcanics and greywackes.

Some recent data support Baragar's concept of a pre-Yellowknife basement. J. Henderson et al. (1971) observed, north of the town of Yellowknife, near Ross Lake on the west bank of Cameron River, boulders of gneissic granodiorite among conglomerates in the greywacke sequence of the Yellowknife Group. Granodiorites are present in a large massif ("Ross Lake granodiorite") northeast of the lake. The geological map compiled by Davidson (1971) suggests that the Ross Lake granodiorites, granite gneisses and migmatites form the core of a large gneiss dome, on the margins of which the greenstone volcano-sedimentary strata of the Paleoprotozoic Yellowknife Group are preserved (Fig.10). On the west bank of the Cameron River the metavolcanics possibly unconformably overlie basement gneisses.

The presence of gneissic inclusions in many older granites suggests the existence of older, highly metamorphosed, supracrustal rocks. In many cases, only slightly metamorphosed rocks (of the greenstone belts) are found adjacent to highly metamorphosed rocks such as gneisses, migmatites, hypersthene plagiogneisses, granulites and charnockites. These units are generally regarded as products of extreme metamorphism of the Keewatin rocks, but this has not been definitively proved.

The Keewatin-type rocks are commonly altered to amphibolite facies, and are even locally granitized, but a transition from such rocks to rocks at granulite facies is nowhere recorded. Such rocks always occur as isolated units, apparently underlying the Keewatin-type rocks. It has been suggested that these high-grade rocks are in the lower part of the Keewatin-type successions, but on account of the abrupt change in metamorphic grade, it is more likely that the gneiss—granulite terrains and the Keewatin strata are separated by a period of orogeny (metamorphism and plutonism). This is indicated by the presence of hypersthene-granulite clasts in conglomerate interbedded with Keewatin-type lavas near the western shore (Caribou River basin) of Hudson Bay (Davison, 1966).

The rocks of the gneiss—granulite complexes are scattered throughout the shield as small linear zones among the gneissic granites. In many localities they are associated with Keewatin-type deposits (overlying them), but in some places they occupy large areas of the older shield. The northwestern part of the Labrador Peninsula is an example of such an area. This is probably the oldest part of the Canadian—Greenland shield and was not extensively reworked by the post-Keewatin (Kenoran) and younger thermo-tectonic events.

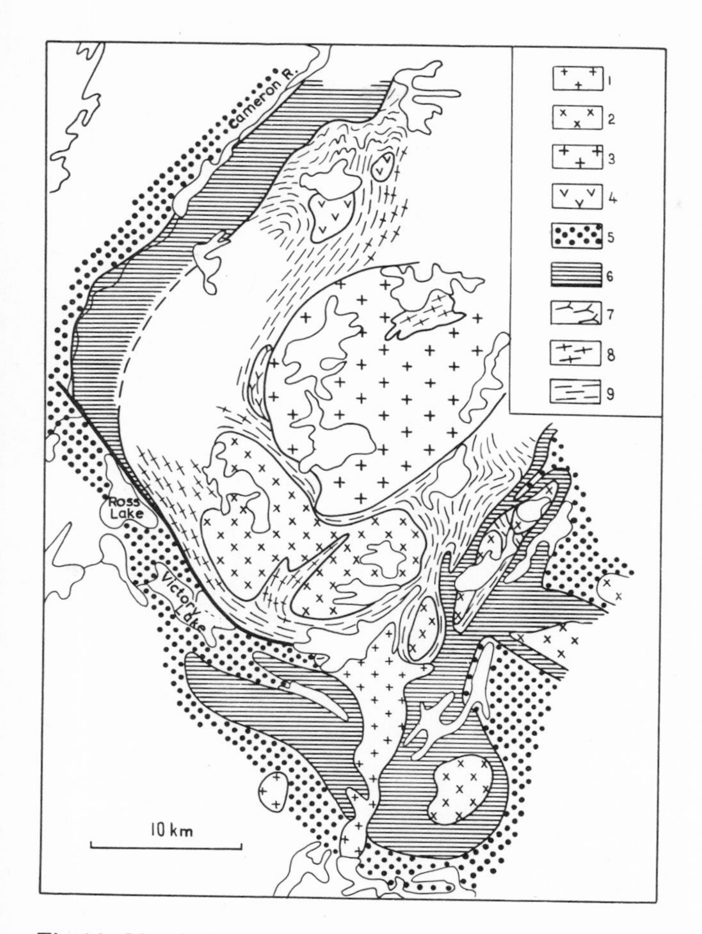

Fig.10. Mantled gneiss dome of the Yellowknife area after a map by Davidson (1972). Salop's interpretation on the basis of data by J.B. Henderson et al. (1972).
Paleoprotozoic: Paleoprotozoic post-Yellowknife intrusive rocks (*1* = muscovite granite; *2* = biotite granite; *3* = bimicaceous adamellite; *4* = diorite); *5* = greywacke, schist and conglomerate of the Yellowknife Group; *6* = metavolcanite of the Yellowknife Group; *Archean*: *7* = gneissose diorite; *8* = gneissose granodiorite and granite; *9* = gneiss (migmatite).

It forms part of the older Ungava protoplatform (Fig.11). The gneisses and granulites which are cut by many granitic bodies, form a complex folded structure of oval shape, dissected and overlain by the linear Mesoprotozoic fold belts on the west, north and east, and on the south by the Abitibi greenstone belt. The shape, size and origin of this large structure are those of a typical gneiss fold oval. This structure differs markedly from the volcanic terrains of the Abitibi belt and similar belts in some other parts of the world which are characterized by the presence of relatively small dome-shaped

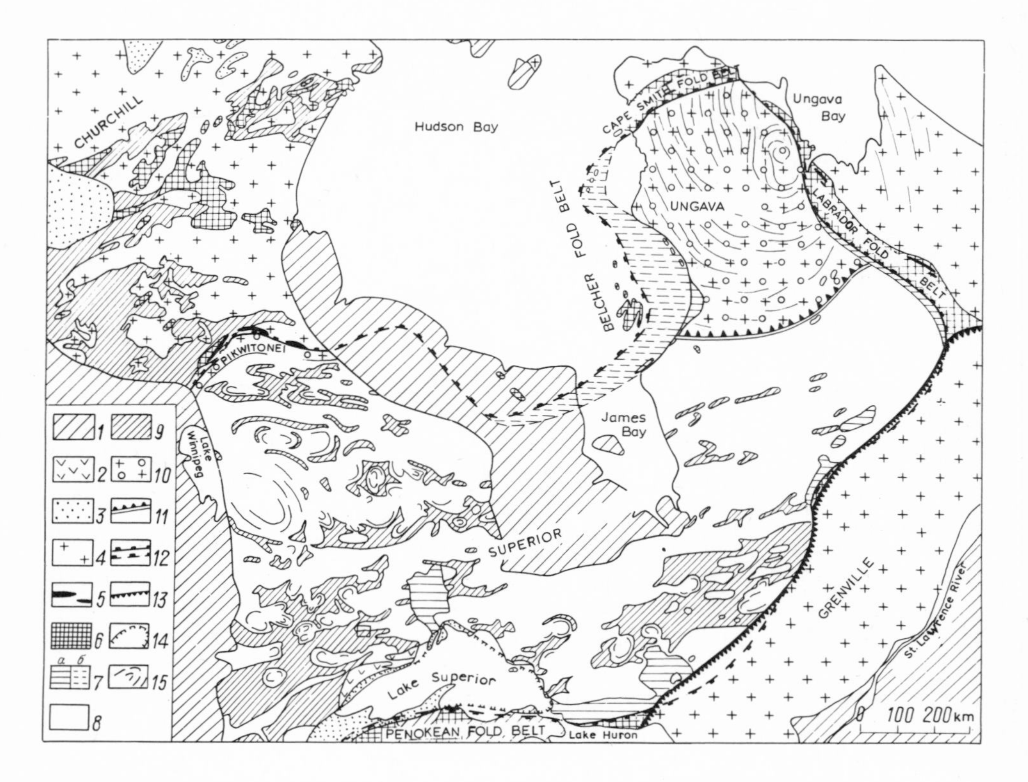

Fig. 11. Generalized geology of the southeastern part of the Canadian shield.
1 = Paleozoic and Mesozoic rocks; *2* = Neoprotozoic gabbroid (Duluth Massif); *3* = Neoprotozoic platform rocks and lavas (in the area of Lake Superior these are partly Epiprotozoic); *4* = Mesoprotozoic and Paleoprotozoic granitoids (partly formed by regeneration of Archean granites and gneiss), together with the Archean granite—gneiss—granulite complex, variously altered by Paleoprotozoic and Mesoprotozoic tectonics and metamorphism (predominantly in the Grenville province and in the eastern part of the Labrador Peninsula); *5* = Mesoprotozoic ultramafics; *6* = Mesoprotozoic miogeosynclinal rocks; *7a* = Mesoprotozoic platform or subplatform rocks; *7b* = the same, but under a cover of younger strata or under water; *8* = Paleoprotozoic granitoids (partly formed by regeneration of Archean granites and gneiss), together with rocks of the Archean crystalline complex, reworked during the Paleoprotozoic orogeny; *9* = Paleoprotozoic sedimentary and volcanic strata (in the Superior province these are the Keewatin-type rocks, which are mostly volcanic); *10* = Archean granite—gneiss—granulite complex not strongly affected by younger movements or metamorphism (Ungava "core" and the Pikwitonei block); *11* = boundary of Paleoprotozoic fold belts and of the area of Archean stabilization (Ungava "core"); *12* = boundaries of Mesoprotozoic fold belts and Mesoprotozoic platforms; *13* = "Grenville front"; *14* = outlines of area of Lower and Upper Keweenawan volcano-sedimentary assemblage; *15* = strikes of schistosity and fold axes.

structures with predominant linear trends of the greenstone belts (Salop, 1971b).

Another, even larger region of the gneisses, and granulites lies in the Grenville tectonic province of Canada. This region suffered Paleo- and Mesoprotozoic folding, and strong uplift at the end of the Neoprotozoic, which, following deep erosion, resulted in exposure of basement rocks of the fold belt.

Gneisses and granulites also occupy large areas in the northern part of the shield, in the Boothia and Melville Peninsulas, in Baffin Island and in other regions which probably earlier formed part of the Ungava protoplatform.

In the Churchill and Slave tectonic provinces granulite gneisses are exposed in the core of Paleo- and Mesoprotozoic fold structures where they acted as rigid massifs. The block of granulite and anorthositic rocks named Pikwitonei subprovince by C.K. Bell (1966), is in a major fault zone at the boundary of the Superior and Churchill provinces. Bell considers the rocks of this subprovince to be older than the Keewatin-type greenstone belts.

The existence of ancient sialic rocks has also been suggested by others such as Gross and Ferguson (1965), and Goodwin (1968).

Radiometric dating of these older (pre-Keewatin-type) rocks, the Keewatin-type assemblages themselves, and the post-greenstone-belt granites usually gives similar results in the range of 2,400—2,800 m.y. (D.H. Anderson, 1965; Davis and Tilton, 1965; K.R. Dawson, 1966; Stockwell, 1968; Hanson et al., 1971). These dates are clearly associated with the Paleoprotozoic (Kenoran) thermal events which affected all the older rocks of the shield. Kenoran granites are widespread in Canada. They formed as a result of remobilization of the pre-Keewatin granitic—gneissic basement rocks. This thermotectonic event also caused widespread development of gneiss domes, made up partly of remobilized basement rocks and partly of metamorphic rocks of the Keewatin—Timiskaming type assemblages. Thus the Kenoran event was widespread and extremely important.

Some Pb-isotope analyses on zircon gave an age of 3,550 m.y. (Catanzaro, 1963). This age is also confirmed by Rb—Sr isochron analysis (Goldich, 1968; Goldich et al., 1970) of the granitic and other gneisses in the southern part of the Canadian—Greenland shield (Minnesota). Himmelberg (1968) showed that these rocks (Morton gneisses) form part of a plutonic and supracrustal gneiss—granite complex (Montevideo) made up of various amphibole—pyroxene, hypersthene and two-pyroxene plagiogneisses, garnet—biotite gneisses, migmatites, and gneiss granites. Just north of this region there are extensive areas of relatively slightly metamorphosed Paleoprotozoic sedimentary—volcanogenic rocks of the Ely (=Keewatin) and Rice Lake Groups. Although the contacts between the gneiss—granulite complex and the sedimentary—volcanogenic terrains are not observed, the sharp decrease in metamorphic grade of the rocks, suggests that rocks of two ages are present.

Gneisses from the gneiss—granulite complex of Labrador Peninsula, on the south of Saglek Bay, gave an age of 3,400—3,600 m.y. on zircon by the Rb—

Sr isochron method (Bridgwater et al., 1973; Hurst, 1973)*.

In Southeastern Manitoba (Lake Winnipeg area) radiometric age determinations suggest that the Kenoran orogeny, though strongly developed there, did not completely "rejuvenate" the basement gneiss. Zircon from the basement gneisses gave an age of 2,950—3,050 m.y., whereas the overlying Keewatin-type assemblage in the Rice Lake area gave an age of about 2,720 m.y. (Ermanovics, 1973).

Isotopic studies of galena from the Cobalt—Noranda area (Ontario—Quebec border) and from some other areas of Canada by Slawson et al. (1963) showed that the original separation of lead from homogenous sources took place about 3,200—3,500 m.y. ago. They considered the ages of 2,400—2,700 m.y., obtained by K—Ar and Rb—Sr analyses on mineral separates, to be the result of late-stage metamorphic events. Lead isotopes indicate the existence of older tectonic and metamorphic events. Thus both types of analysis, Pb-model and Pb—U show that the Precambrian geology of North America began more than 3,200 m.y. ago. These very old rocks cover large areas of North America (Slawson et al., 1963, p.413).

Goodwin (1968) referred to the strong influence of Kenoran metamorphism on older rocks and was pessimistic about the possibility of establishing their true age. However, the data mentioned above, together with data from Archean rocks of Greenland, suggest that the problem can be solved.

Archean rocks are widespread in *Greenland*. The Archean rocks of the Godthaab, Nordland, Finnefeld and Isortoq gneiss—granulite complexes occupy the greater part of the west coast of Greenland. The pre-Ketilidian complex (including the Cape Farewell complex) at the southern extremity of the island is also Archean. All these complexes contain supracrustal rocks that have been highly metamorphosed under granulite and amphibolite facies. These consist largely of gneisses of variable composition including hypersthene-bearing gneisses and less abundant amphibolites, pyroxene amphibolites, granulites, locally marbles and calc-silicate rocks. In some areas meta-anorthosite bodies are present in the gneisses and granulites (in some cases they are chromite-bearing). They are folded, and metamorphosed together with the stratified host rocks (Windley, 1969a). All of these rocks are strongly migmatized, and are commonly transformed into granite gneisses and charnockites.

K—Ar dates from these rocks (and also from other areas) usually give

---

*Some new data became available since the manuscript went to press. These data concern Rb—Sr isochron determinations of Archean rocks of the Canadian shield. Tonalite gneiss, which is widely distributed throughout the Grenville province (southeast of the town of Chibougamau), gave a minimum of 3,000 m.y. (Frith and Doig, 1975). Tonalite gneiss from the pre-Yellowknife basement in the western part of the Slave province is about 3,000 m.y. old (Frith et al., 1974). Thus wider application of the Rb—Sr isochron method recently provided evidence of the presence of old rocks in younger orogenic belts. Such evidence was not obtainable by the K—Ar method.

"rejuvenated" values (1,900—2,800 m.y.), but dates as old as 3,400 (3,210) m.y. were obtained by this method from rocks of Southwestern Greenland (Windley, 1969b). Pb-isotope dating of zircon from Archean granite gneiss of the same area gave an age of 3,000 m.y.

Rb—Sr analyses of many samples (some scores) from the Amitsoq gneiss and granite gneisses of the Godthaab complex yielded an age of 3,700—3,760 m.y., with an initial ratio of $^{87}Sr/^{86}Sr = 0.7009—0.7015$ (Moorbath et al., 1972; Bridgwater, 1973; Pankhurst et al., 1973). Pb-isotope analysis on zircon gave an age of 3,650—3,680 m.y. (Black et al., 1971; Baadsgaard, 1973). The same gneiss was dated at 3,620±100 m.y. (Black et al., 1971) by the Pb-isochron analysis. These values are the most reliable for any Archean complex in the world. Even the oldest values do not give the age of the Archean supracrustals, but the time of their granitization and first metamorphism, for the isochron method indicates the time of isotopic homogenization after recrystallization. Earlier age determinations from the Amitsoq gneisses gave a figure of 3,980±170 m.y. (Black et al., 1971). This result was too high because of erroneous selection of samples (Moorbath et al., 1972).

The Isortoq granitic gneiss has many inclusions (up to some hundreds of metres long) of various well-banded rocks, predominantly amphibolites with rarer metabasites with ferruginous clinopyroxene, rocks rich in quartz with magnetite, rhombic pyroxene, garnet, grünerite, and various gneisses. Most of these rocks are undoubtedly of supracrustal origin and include some highly metamorphosed sedimentary iron formations (McGregor and Bridgwater, 1973). The Isortoq granite gneisses are cut by the numerous sheared and locally granitized amphibolite Ameralik dikes, which do not appear to penetrate the metavolcanic Malene rocks which are presumed to unconformably overlie a basement of granitic gneiss. The problem of the age of the Malene supracrustals is discussed briefly in the next chapter.

The Archean gneiss complex of Southwestern Greenland contains folded, layered, chromite-bearing intrusions of the Fiskenaesset anorthosite, which is probably Archean too. The Rb—Sr isochron age of these rocks is about 2,900 m.y. (Bridgwater, 1973), but it is probable that this age is "rejuvenated" due to the effects of a later metamorphism. There are some xenoliths of meta-anorthosite in the Ameralik dikes (McGregor, 1968).

In the *North American platform*, Archean rocks, overlain by platform deposits, occur in the basement of the Eastern Rocky Mountains and in the Colorado Plateau. These rocks are biotite, sillimanite, and cordierite (in some cases hypersthene-bearing) gneisses, amphibolites, quartzites and marbles. The gneisses are usually migmatized and are cut by bodies of banded granitic gneiss. Pb-isotope analysis on zircons gave an age of 3,200 m.y. for metamorphism of the gneissic basement complex of the Big Horn Mountains (Heimlich and Banks, 1968). K—Ar dating yielded younger ("rejuvenated") values of 2,890 (2,900) m.y. The Rb—Sr age (2,410 m.y.) is also "rejuvenated".

Widely distributed granitized high-grade metamorphic rocks are present in some parts of the *North American Cordillera* near the Eastern Rocky Mountains, but it is only recently that isotopic dating has shown their old age. Rb—Sr isochron analysis gave an age of 3,300 m.y. for the granulite—gneiss basement complex of Southwestern Montana (the Pony complex) (Brookins, 1968), and a Pb-isotope analysis on zircon gave an age of 3,500 m.y. for a similar complex in the Montana—Wyoming area (Beartooth Mountains) (Catanzaro and Kulp, 1964). In Idaho (Albion Mountains) dating by the Rb—Sr isochron method gave an age of 3,700 m.y. for granulite gneisses exposed in the cores of mantled gneiss domes (Sayyah, 1965; Armstrong, 1968). Precise isotopic techniques therefore reveal the Archean age of granulite—gneiss basement rocks in a number of areas of the North American Cordillera. K—Ar dating of the same rocks gave "rejuvenated" values not greater than 2,890—2,900 m.y.

There may be Archean basement rocks in the *Appalachian fold belts* (Blue Ridge and Piedmont areas). The Baltimore gneiss—granulite complex and the Cranberry gneisses are examples of possible Archean rocks. These complexes are exposed in the cores of Upper Precambrian and Paleozoic fold structures. In addition to widely distributed gneisses and migmatites, the Baltimore gneiss complex includes stratiform anorthosites and charnockites which are characteristic of many Archean complexes throughout the world. Isotopic dating of these rocks by different methods invariably gives "rejuvenated" values that range from 350 to 1,300 m.y. These dates are undoubtedly due to superposition of various thermal processes in the late Precambrian and Paleozoic.

Archean basement rocks are probably also present in the *East Greenland fold belt*, in the Kong Oskars Fjord and Scoresby Sound areas. In these areas the older rocks comprise various gneisses, amphibolites, migmatites and gneissic granites. Rb—Sr isochron analysis yielded an age of 3,000±250 m.y. for augen granite gneiss although these rocks were strongly reworked by the Caledonian movements. The Rb—Sr isochron and Pb-isotope age of zircon from gneiss of the Danmarkshaven area also yielded about 3,000 m.y., and Rb—Sr and K—Ar ages on minerals fall into the 320—380 m.y. range, which is undoubtedly related to strong isotopic "rejuvenation" of the rocks (Bridgwater, 1973).

In most cases the stratigraphy of the Archean complexes of North America and Greenland is either completely unknown or poorly studied. One exception is the Grenville Archean complex in Southeastern Canada (Grenville province and the Adirondacks in the U.S.A.). This complex may be divided into two parts according to Wynne-Edwards (Wynne-Edwards et al., 1966; Wynne-Edwards, 1967). The lower part consists of various gneisses, largely hypersthene and two-pyroxene varieties, granulites, amphibolites and also charnockites formed by granitization of hypersthene gneisses. Locally folded, sheet-like anorthosite bodies are present among these rocks (particularly in the Adirondack Mountains). These anorthosites are usually considered (erroneously) together with the younger (Neoprotozoic) anorthosites which form large intrusive cross-cutting massifs.

The upper part of the complex, known as the Grenville Group, is characterized by interbanded gneisses, marbles (calc-silicate rocks), and quartzite bands. According to Wynne-Edwards, the following generalized sequence (in ascending order) is characteristic of the southern part of the Grenville province: (1) gneiss and granulite; (2) marble; (3) gneiss; (4) marble; (5) quartzite; and (6) marble. However, in the Adirondacks, according to De Waard and Walton (1967), this sequence seems to be reversed. This may be explained by different interpretation of rather complex fold structures. There is no common opinion as to the relationships between the lower (gneiss—granulite) part of the complex and the upper gneiss—marble—quartzite (Grenville Group) part. Wynne-Edwards proposed an unconformity between them so that the lower part would be Archean and the upper part, Proterozoic (Aphebian).

Others do not believe that this unconformity exists and consider all these units as a single complex. There is an equal lack of agreement among American geologists concerning relationships in the Adirondacks. De Waard and Walton consider the Grenville Group to have been deposited on top of the eroded surface of a folded gneiss—charnockite—anorthosite basement. Others, including Buddington, strongly opposed this interpretation. There appears to be no reliable evidence for the existence of an unconformity within the complex.

Problems concerning the structure and age of rocks of the Grenville complex have been debated for many years. Recent studies suggest that the metamorphic rocks of the Hermon and Flinton Groups are Protozoic (see Chapters 5 and 6) whereas they were previously assigned to the Grenville complex. In the following discussion these rocks are excluded. Determination of the age of this complex is difficult because of the fact that it has undergone polyphase deformation and metamorphism and the radiometric datings are accordingly "rejuvenated". The same phenomenon is characteristic of many older fold belts such as the Stanovoy belt in Siberia and the Mozambique and Nigerian belts of Africa. The most common dates for Precambrian rocks of the Grenville province (K—Ar method in particular) are in the 950—1,100 m.y. range. These have been interpreted as indicating the time of original metamorphism of the Grenville complex. However, recent work, using different methods, has resulted in much older values (2,500 m.y. by Rb—Sr analysis and 2,700—2,800 m.y. by Pb-isotope analysis (Krogh and Brooks, 1968—1969) from an area close to the so-called Grenville Front which separates the Grenville province from the stable Superior province (Fig.11)*. K—Ar dates from the Grenville province appear to be older in

*See footnote on p. 74

areas closer to the Superior province — from 1,000 (950) m.y. in the southeastern part of the belt to 2,600 (2,480) m.y. in the northwestern part (Chadwick and Coe, 1973). K—Ar dates from Precambrian rocks appear to be younger (from 1,000 to 360 m.y.) in a southeasterly direction from the Grenville province towards the Appalachians (Long, 1961). In the southern part of the Grenville province a sheet-like granite body in paragneisses gave a date of 1,725 m.y. (Rb—Sr isochron) and zircon from the same granite gave a date of 1,660 m.y. by Pb-isotope analysis. Zircon from host paragneiss gave an age of 1,300 m.y. (Krogh and Davis, 1972). This discrepancy is explained as being due to later metamorphic effects. Probably all of these values also reflect later thermal events.

Canadian geologists commonly compare the Grenville complex with the Mesoprotozoic (Aphebian) Huronian Supergroup in the Superior province. However, there is a very abrupt change in metamorphic grade of the correlated units over a very short distance and, more importantly, there are significant lithological differences. In a southeasterly direction from Sudbury, at the boundary between the Grenville and Southern provinces, relatively unmetamorphosed Huronian quartzitic sandstones are juxtaposed, across a 150 m thick mylonite zone, with highly metamorphosed dark biotitic greywackes which are interlayered with metavolcanics (?) (amphibolites). These rocks have been compared to the Huronian, but they are perhaps more similar to sedimentary and volcanic rocks of Keewatin-type (Paleoprotozoic). This comparison is all the more reasonable, considering that an age of about 2,700 m.y. (Krogh and Brooks, 1968—1969; Krogh and Hurley, 1968) was reported from granite gneiss that cuts these metamorphic strata (zircon from pegmatite). A few hundred metres to the southeast across the strike, and possibly separated by another large fault zone, there are strongly folded migmatitic gneisses with interbands of calc-silicate (diopside-bearing) rocks of "Grenville-type".

The lithology and sequence of rocks in the Grenville complex are similar to those of the middle part of the Archean section in the Aldan area (the Aldan Group) which has been suggested as the world stratotype for the era.

The lower part is similar to the lower part of the Timpton Subgroup in the Aldan area which is largely composed of basic schists and gneisses (metabasites), and the upper part of the Grenville complex is similar to the upper part of this subgroup, in that they both contain some carbonates among metabasites. Stratiform anhydrite deposits are present in the carbonate rocks of both complexes. The Grenville complex is even more similar to the Archean rocks in the western part of the Ukrainian shield (Bug Group) where quartzites are commonly present in a metabasite—carbonate sequence which overlies metabasites.

The Grenville complex might well serve as a stratotype for the Archean in North America and Greenland, but later thermo-tectonic events present problems in establishing the radiometric age of these rocks. The gneiss—granulite complex of the Minnesota River valley region (Montevideo and Morton gneisses) which is reliably dated by various isotopic methods is proposed as a temporary stratotype.

General Characteristics

The Archean supracrustal units are generally indivisible complexes without recognizable internal unconformities. This apparent stratigraphic simplicity is probably due to the high-grade metamorphism and complex folding undergone by these rocks. Published data on the existence of such unconformities in some of these complexes are not certain.

All Archean complexes are characterized by rocks metamorphosed to granulite and amphibolite facies. The different facies appear to be irregularly distributed and there is no regular linear metamorphic zonation. Granulite-facies rocks are extensively developed and rocks of progressive amphibolite facies (upper) are locally developed. Rocks of transitional subfacies (hornblende granulites) are abundant.

In the fold zones and in stable shield areas adjacent to them the Archean rocks have, in some cases, undergone retrograde metamorphism with development of minerals indicating lower-temperature facies. Retrograde metamorphism is particularly common in Archean rocks forming the basement of Protozoic and younger mobile belts (geosynclinal systems). The strongly eroded Stanovoy Range fold belt is a good example. In that region the basement of a Paleoprotozoic geosyncline displays Archean rocks retrograded to amphibolite or epidote-amphibolite facies. Some workers contend that the amphibolite facies in the Archean rocks is always secondary, superimposed by retrograde processes on rocks of granulite facies. Data from the Aldan shield indicate that this view is incorrect (Salop and Travin, 1974). Archean rocks are almost everywhere migmatized and granitized, but the intensity of these events varies in different areas and in different rock types.

Slightly metamorphosed rocks at greenschist facies or epidote-amphibolite facies of progressive metamorphism are not known in Archean sequences (older than 3,500 m.y.) of the Northern Hemisphere. Ferruginous strata of the Isua complex in Southwest Greenland may be an exception. However, these rocks may not be Archean (see Chapter 5). A preliminary study suggests that they are also absent from Archean sequences of the Southern Hemisphere. Relatively unmetamorphosed sedimentary and volcanic rocks of the Swaziland Supergroup (South Africa), considered by many to be the oldest rocks in Africa, are here assigned to the Paleoprotozoic. Pb-isochron dating gave an age of 3,360 m.y. (Van Niekerk and Burger, 1969) on sulphides and zircon from quartz porphyries of the Onverwacht Group. This date probably indicates the time of formation of the lava, and not the age of metamorphism. Rb—Sr whole-rock dating of the phyllitic argillaceous shales of the Middle (Fig Tree) Group gave 2,980 m.y. This date, and ages obtained from the Consort pegmatites (Allsopp et al., 1968) which cut the youngest (Moodies) group (3,000 m.y.) support a Paleoprotozoic age for this group. Rocks of the Swaziland Supergroup contain microscopic remains of blue-green algae and its rocks lie unconformably on older granites, gneisses and granulites of the basement (Allsopp et al., 1968; Mattews and Scharrer, 1968; Condie et al., 1970). It is also similar to other Paleoprotozoic complexes of South Africa. There appears to be an unusually high initial ratio of $^{87}Sr/^{86}Sr$ in the Consort pegmatites (0.770), and also in the Fig Tree schists. Thus, there may have been an excess of radiogenic ^{87}Sr, which originated in the older basement rocks, resulting in a slightly "older" isotopic age. Details of the Swaziland Supergroup and its relationships with the underlying rocks will be published in another work dealing with the southern continents. Additional older age values (as old as 3,000—3,100 m.y.) are reported for granites of Swaziland and Rhodesia associated with Paleoprotozoic volcanics. The oldest values obtained so far (>3,500 m.y.) were obtained from gneissic granites in South Africa which are unconformably overlain by a Paleoprotozoic sedimentary and volcanic assemblage.

In Archean complexes primary structures are rare to absent. However, various kinds of bedding are preserved in most cases.

The granulite facies (regional) is a specific characteristic of Archean complexes and is absent or only locally developed in younger complexes. This idea has been opposed by many authors, but the examples they cite of extensive younger granulites, are not convincing. In most cases the post-Archean age of the granulites is not certain, or is based on misinterpretation of geological observations or misunderstanding of isotopic dating. This is particularly true in the case of a recently published book dealing specifically with the granulite facies of metamorphism (Drugova and Glebovitsky, 1971). While charnockite-like hypersthene-bearing granites (and granodiorites) with some of the characteristics of intrusive bodies are present in some younger (Protozoic) Precambrian complexes (Shemyakin, 1972), they are mostly attributable to partial or complete remelting of Archean hypersthene plagiogneisses and metasomatic charnockites. Charnockites and granulite-facies gneisses in the Ladoga region of Karelia have been shown to be Archean basement overlain by metasedimentary rocks of the Mesoprotozoic Ladoga Group, and not highly metamorphosed correlatives of this group as was previously thought to be the case.

The problem of the age of granulite-facies metamorphism in Archean rocks of Western Greenland is very complex. Some geologists (McGregor, 1968; Black et al., 1973; Pankhurst et al., 1973) consider this metamorphism to have been superimposed (2,600—2,850 m.y. ago) on rocks that were originally metamorphosed under amphibolite facies 3,500—3,700 m.y. ago. This conclusion was reached from the fact that granulite-facies rocks of the Godthaab area gave a Rb—Sr isochron age of 2,850 m.y., and no older values are known for them. However, in the same region, the Amitsoq gneiss with mineral assemblages of the amphibolite facies, gave an age of about 3,700 m.y. The concept of superimposed granulite-facies metamorphism seems unlikely because the Paleoprotozoic supracrustals of the Malene and Isua(?) Groups (>2,900 m.y. old) which unconformably overlie gneisses, appear to only have suffered amphibolite-facies metamorphism (in the case of the Isua rocks greenschist facies is locally developed). It is probable that progressive metamorphism to both amphibolite and granulite facies in the Archean complex of Greenland developed simultaneously, though the amphibolite facies may be locally retrograde. It is possible that thermal processes were superimposed on granulite facies rocks 2,600—2,850 m.y. ago. They did not, however, cause strong retrograde metamorphic effects, but caused isotope homogenization ("cryptometamorphism"). From descriptions of granulites in Greenland there is some evidence of retrograde changes (replacement of pyroxene by amphibole and mica, formation of epidote, green biotite and sphene, etc.). The Amitsoq gneiss in the area of granulite-facies metamorphism (in the Buksefjord region for example) is charnockitic, and the Ameralik dikes (which have undergone amphibolite-facies metamorphism) cut not only the Amitsoq gneiss, but also the granulites (Chadwick and Coe, 1973) which suggests very early and coeval metamorphism to both granulite and amphibolite facies. The Ameralik dikes are supposed to be older than the Malene supracrustals which are cut by the Nuk granite (about 3,000 m.y. old). Thus, the granulite-facies metamorphism must be older than the indicated age.

All the Archean complexes of the Northern Hemisphere are characterized by abundant melanocratic amphibole, amphibole—pyroxene, and pyroxene plagiogneisses, gneisses, crystalline schists and amphibolites. Their chemical and mineralogical composition, and mode of occurrence suggest that they are highly metamorphosed basic and ultrabasic lavas and tuffs, and in some cases, sheet-like intrusive gabbroid bodies. In some gneiss complexes, stratiform meta-anorthosite bodies occur (under amphibolite or granulite facies). In many areas metabasites are predominant, but they are commonly strongly granitized and migmatized gneissic granites with relics of metabasites and hypersthene granites (charnockites and enderbites). During granitization, pyroxene and amphibole may be replaced by biotite, so that mesocratic, and even leucocratic biotite or biotite—amphibole gneisses are formed after metabasites. These rocks contain only relics of pyroxene. This fact should be taken into consideration in evaluating the proportion of metabasites in Archean sequences.

Metamorphosed acid volcanics are not recognized in Archean complexes, but this does not rule out the possibility that acid volcanics were present in the Archean. Their apparent absence may be due to a lack of reliable criteria for recognizing such rocks when they have been strongly metamorphosed. Many porphyries which have developed a strong schistosity and are recrystallized (even under greenschist-facies conditions) are transformed into sericite—chlorite schists which can hardly be distinguished from para-schists. Only the relics of blastoporphyritic texture reveal their origin. It has been suggested that some biotite (leptite-like) gneisses in a number of Archean complexes are altered acid volcanics.

In some strata garnet—biotite, sillimanite- and cordierite-bearing gneisses and quartzites are closely associated with metabasites. Rocks with abundant sillimanite, corundum and spinel, in some cases rich in magnetite, are also present. Judging by their composition, these may be metamorphosed high-alumina sedimentary rocks, formed by reworking of material from fossil soils. In post-Archean times, especially in Upper Proterozoic and Phanerozoic geosynclinal complexes, the association of volcanics, quartzites and high-alumina rocks is rare-to-absent. The Archean rock association is probably unique. The problem of origin of Archean rocks in general and of Archean quartzites in particular is discussed in detail in Chapter 10.

The marbles and graphitic gneisses are undoubtedly of sedimentary origin. Graphite genesis in Archean gneisses and marbles is a contentious issue. Some people suggest that carbon appeared from ammonia-rich water and carbon dioxide, as a result of radiogenic synthesis under the influence of cosmic radiation as is thought to be the case for complex hydrocarbons in meteorites. Others have suggested that the graphite is biogenic, on the basis of the isotopic composition of the carbon. The latter point of view seems more reasonable. Radiogenic synthesis may have been important locally, but it could have taken place only in the outer layers of the dense Archean atmosphere. However, graphitic gneisses (and marbles) are widespread in the Archean of some areas, and even include large economic deposits of graphite. It is probable that these deposits indicate the existence of primitive biological systems as early as the Archean (see Chapter 10). There are supposed remains of stromatolites in calc-silicate bands in the Archean granulite complex of the Kola Peninsula (Ivliev, 1971). These structures are probably non-biogenic. They appear to be the result of boudinage in carbonate layers included in rocks of different competence. However, because of the complex structural setting of these rocks, the possibility that the carbonate rocks are Mesoprotozoic, and that the granulite complex has been thrust over them, cannot be ruled out.

Many Archean complexes include bands and lenses of syngenetic metamorphosed iron formation. They are almost invariably associated with amphibolites and basic (amphibole and pyroxene) crystalline schists. Magnetite, in some cases partially or completely hematitized (martite) is an important component of these units. Three types of ore bodies are recognized. The first type is banded; the banding is commonly coarse or poorly expressed. In rare cases finely banded rocks (layers rich in magnetite, intercalated with layers with little or no magnetite) occur. All of these layers contain abundant amphibole and pyroxene (monoclinic and rhombic), together with garnet, biotite, quartz and feldspar. Quartz mainly occurs in layers without magnetite; in some cases it is so abundant that the rock has the appearance of quartzite or quartzite schist. This type of iron formation has a general resemblance to jaspilitic ores but it differs from the characteristic, younger Precambrian ores in having poorly expressed and thicker banding, by having abundant silicates and aluminosilicates, a lower quartz content and in the absence of carbonates.

Another type of iron ore that is perhaps more characteristic of the Archean, is aluminosilicate and silicate—magnetite ore which is included in the rock as interbands, lenses, isolated pods, and disseminations of magnetite in amphibolites and amphibole—pyroxene crystalline schists. These two types of ore commonly occur together, associated with metabasites. They form a characteristic Archean ore association for which the name "Priazov type" (after the Priazov deposits) is proposed. Their close association with metabasites strongly suggests that their formation is related to volcanic processes.

The third kind of Archean iron ore is of limited distribution. It occurs as magnetite (martite) disseminations in sillimanite-rich quartzite bands. These ores are probably formed by reworking of materials derived from fossil soils.

One striking feature of the Archean is the absence of conglomerates and coarse sandstones. This cannot be explained as being due to the high grade of metamorphism, because conglomeratic textures tend to be very stable, survive any degree of recrystallization and even the first stages of metasomatic granitization (Salop, 1964—1967). It is possible that coarse clastics were absent or rare in the Archean. Although there are many reports of conglomerates in gneiss complexes of Archean age, they have all proved to be tectonites ("pseudoconglomerates") or agmatite breccias (Krylova and Neyelov, 1960; Ravich, 1963; Bogdanov, 1971a; Drugova, 1971). In particular, the metaconglomerates of the Kola—Belomorsky complex in the Volchyi tundra of the Kola Peninsula (recently described by S.I. Makiyevsky and K.A. Nikolayeva) are considered by Bogdanov (1971a) as typical tectonites. In a number of cases younger Precambrian units were considered to be Archean conglomerates. They unconformably overlie Archean gneisses or have tectonic contacts with them.

A striking feature of the Archean rocks is their uniform composition over very large areas. Abrupt facies changes are recorded only in the metavolcanics (orthoamphibolites and some pyroxene orthoschists). In many cases tectonic thickening and thinning of beds, or even total loss of units, particularly in successions that contain relatively incompetent carbonate rocks, have been interpreted as facies changes. Primary facies zonation is not known.

One of the most noteworthy features of the Archean complexes of Northern Eurasia and possibly of the world, is the striking similarity in lithology and succession of strata even in widely separated regions (Salop, 1968a). The most complete Archean sections are those of the Aldan shield (Aldan Group which is the Archean stratotype of Siberia and proposed as the world stratotype of the era) which are characterized by the presence in the lower part of the gneiss complex of predominant metavolcanics (orthoamphibolite or pyroxene orthogneiss), and quartzites with interbanded alumina-rich gneisses, crystalline schists, and locally, magnetite quartzites. Up-section there is a thick, monotonous sequence of basic crystalline schist (and plagiogneiss) or amphibolite derived from basic volcanics. Still higher in the section there are similar metavolcanic units, but which are interbanded with thick carbonate units. These sections are capped by garnet gneisses, orthoamphibolites, pyroxene schists, quartzites and minor amounts of other rock types.

The same general sequence is noted in incomplete sections in other areas.

The origin of such a uniform sequence is not certain. It is possible that the change from one facies to another was a result of a normal sequence characteristic of volcano-sedimentary assemblages of different ages. Perhaps the most reasonable explanation of this peculiarity of the Archean sequences is that it reflects the evolution of the geochemical environment during sedimentation and also sedimentary (mainly chemical) differentiation.

The question arises as to whether these similar sequences were formed approximately simultaneously. In the case of Phanerozoic complexes such an interpretation would be most unlikely. The increasing heterogeneity of the crustal structure of the earth with time, and the non-synchronous development of its various parts did not favour widespread development of one facies type at any one time. However, the analysis of composition and paragenesis of the oldest supracrustal rocks show little evidence of differential tectonic movements. The following features were the most characteristic of the Archean: a lack of any definite facies zonation, lack of high relief or any long-lived source areas. The greater part of earth's surface was probably covered by ocean. The land areas were low peninsulas or emergent shoals of irregular shape (Salop, 1964).

Correlation of Archean complexes on the basis of hypothetical conditions is at best speculative. However, the present level of knowledge does not permit any other possibility for correlation of Archean rocks of widely separated regions.

In correlation charts of the Precambrian of Siberia and the European part of the U.S.S.R. (Tables I—II) Archean strata (formations, groups) of similar composition are placed between horizontal lines. The oldest supracrustal Archean rocks in the lower part of the tables (A^I) are characterized by the association of quartzite, amphibolite or high-alumina (sillimanite, garnet—corundum, cordierite) schist and gneiss, with magnetite quartzite, and other magnetite-rich rocks. As stated above this rock association is characteristic of Archean complexes throughout the world. This complex is called the ferallite—amphibolite complex because of the characteristic formations (Salop, 1968a), but it is better named "ferallite—amphibolite—quartzite". Judging from the composition of the rocks an association of metavolcanics and residual products of chemical weathering is characteristic of Lower Archean successions.

In Siberia, the following stratigraphic units are assigned to this part of the Archean: the Iyengra Subgroup of the Aldan shield, the Daldyn Subgroup of the Anabar Massif, the lower parts of the Kurulta Group (Ungra and Chaynyt Formations) and the Zverevo Group of the Stanovoy Range area, possibly part of the Yerma Formation of the Sayan region and the Kuzeyevo Formation (?) of the Yenisei Ridge. Rocks of this kind are poorly developed in Europe. The lowermost part of the Pinkel'yavr Formation of the Kola Peninsula, the Keret' Formation of the White Sea area and the lowest part of the Ingul, and Orekhov—Pavlograd Group of the Ukraine, all of which are characterized by the presence of magnetite quartzites and aluminous rocks, may be part of this complex or, to be more exact, of its topmost part, transitional with the second (metabasite) complex. Abundant quartzite units in the gneiss—granulite complex of Southern Sweden suggest that these rocks may belong to the lower part of the Archean succession. The lower complex is not known in America.

Economic deposits of rock crystal of the lateral-secretion type are localized in quartzite horizons of this complex in the Aldan shield (Kargat'yev, 1970). In the Aldan and Anabar shields, in the Stanovoy Range fold belt, in the Kola Peninsula, and in the Ukraine, metamorphosed iron formation (jaspilite) is closely associated with volcanogenic rocks. This complex everywhere contains horizons of high-alumina rocks (sillimanite and corundum gneisses) which may be of economic value as raw material for the production of aluminium, different refractories and abrasive materials.

The overlying complex (A^{II}) is a thick monotonous unit of amphibolite and/or hypersthene, rarely two-pyroxene, basic (sometimes ultrabasic) crystalline schist, or plagioclase may be predominant in these strata. Other gneiss types, and crystalline schists occur in subordinate amounts, quartzite is still less abundant and high-alumina rocks and marble are present as thin bands. Charnockite and enderbite have formed after hypersthene rocks as a result of granitization. This complex includes abundant rocks which many geologists consider to be altered lavas of basic composition and partly intrusive rocks of similar composition. Thus, the complex is characterized by an abundance of metabasites.

The metabasite complex appears to be more extensive than the underlying one. In Siberia the lower part of the Timpton Subgroup (Nimgerkan and Ungra Formations) of the Aldan shield, are assigned to this complex, as also are the Verkhneanabar Subgroup of the Anabar shield, the Kuzeyevo and Atamanovka Formations of the Yenisei Range, most of the Sharyzhalgay Group, and Yerma complex of the Sayan region, the Talanchanskaya Subgroup of the Baikal region, the Ileir strata of Central Vitim, a large part of the Stanovoy complex of the Stanovoy area, and others. In the East European platform this complex is represented by most of the gneiss—granulite terrain of the Baltic shield (in particular, most of the Pinkel'yavr Formation of the Kola Group and the Khetalambino Formation of the Belomorskaya Group). In the Ukrainian shield the Pobujian and Dniester—Bug Formations of the Bug Group, the Volodarsk Formation of the Ros'—Tikich Group, the major part of the Ingul Group, the Novopavlograd Formation of the Orekhov—Pavlograd Group and the Lozovatka Subgroup of the Priazov Group all belong to this complex. Similar rocks are also present in the basement of the Russian platform. Probably most of the gneiss—granulite complexes of Southern Eurasia (China, India, etc.), of Western Europe ("the monotonous sequences" of the Bohemian Massif for example), and of North America (in particular, the lower metabasite part of the Grenville complex) are all part of this same complex.

The most common ore deposits of the syngenetic type in this complex are iron-formations of the Priazov type which are closely associated with volcanic rocks. The iron formations vary from place to place and include the following types: massive and disseminated, sometimes banded magnetite—silicate ore of the Pinkel'yavr Formation (Kola Peninsula), of the Volodarsk Formation, the Ingul Group, the Novopavlograd Formation (the Ukraine), the Oboyan Group (Kursk magnetic anomaly), of the Zoga Formation (Sayan region), of the Mogot Formation (Stanovoy Ridge), metabasites of the Grenville complex (Ontario and Quebec, Canada) of the gneiss—granulite complex on Baffin Island, and in Nevada and Wyoming (U.S.A.), of metabasite in Southern India, etc. The reserves in ore of this type are very low in spite of the numerous deposits. Chromite mineralization is confined to the stratiform Archean anorthosites in Southwestern Greenland which occur among basic crystalline schists and gneiss.

The third Archean unit (A^{III}) is rather complex. It comprises various gneisses, basic crystalline orthoschists and amphibolites, but is most characterized by its carbonate rocks; marble, calc-silicate rocks, diopside, diopside—scapolite and other types of calcic crystalline schists. Bands of graphitic gneiss, marble, high-alumina rocks and quartzites are quite abundant. The content and position in the section of carbonate rocks varies from place to place. In most areas (with the exception of the Baikal region, Pobujian area, Pamir and Southeastern Canada) they are of subordinate importance, and seldom make up discrete units, but rather occur interbanded with other rocks, largely orthoamphibolite and pyroxene plagiogneiss, but sometimes with quartzites. In the Aldan shield, for example, metabasites make up the major part of this complex section and carbonate rocks are present as subordinate scattered bands. This complex, however, may be referred to as metabasite—carbonate for these are the most characteristic rocks.

In Siberia the following units are assigned to this complex; the Fedorov Mines, Seym (Idzhek) and Kyurikan Formations in the Aldan shield, the Khaptasynnakh strata of Anabar, the Elgashet and Reshet Formations of Eastern Sayan, the Slyudyanka Group of the Sayan region and Khamar-

Daban, the Priolkhon complex of the Western Baikal area, the Svyatoy Nos strata of the Eastern Baikal region, the Tuldun strata of the Middle Vitim and many others. In other parts of Asia the following units are involved: the Ruzhinsk Formation of the Ussuri district, the Kossov Formation of the Taygonos Peninsula, the Kara-Kul'dzha Formation of the Takhtalyk Ridge (Tien-Shan), the Vakhan Group of Pamir and some formations in Afghanistan, China and India. In the European part of the U.S.S.R. the upper complex includes the upper part of the Khetalambino Formation of the White Sea area, the Chudzyavr Formation of the Kola Peninsula, the Teterev—Bug Formation of the Pobujian area, the Belotserkovsk Formation of the central part of the Ukrainian shield, the Orekhov Formation of the Dnieper area, the Korsak—Shovkai Subgroup of the Azov area and some others. In Western Europe the rocks most typical of this complex are "the Varied group" of the Bohemian Massif. In North America the Grenville Group (s.str.) and probably, more strata in different areas of the continent, form part of this complex. In the Arctic Islands of Canada and in Greenland the stratigraphic position of similar rocks is not certain.

Economic deposits in rocks of this complex are varied and numerous. They include syngenetic deposits as well as metasomatic ones, confined to carbonate rocks. The latter are of great industrial importance. To them belong, for example, the well-known deposits of phlogopite in the Aldan and Anabar shields, in the Southern Baikal area, in the Ukraine and in other areas of this type. Also important is magnetite ore in the form of large metasomatic-type deposits in the Fedorov Mines Formation of the Aldan shield (Southern Aldan Group of deposits of this type). Metasomatic lazurite deposits are also of great value and occur among the marbles of the Southern Baikal region, Pamir, Afghanistan and on Baffin Island. Almost all the lazurite deposits of the world are confined largely to Archean carbonate rocks. This is possibly related to the fact that the Archean carbonate rocks formed under an atmosphere and hydrosphere deficient in oxygen. In addition to sulphates, other sulphur-rich compounds also precipitated with the carbonates. Probably the original rocks, which later underwent metasomatism, were the sulphate and sulphide-rich limestones (and dolomites). Lazurite is a composite alumino-silicate including Na_2S and Na_2SO_4.

The Fedorov Mines Formation of the Aldan shield contains layers and lenses of anhydrite-bearing diopside—plagioclase crystalline schists with a sulphate content of up to 15—25%. Diopside, diopside—scapolite, and phlogopitic rocks are also present. In these rocks anhydrite is present with other minerals of granulite facies (Kargat'yev, 1970). In granitized rocks the potassium feldspar—perthite is developed after it. According to Brown (1973) the same situation exists in Canada where anhydrite together with diopside, phlogopite, tremolite and serpentine are present as interbands and inclusions among the marbles of the Grenville Group.

The syngenetic deposits of the metabasite—carbonate complex comprise the Priazov-type iron ores which are closely associated with metabasites. These form large deposits in the Azov region (the Mariupol deposits, the Kusungur and others), and some smaller, less well-known deposits of the Pobujian area, in the Kola Peninsula and elsewhere. The graphitic gneisses and marbles are of economic importance. Large deposits of graphite are present in the Pobujian area (the Zaval'evo) and in the Azov region (the Troitsk and Karatyuk deposits), in the Aldan shield and in many other areas. Carbonate rocks forming part of the complex in the southern Baikal area contain bands of quartz—diopside crystalline schist rich in apatite (metamorphosed phosphate—chert—carbonate deposits) which are of economic value because of the apatite. Apatite-bearing rocks are also present in carbonate rocks of many other regions (Priolkhon area, Priazov area etc.). In this complex there are local high-alumina rocks with sillimanite corundum and spinel. The marble is widely used as building material and industrial stone.

The uppermost, that is the fourth, Archean complex (A^{IV}) is known from only a few areas. In the Aldan shield this horizon is represented by the Sutam Formation of the Dzheltula Subgroup, in the Anabar shield by the Billeekh—Tamakh Formation of the Khapchan Subgroup, in the Eastern Sayan by the Golumbeyka Formation of the Biryusa Group, in the Far East by the Matveyevsk Formation of the Iman Group, in Kazakhstan by the upper part of the Zerenda Group, in Karelia by the Loukhi Formation of the Belomorskaya Group, in the Kola Peninsula by the Volshpakh Formation of the Kola Group, in the Ukraine by the upper part of the Bug Group and the Karatysh Subgroup of the Priazov Group. Correlative rocks may be present in the Archean complexes of North America and Southern Asia.

This complex comprises mainly garnet-bearing biotite gneiss, crystalline schist, rare garnet, sillimanite, cordierite-bearing gneiss, quartzites and some pyroxene and amphibole gneiss. Peculiar garnet-bearing granulite was formed under granulite-facies metamorphism by granitization of garnet—biotite gneiss. Prior to metamorphism the gneisses of this complex may have been pelitic rocks, and probably in part, acid volcanics.

No mineral deposits are present in this complex, but probably some of the garnet-rich gneisses may be used as raw material in the abrasives industry.

The proposed grouping of Archean units into four successive complexes provides the possibility of predicting the kinds of mineral deposits likely to be found in a given succession. However, the stratigraphic (correlational) aspect of some of the proposed subdivisions needs further substantiation.

The Archean supracrustal strata are everywhere intensely folded. They have a diagnostic tectonic style in large areas which have not undergone secondary deformation. For example, in the Aldan and Anabar shields the Archean tectonic style is distinctly different from that of younger Precambrian and Phanerozoic fold belts. These older gneiss complexes are characteristically made up of isometric or elongated, irregular ("amoeba-like") oval-

and roundshaped structures. They are somewhat similar to gneiss domes, but they differ from them in their larger size (from 100 to 800 km in diameter), and by having a rather complex internal structure. Such structures, which are here called gneiss fold ovals (Salop, 1971b) represent fold systems characterized by concentric grouping of folds which are predominantly linear and isoclinal with distinct centripetal vergence (mass movement to the centre of the ovals). The larger folds are complicated by many folds of lower order down to microfolds. The axial surfaces of the folds are themselves sometimes curved about horizontal axes.

The intensity of plastic deformation is directly related to the abundance of granitic material present (degree of granitization). In the cores of the fold ovals there are bodies of pink alaskite granite. These are largely conformable (partly cross-cutting) bodies formed by rheomorphism of a granitized mass and selective anatexis of the ancient gneissic basement during late stages of a tectono-plutonic cycle. Earlier autochthonous metasomatic plagioclase (or microcline—plagioclase) gneissic granite is commonly developed and has also undergone complex folding, together with the gneiss. The biggest occurrences of such rocks are in the inner part of the ovals. Between the fold ovals there are areas characterized by chaotic orientation of folds, lacking distinct vergence, and with development of brachyform and small dome structures.

The tectonic style of the Archean rocks clearly indicates the high plasticity of the material and also a close relationship between folding, granitization and anatexis, which were accompanied by the uplift of the granitized mass.

The tectonics of such gneiss fold ovals and of the interoval areas is well studied in the Aldan shield area where more than ten such oval systems have been recognized. These systems range from simple ovals to some that are amoeba-shaped (Fig.12). Description of these structures and discussion of their genesis has been reported elsewhere (Salop, 1971b; Salop and Travin, 1974). Similar structures are known from many areas of different continents. However, they are still poorly known, so that a few examples are described from the Northern Hemisphere. In Siberia they are present, in addition to those of the Aldan shield, in the Baikal mountain area, where parts of fold ovals are present in median massifs (blocks) composed of Archean rocks (Salop, 1964—1967). According to geophysical data (mainly magnetic surveying) concentric structures commensurate with the fold ovals may be outlined in the basement of the Siberian platform. They are also present in the area of the Irkutsk amphitheatre and in the region north of the Aldan shield, and in the northern part of the platform.

Such structures have also been recognized in air photographs of the Anabar shield. Within this shield area parts of three fold ovals of different size may be seen. The largest of these is also the most easterly. It appears to be about 750 km along the long axis. This estimate is based on studies of the orientation and shape of folds. The median oval has a complicated lobate shape, with gneiss domes confined to its central part. It resembles in this

Fig. 12. Gneiss fold ovals in the Archean complex of the Aldan shield.
1 = gneiss fold ovals; *2* = intra-oval regions; *3* = post-Archean rocks of the platform cover; *4* = strike of fold axes in fold ovals; *5* = strike of fold axes and partly of schistosity in intra-oval regions; *6* = synclinal zones; *7* = gneiss domes; *8* = norhtern boundary of Stanovoy Range fold belt. *Gneiss fold ovals*: *I* = Chara; *II* = Nelyuka; *III* = Verkhnealdan; *IV* = Verkhnetimpton; *V* = Nizhnetimpton, and Sunnagin fold ovals were presented in earlier articles by Salop (1971b; 1974).

respect the Verkhnealdan "amoeboid". Preferred orientation of folds close to the margins of the ovals caused a predominant northwestern strike of the Archean structures of the massif. Complicated chaotic (non-oriented) folding is typical of the interoval areas. The shape of the ovals under the cover of platform deposits on the Anabar Massif may be approximated from the arch-like curved linear magnetic anomalies (Fig.13).

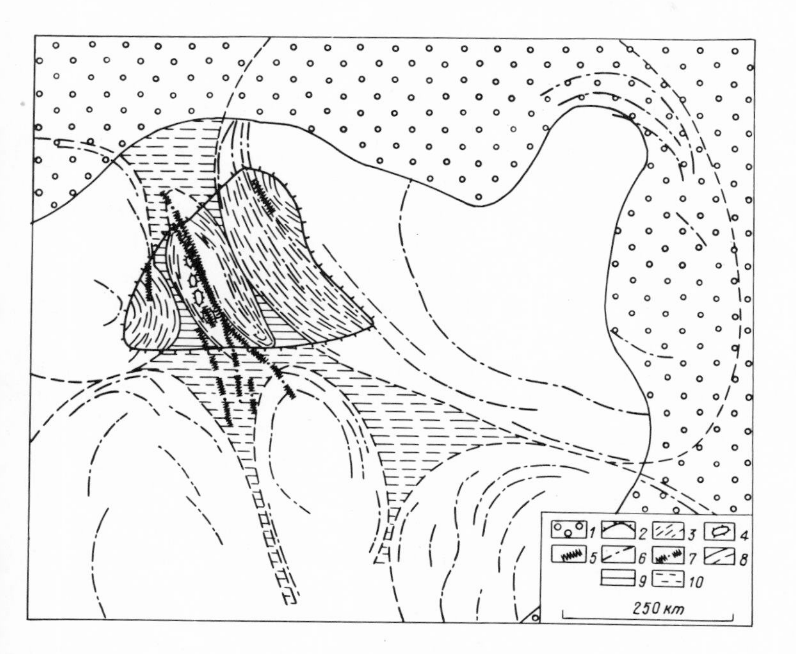

Fig.13. Gneiss fold ovals in the Archean basement of the northern part of the Siberian platform. Fold trends on the Anabar shield are based on detailed air-photo interpretation, magnetic anomaly trends are based on data of L.V. Bulina.
1 = marginal trough strata; *2* = outlines of the Anabar shield; *3* = fold strikes in Archean gneiss; *4* = gneiss domes; *5* = fault zones (according to aerosurvey interpretations and to geological surveys done at the Scientific Research Institute of Geology of the Arctic); *6* = magnetic anomalies interpreted as folded structures of the Archean basement; *7* = magnetic anomalies interpreted as zones of tectonic dislocation of the basement; *8* = bound-areas of gneiss fold ovals (observed and presumed); *9* = intra-oval regions with typical structures in the Anabar Massif; *10* = intra-oval regions under platform cover (presumed).

Oval fold systems are also characteristic of the Archean basement in the East European platform. In many parts of the shields they are strongly affected by younger Protozoic faults and separated by large granitoid intrusions so that they exist only as fragments (the Ukraine, Karelia). In the Baltic shield the Central Finland granite—gneiss massif was probably originally a gneiss oval. Hypersthene-bearing granite is reported in its inner part. It is surrounded by gneiss (sometimes hypersthene-bearing), migmatite and gneissic granite which make up a concentric fold system up to 400 km in diameter. All these rocks are usually considered to be Svecofennian, but many data suggest that the Central Finland massif is a block (median massif) of Archean rocks in the Svecofennian eugeosynclinal belt, surrounded by Mesoprotozoic (Bothnian) metamorphic strata (Salop, 1971a). During the Svecofennian (Karelian) orogeny Archean rocks suffered selective anatectic mobilization which led to the formation of the allochthonous granite. They also underwent intensive tectonic reworking which obscured the original relationships.

On the basis of geophysical data gneiss fold ovals are clearly developed in the basement of the Russian plate. These anomalies correspond to places where drilling has revealed large areas of gneiss (migmatite) and granulite. Fig.14 shows the structures of fold ovals established in the platform basement according to linear magnetic anomalies of variable intensity (map of magnetic rocks of the basement of the Russian plate, compiled by A.N. Berkovsky et al. and edited by V.A. Dedeyev, 1970). One of the fold ovals plotted on Fig.14 (the Mazovets oval in the eastern part of Poland) is based on information supplied by W. Ryka (1968, 1970) from drilling and geophysical data. Most of the ovals fall between 250 and 600 km. Similar groups of folds in the shape of gneiss fold ovals are present in the Archean rocks of the Canadian shield (the Ungava craton) and in the Bohemian Massif (see Figs.8 and 11).

Precambrian tectonics is not the subject of this work. Special attention was given to the structures of the Archean complexes because of their unique style. The most outstanding feature of Archean tectonics is the grouping of folds in the shape of large closed systems (ovals, "amoeboids") scattered irregularly throughout all areas where gneiss complexes are developed.

These fold groupings suggest that there was no rigid tectonic framework — no bounding by cratonic blocks (platforms). This in turn suggests widespread mobility of the earth's crust during the Archean and characterizes the older permobile stage in its development.

The older gneiss complexes of the Archean are everywhere separated by a great structural unconformity from the overlying Paleoprotozoic complexes. The widespread occurrence of this unconformity suggests that the "Epiarchean" gap was of long duration and that the geological history of formation of the unconformity was long and complex with diachronous migration of uplift and submergence of the earth's crust in different places (see Fig.3).

Radiometric dating (mainly Pb-isotope, Pb-isochron and Rb—Sr isochron analyses) of Archean plutonic and metamorphic rocks yields ages of about 3,500 m.y., sometimes as old as 3,700 m.y., for the time of the earliest thermal processes. This is probably the age of the most intensive metamorphic and plutonic processes of the Saamian diastrophic cycle which closed the Archean Era and the permobile stage in the development of the earth's crust (corresponding to the Cryptozoic Eon). Thus, 3,500 m.y. is to be taken

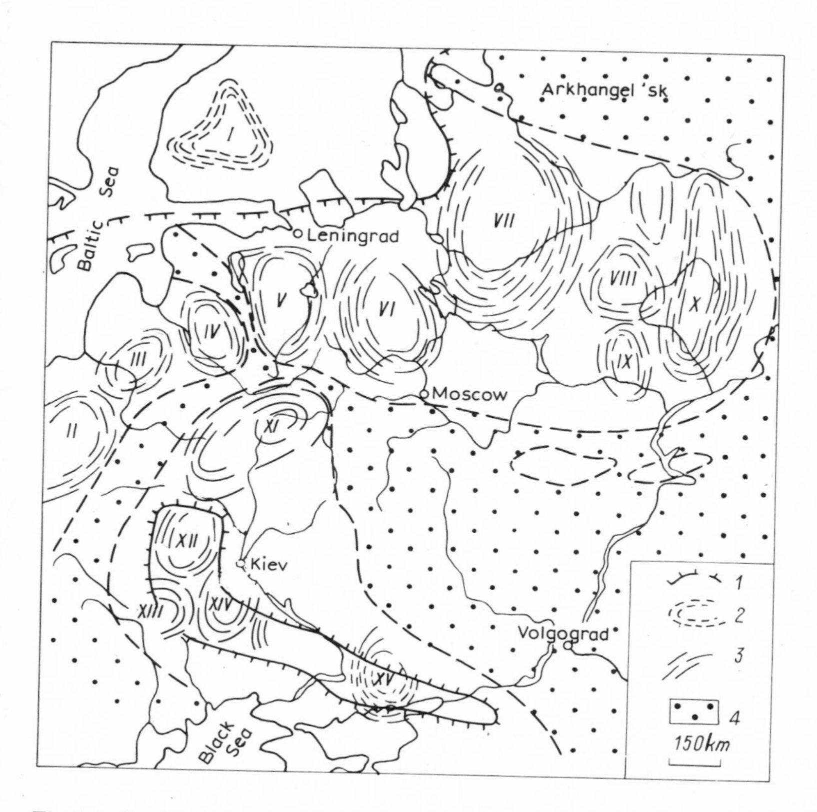

Fig. 14. Gneiss fold ovals in the Archean basement of the East European platform. In the Russian plate the structures are shown according to a magnetic map edited by Dedeyev (1970), in Byelorussia according to data of Dominikovsky and Medushevskaya (1973), in the western part of the Ukrainian shield according to Gintov (1973), and for Poland according to Ryka (1970).
1 = boundary of shields; *2* = fold axes of Archean rocks in ovals on the shields; *3* = linear magnetic anomaly; *4* = fields of linear magnetic anomalies of "belt-type" in the basement. *Gneiss fold ovals*: *I* = Central Finland; *II* = Mazovets; *III* = Kaunas; *IV* = Resekne; *V* = Novgorod; *VI* = Bologoye; *VII* = Vologda; *VIII* = Moloma; *IX* = Sanchura; *X* = Vyatka; *XI* = Byelorussia; *XII* = Zhitomir; *XIII* = Khmelnitsk; *XIV* = Uman; *XV* = Priazov(?).

as the upper geochronological boundary of the era (and eon).

The lower Archean boundary is not certain because its base is nowhere seen. However, later on (Chapter 10) some arguments will be presented in favour of the idea that supracrustal rocks of basal aspect are present in the lower part of the most complete Archean sections such as that of the Aldan shield. Thus, the possibility of finding pre-Archean (Katarchean) basement in some places cannot be ruled out.

CHAPTER 5

THE PALEOPROTOZOIC

Rocks formed after the Saamian diastrophism and before the end of the Kenoran diastrophism are attributed to the Paleoprotozoic Era (between 3,500 and 2,600 m.y. ago). Supracrustal strata of this era generally lie unconformably on Archean rocks. In some areas relationships with the underlying rocks are obscured by faulting or later folding which remobilized the older tectonic grain. Original relationships can, however, be deciphered in some areas. The main lines of evidence are as follows: presence of clasts of Archean metamorphic and plutonic rocks in Paleoprotozoic terrigenous deposits, marked changes in degree of metamorphism and different tectonic styles of both complexes, etc.

Paleoprotozoic rocks are widespread in many parts of the world. Usually they form a large structural stage which is quite different from the underlying Archean and the overlying Mesoprotozoic. They appear to form single complexes together with Mesoprotozoic strata only in some areas of Asia. The Paleoprotozoic rocks are commonly only slightly metamorphosed in contrast to the Archean rocks. Original structures and textures of both sedimentary and igneous rocks are commonly preserved in the Paleoprotozoic sequences. In many cases these formations are well dated by various isotopic methods and they are well defined among other Precambrian strata.

Regional Review and Principal Rock Sequences

Europe

Metamorphic rocks belonging to the Paleoprotozoic Era are widely developed in the *East European platform* where they outcrop in the Baltic and Ukrainian shields and are reported from drillings in a number of locations on the Russian plate where, together with Archean rocks, they form the basement of the platform. This era is characterized in many regions by strong development of eugeosynclinal sedimentary—volcanic rocks with associated jaspilite. The Gimola Group of Western Karelia is suggested as the European stratotype for these deposits. It lies unconformably on Archean gneisses and granites and is cut by granite and granodiorite with an age of 2,750—2,770 m.y. by Pb-isotope analysis on accessory zircon and sphene (Kouvo and Tilton, 1966; Tilton, 1968). These dates are from granitoids that cut the Gimola Group in Finland near the boundary with the U.S.S.R. In that region rocks of the Gimola Group are known as the Ilomanti strata. In the U.S.S.R. schist of the Gimola Group gave an age of 2,750—2,830 m.y. by K—Ar analysis of amphibole (Gerling et al., 1965b), and sulphide ores in rocks of this group have an age of 2,480—2,530 m.y. by the Pb-model method (Tugarinov and Voytkevich, 1970).

According to Chernov et al. (1970) the Gimola Group is composed of four formations which, in ascending sequence, are: (1) Sukkozero (150 m) comprised of metaconglomerate and tuff conglomerate with interbands of biotite, amphibole schist and meta-andesite; (2) Kostamuksha (40—350 m) composed of quartz—mica and chlorite schist, and leptite with bands of magnetite quartzite (jaspilite); (3) Mezhozero (40—500 m) made up of leptite gneiss, tuff breccia and tuff schist; and (4) Kadiozero (80—620 m) similar to the underlying formation, but containing interbeds of magnetite-rich rocks. According to Kharitonov et al. (1964) the Sukkozero conglomerate is in the upper part of the group, and it is underlain by schist and gneissose meta-arkose beneath the Kostamuksha Formation.

In other areas of Karelia the following sedimentary—volcanogenic groups are correlative with the Gimola Group: the Khautovara, Yalonvara, Bergaul and Parandova. These groups are characterized by greater abundance of volcanic rocks and fewer interbeds of jaspilite ores. All these groups are in a similar stratigraphic position to those of the Gimola Group (between Archean and Mesoprotozoic rocks), and are of the same age (2,600—2,800 m.y.) according to different isotopic analyses (Salop, 1971a). The Pb-isotope age of zircon from granodiorite cutting the Khautovara Group is 2,800 m.y. (Bibikova et al., 1964).

In the Kola Peninsula there are sedimentary and volcanic groups analogous to the stratotype. These include the Tundra, Voronya—Kolmozero and Lebyazh'ya Groups. The Lebyazh'ya Group, until recently, was considered to be a formation, and was attributed to the lowest part of the Keyvy Group. However, the Keyvy Group comprises a variety of strata that are probably not all coeval (see below), and the Lebyazh'ya Formation is a discrete sedimentary—volcanic group (Golovenok, 1971; Mirskaya, 1971). The Voronya—Kolmozero Group is here considered to exclude the upper Chervurt Formation which unconformably overlies the metavolcanic rocks. The age of metamorphism of the rocks of the Voronya—Kolmozero and Tundra Groups is 2,600—2,800 m.y. (K—Ar analysis on mica and amphibole; Garifulin, 1971).

Correlatives of the Paleoprotozoic rocks of Karelia and the Kola Peninsula are widely developed in the western part of the Baltic shield outside of the U.S.S.R. These rocks include greenstones associated with jaspilite and known as the Koli—Pielisjarvi, Nurmes, Ilomanti, Sodankila and Vihdi strata, and some others in Finland; leptite formations of Central Sweden, Sør—Varanger (Bjornevatn) strata in Northern Norway, and possibly the Rjukan Group in Southern Norway. The radiometric age of most of these groups is "rejuvenated" by later metamorphic processes (predominantly Mesoprotozoic — Karelian—Svecofennian) but their stratigraphic position in the Precambrian section is quite analogous to that of Paleoprotozoic strata in the eastern part of the shield. Their age, correlation and comparison with corresponding groups of the Karelian Kola region have been described in an earlier paper by the author (Salop, 1971a).

Rb—Sr dating of the plutonic Haparanda complex, which intrudes sedimentary and volcanic jaspilite-rich rocks of the Lapponian (s.str.) in Northern Sweden and adjacent areas of Finland, produced some interesting results concerning the age of Paleoprotozoic strata in Scandinavia. The Haparanda complex is a series of intrusive rocks including rare anorthosite and hornblendite and widely distributed gabbro, diorite and granite. The granites cut the basic rocks, and are grey and coarse-grained, with porphyroblasts of microcline which resemble the well-known Revsund granite of the Norbotten province in Sweden. There is also a younger red granite of the Karelian (Mesoprotozoic) cycle. Rb—Sr datings of microcline-bearing rocks of the Haparanda complex by Brotzen (1973) gave points that lie on a line between two isochrons (2,680 and 1,820 m.y., respectively). According to Brotzen "the greatest value (2,680 m.y.) is the best indication of the age of emplacement of the complex" (p.22). The lowest values indicate the time of later biotitization and microclinization, corresponding to the Karelian cycle.

The position of the sedimentary and volcanic rocks of the Kiruna complex (Kiruna—Arvidsjaur) in the Precambrian section of Northern Sweden is uncertain. Swedish geologists such as Ödman (1972) assigned it to the Early Svecofennian. However, recently a Rb—Sr isochron age in the 1,600—1,725 m.y. range was reported for volcanics of this group. Because of these dates, and also because of the low metamorphic grade of this group, many Swedish geologists (E. Welin, G. Erikson, G. Kautsky) suggested separation of this complex from other sedimentary, volcanic and ferruginous strata of the Svecofennian area and that it should be considered "Upper Karelian" (sub-Jotnian). Many geological data cited by Ödman (1972) tend to contradict this view. Rb—Sr isochron dates from acid volcanics of the Kiruna Group are close to, or slightly lower than, dates from the younger Revsund granite (1,785 m.y.). These older dates coincide with Rb—Sr dates from the Archean (?) Vuolosjärvi—Vakkojäd granite gneiss which underlies the Kiruna Group. Pb-isotope dates on zircon and sphene from the granite gneisses gave an age of 2,750—2,800 m.y. (Kouvo and Tilton, 1966). The Karelian (Mesoprotozoic) Lina granite, in the same area of Northern Sweden, gave a Rb—Sr isochron age of 1,565—1,820 m.y. (Welin et al., 1971). All of these facts indicate a general isotopic "rejuvenation" due to the effects of the Karelian orogeny. Probably the basement granite gneiss was already mobilized at the end of the Paleoprotozoic (during the Kenoran orogeny).

In the *Ukrainian shield*, greenstone-type volcano-sedimentary assemblages (with jaspilite) in the Dnieper region and Krivoy Rog area provide examples of eugeosynclinal Paleoprotozoic sequences. They are known under different names, but the ones commonly used (Konksko—Verkhovtsevo or Konksko—Belozersk Groups) are poor ones because, in the areas of the Konksk and Belozersk magnetic anomalies, there are also younger metamorphic rocks which are erroneously included in these groups. The names Verkhovtsevo or Bazavluk Groups are suggested after the magnetic anomalies of the same names where these rocks are widely developed and studied in detail.

Sedimentary and volcanic rocks form composite synclines, the shape of which is determined by the gneiss—granite domes separating them. At the boundary between the gneissic granites and greenstone belts there is commonly a narrow zone in which the volcanic rocks are feldspathized and migmatized. The greenstone assemblage tends to be metamorphosed under amphibolite or epidote-amphibolite facies condition. The metamorphic grade of the rocks decreases rapidly to greenschist facies away from the granites. Most of the rocks of the group are only slightly metamorphosed, and many primary structures and textures are well preserved. Amygdaloidal lavas with porphyritic and spilitic textures are common among the basic volcanics. Zonal metamorphism developed around the domes appears to be related to thermal and tectonic remobilization of the gneiss and granite basement during formation of the domal structures. Deposition of the volcano-sedimentary assemblages probably took place on the eroded surface of older crystalline rocks. Such unconformable relationships are established only in the Krivoy Rog area where conglomerates and breccias are present at the base of the metabasite succession (Kukhareva, 1972), and quartzites overlie a fossil soil developed on gneisses and granites (Dobrokhotov, 1969).

The greenstones of the Dnieper area have yielded a K—Ar age (amphibole) of 2,600—2,800 m.y. (Ladieva, 1965). The age of the basement gneiss granite is 3,500 m.y. Iron-rich rocks of the Krivoy Rog Group unconformably overlie different metabasite units and, in some cases, lie directly on the basement gneissic granites.

In central and western parts of the Ukrainian shield the thick Teterev (Zvenigorod) Group probably lies unconformably on the Archean Bug and Ros'—Tikich Groups. The stratigraphy of this Paleoprotozoic group is poorly known, but the lower part includes metaconglomerate, biotite- and sillimanite-bearing gneiss and amphibolite (Stanishov, Gorodok Formation). The overlying Kocherov Formation is made up of metabasite (amphibolite and amphibole gneiss) with interbands of marble (Shotsky, 1967; Laz'ko et al., 1970). The Pb-isochron age of metamorphism of biotite gneisses of the Teterev Group is 2,750—2,800 m.y. (material of A.D. Dashkova and T.V. Bilibina of the V.S.E.G.E.I. Laboratory).

Metamorphosed sedimentary and volcanic rocks, which probably correspond to the Teterev Group, are reported in the basement of the Russian plate in the Volyn'—Podolia areas adjacent to the western part of the Ukrainian shield (Yatsenko et al., 1969).

In the *central part of the Russian plate* in the area of the Kursk magnetic anomaly (Voronezh Massif), the Mikhaylovsk (Staryoskol) Group is Paleoprotozoic. It is composed of metabasite and altered acid volcanics with interbeds of magnetite-silicate ores. At the base of the Mikhaylovsk Group the

Archean gneisses locally have a weathered zone (paleosol) containing dickite, metagalluasite and other aluminous minerals (Bobrov et al., 1973). The Mesoprotozoic Kursk Group, with conglomerate at its base, unconformably overlies rocks of this group and the underlying gneissic granites. The metamorphic age of the Mikhaylovsk Group volcanics is about 2,700 m.y. according to K—Ar dating.

In Eastern Byelorussia the iron-rich Okolov Group is probably correlative with the Mikhaylovsk Group. It is known from drill holes in the platform basement.

In the fold belts around the East European platform Paleoprotozoic strata are present in the *Urals* and, probably, in the Caucasus. In the western slopes of the Urals, sedimentary and volcanic rocks of the Tukmalin Group in the Taratash complex unconformably overlie gneisses of the Archean Arshin Group and are transgressively overlain by the Mesoprotozoic Burzyan Group. They are accordingly assigned to the Paleoprotozoic. According to Smirnov (Smirnov, 1964; Abdulin and Smirnov, 1971) this group is made up of micaceous metamorphic quartzites, schists and gneisses, with thick metadiabase, metaporphyrite and amphibolite units associated with banded iron formations (jaspilites). The metamorphic age of the Tukmalin Group gneiss is 2,700 m.y. by α-Pb analysis on zircon.

Paleoprotozoic strata are probably widely distributed in the axial part of the Urals fold belt, but their radiometric age is difficult to determine because of several superimposed metamorphic episodes. The Ufaley, Sysert and Ilmenogorsk metamorphic complexes are the most reliably identified as Paleoprotozoic. They occur in the central Uraltau zone of the Central and Southern Urals. The Kharbey metamorphic complex of the Polar Urals is probably also Paleoprotozoic. The Rb—Sr isochron age of gneiss in the Ilmenogorsk complex is 2,500 m.y. (Ovchinnikov et al., 1968), and granite cutting the Kharbey complex gave ages from 1,670 to 2,490 m.y. by Pb-isotope analysis on zircon (Pronin, 1965). The Shatmaga strata of the sub-Polar Urals and the Taldyk Group of Mugodzhary unconformably overlie the Archean (?) Kainda gneiss complex and are probably also Paleoprotozoic. All of them are thick, strongly metamorphosed, volcano-sedimentary complexes. Some of them (the Ufaley complex, for instance) contain banded magnetite quartzites and magnetite schists associated with metavolcanics.

In the Caucasus amphibolitic gneisses of the Zelenchuk Group in the Main Caucasus Ridge lie on an older Precambrian basement (Potapenko, 1969). These are tentatively assigned to the Paleoprotozoic.

Perhaps Paleoprotozoic rocks are also present in the *Caledonides and Hercynides of Western Europe*, but they are not identified with any degree of certainty. It is highly probable that they are present in the Highlands of Scotland where granitoid intrusions dated at about 2,600 m.y. are reported and retrograde metamorphism of the same age (Scourian) is superimposed on Archean rocks.

The so-called Pentevrian complex of the Armorican Massif of France is made up of various gneissoid granodiorites, among which areas of crystalline schist, amphibolite and migmatite are reported. These rocks may also be Paleoprotozoic. According to new geological and radiometric (Rb—Sr isochron) data this complex suffered three periods of metamorphism (Roach et al., 1972). The first occurred at the end of the Paleoprotozoic about 2,800—2,700 m.y. ago (the Icartian=Kenoran cycle), and later ones took place at the end of the Mesoprotozoic — 2,000—1,900 m.y. ago (the Lihouan=Karelian cycle), and at the end of the Neoprotozoic — 1,100—900 m.y. ago (the Late Pentevrian=Grenville cycle). The Neoprotozoic strata (Brioverian) lie unconformably on the "basement complex" described above.

Asia

In Asia Paleoprotozoic supracrustal and plutonic complexes are best studied in Siberia, especially in the fold belts on the southern margin of the Siberian platform. They everywhere include variously metamorphosed volcano-sedimentary complexes lying unconformably on Archean rocks. Two types of sections are recognized on the basis of the nature of their relationships with the overlying Mesoprotozoic rocks. In the first type, Paleoprotozoic rocks (and intrusive rocks cutting them) are overlain unconformably by Mesoprotozoic rocks. The second situation is characterized by conformable relationships between the rocks of both complexes. The second type is mostly typical of the Baikal and East Sayan fold zones where the Paleoprotozoic and Mesoprotozoic strata form a single structural stage, though the rocks of both eras are distinctly different in lithology and are sometimes separated by local stratigraphic unconformities.

Formerly such continuous sections were assigned as a whole to the Paleoprotozoic (or Lower Proterozoic), but now their age needs revision. Their stratigraphic position is well established for they lie unconformably on gneisses of the Archean basement, and are cut by Mesoprotozoic granites, the age of which is reliably placed between 1,900 and 2,000 m.y. by different isotopic methods. Neoprotozoic strata (including the Lower Neoprotozoic) unconformably overlie them and the granites. These complexes therefore formed in the time interval between the end of the Archean and the beginning of the Neoprotozoic. The upper parts of these complexes include rocks with stromatolites and oncolites typical of the Mesoprotozoic. It seems unlikely, however, that these complexes are totally Mesoprotozoic, and that Paleoprotozoic strata are completely missing from these regions. Many data indicate that rocks of both eras are present in these complexes. Perhaps the best evidence is the good correlation of the lower and upper parts of such complexes with Paleoprotozoic and Mesoprotozoic strata, respectively, of adjacent areas where they are separated by an angular unconformity and dated by isotopic methods. Locally, along strike, there is a gradual transition of the upper part of the continuous complex into Mesoprotozoic strata. The latter unconformably overlie Paleoprotozoic strata which are equivalent to the lower part of this complex.

The continuous complexes, in almost every case, possess a twofold structure, which is shown either by the existence of two incomplete sedimentation cycles, each of which starts with terrigenous rocks and finishes with carbonates or by a rather abrupt change (up-section) from volcanic rocks to sedimentary or volcano-sedimentary (mostly terrigenous) rocks. In this case a local unconformity is developed at the boundary of the two units.

In the eugeosynclinal parts of fold belts in Southern Siberia the lower part of these sequences, comprising the volcanic and sedimentary strata, includes finely banded ferruginous and silica-rich rocks (jaspilites). This association of jaspilite with volcano-sedimentary assemblages is typical of the Paleoprotozoic in Europe. Volcanogenic rocks of spilite—keratophyre type, similar to those in the lower parts of eugeosynclinal complexes of Siberia, are also characteristic of many other areas of the world.

The Paleoprotozoic stratotype for Siberia is best placed in the well-studied sections of the *Baikal fold belt* (Salop, 1964—1967). The lower, essentially volcanic, part of the Muya Group (including the Parama and Kilyana Subgroups), is chosen as a stratotype. Its general thickness is more than 6,000 m. Its relationships with underlying and overlying rocks is certain. It overlies Archean gneiss with angular unconformity and is itself conformably overlain by the volcano-sedimentary Zama Subgroup, at the base of which a thick quartzite formation is everywhere reported. Possibly this unit represents the beginning of the new Mesoprotozoic cycle of sedimentation (Fig. 15).

The base of the Parama (lower subgroup of the Muya Group) is made up of metamorphosed conglomerates and sandstones with some metavolcanic horizons (Samokut Formation). Above these rocks there are marbles, interbedded with metavolcanics and highly altered tuffs (Bulunda Formation). The Kilyana Subgroup which makes up the greater part of the Paleoprotozoic section, is largely composed of acid and basic volcanics and tuffs of the spilite—keratophyre association which typically occurs at the base of eugeosynclinal sections of various ages. Its diagnostic feature is the occurrence of jaspilite bands in certain facies.

The Paleoprotozoic age of the lower part of the Muya Group (Parama and Kilyana Subgroups) is shown by the fact that it may be correlated with volcano-sedimentary Paleoprotozoic rocks developed on the northeastern flank of the Baikal fold belt where it is juxtaposed against the Stanovoy fold belt, near the Chara block of the older craton (protoplatform). It has been established (Salop, 1964—1967) that the upper part of the Muya Group (Zama Subgroup) is a facies equivalent of the terrigenous Mesoprotozoic Udokan Group. The latter unconformably overlies the volcano-sedimentary complex (see Chapter 6).

The volcano-sedimentary rocks have recently been studied by Fedorovsky

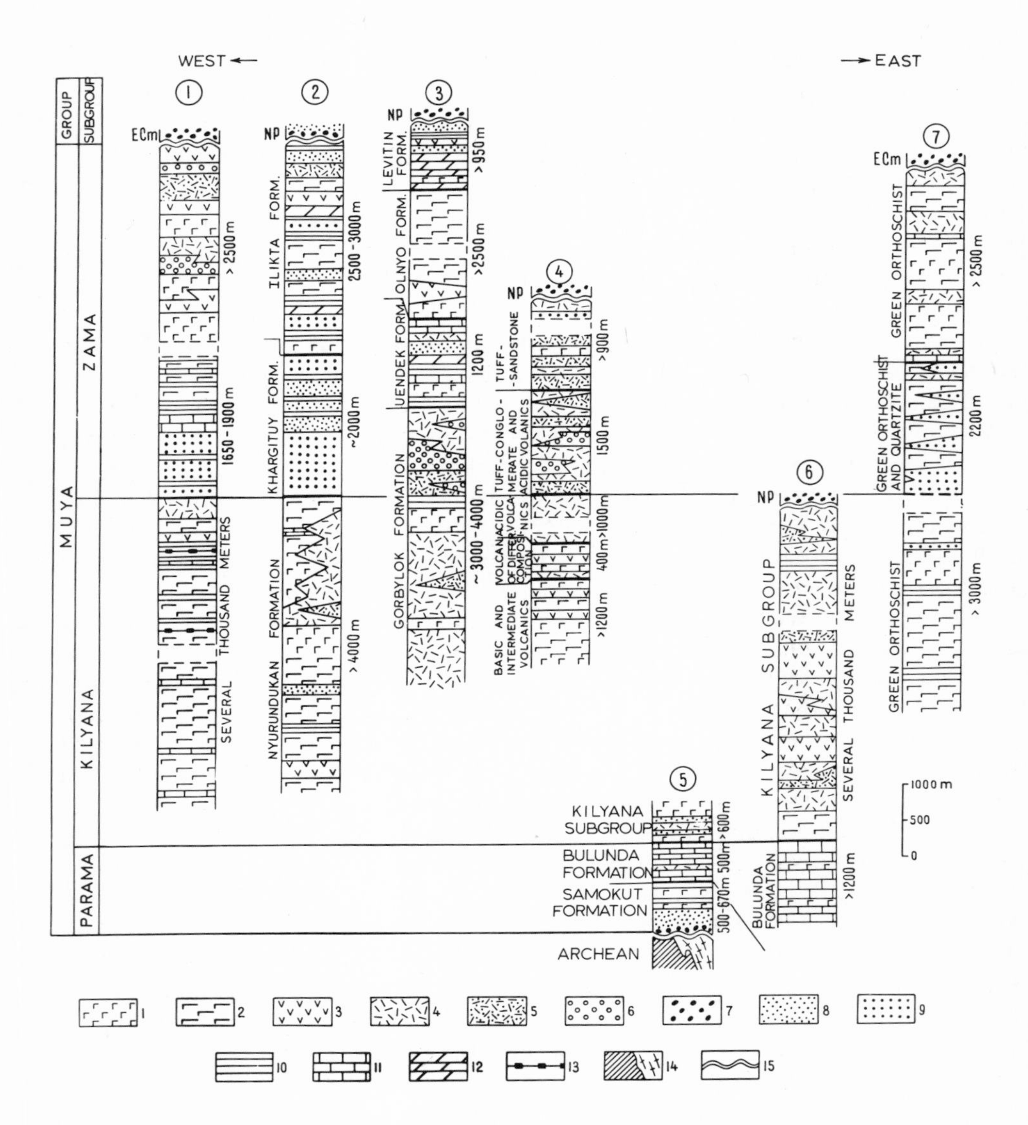

Fig. 15. Correlation of the Muya Group sections in the Baikal fold belt.
1 = basic volcanics; *2* = green orthoschist and orthoamphibolite; *3* = intermediate volcanics; *4* = acid volcanics; *5* = tuff of predominantly acid and intermediate composition; *6* = tuff—conglomerate; *7* = conglomerate; *8* = lithic sandstone; *9* = quartzite; *10* = paraschist; *11* = crystalline limestone; *12* = crystalline dolomite; *13* = jaspilite; *14* = Archean gneiss, marble and granite gneiss; *15* = angular unconformity; *NP* = Neoprotozoic; *ECm* = Eocambrian. *Numbers in columns*: *1* = Northwestern Baikal region (Tyya and Kholodnaya Rivers); *2* = Western Baikal region (Baikal and Primorian Ranges); *3* = South Muya Range (Gorbylok River); *4* = North Muya Range (Kilyana River); *5* = North Muya Range (Samokut River); *6* = Delyun—Uran Range (Yanguda River); *7* = western part of the Kodar Range (Taloi River).

and Leytes (1968), Nuzhnov et al. (1968), Mironyuk et al. (1971) and also by V.A. Kudryavtsev and L.V. Travin. The thick volcano-sedimentary Olondo Group (up to 2,500 m) is typical. It is locally developed in the Olondo and Khani River regions in the Khani—Olondo graben. Nuzhnov et al. (1968) called it the Olondo Formation and it was considered to be part of the Sakhabor Group. Recent work,however, showed that some younger (Mesoprotozoic) strata of the Udokan Group were erroneously included with it. The Olondo Group is made up of metavolcanics of both basic and acid composition, with subordinate interlayers of metasandstone, metatuffite and micaceous schist under greenschist and epidote-amphibolite facies. These rocks are cut by amphibolitized gabbro diabase and also by granite.

In the Khani—Olondo graben the Olondo Group is infolded among Archean gneisses and its relationships with (overlying) Mesoprotozoic strata of the Udokan Group are not clear. Some workers (Fedorovsky and Leytes, 1968) believe that the rocks here assigned to the Olondo Group, in fact conformably overlie the Udokan Group. New studies made by V.A. Kudryavtsev, however, showed that the Olondo Group overlies Archean gneisses, and that the Udokan Group overlies the Olondo with angular unconformity.

Sedimentary and volcanic rocks in the southern part of the bald mountains of Burpala, and to the south, in a large region of the Kalar River basin, and also in some other areas of the Olekma—Vitim mountain-land are very similar to the Olondo Group (data of V.S. Fedorovsky and A.M. Leytes). These rocks are metamorphosed volcanics, psammites and pelites with bands of magnetite—cummingtonite schist, and banded iron formation (jaspilite) associated with basic metavolcanics (amphibolites). There are also some units of marble and conglomerate. Metamorphism of these rocks commonly reaches the amphibolite facies. All these strata are undoubtedly younger than the Archean gneisses, but their relationships with the Udokan Group are not certain, because the contacts are usually tectonic. There is a significant change in degree of metamorphism between these rocks and those of the Udokan Group. It is commonly believed that the Udokan Group is younger. The same relationship is indicated by isotopic age data.

Mironyuk et al. (1971) included all these rocks in the Borsala Group (all the metamorphic rocks of the Olekma—Vitim mountain-land and those of the western part of the Aldan shield that contain iron formations). The Borsala Group appears to be heterogenous, comprising both Paleoprotozoic volcano-sedimentary strata and Archean gneiss with interbedded magnetite-rich crystalline schists and quartzites. In addition to these strata, the Borsala Group includes many other locally developed Paleoprotozoic volcano-sedimentary iron-bearing rocks occurring in grabens among Archean gneisses of the Chara block of the Aldan shield (e.g. in Tyany and Oryus—Miele River basins).

According to V.S. Fedorovsky and A.M. Leytes these rocks were formed in narrow geosynclinal troughs bounded by faults formed at the time of initiation of the mobile belt. These authors pointed out some troughs of this kind in the Olekma—Vitim mountain-land. However, some strata attributed to such "trough complexes" are in fact developed within the limits of the broad Baikal eugeosynclinal belt and belong to the Muya Group. This refers particularly to rocks of the so-called "Tallai trough", and the volcano-sedimentary assemblage that occupies large areas in the Kalar River basin ("the Kalar trough"). These facts underline once more the close similarity of strata developed in "troughs" to rocks of the Muya Group.

The Paleoprotozoic age of the volcano-sedimentary assemblage of the Olekma—Vitim mountain-land is clearly shown by isotopic analysis. K—Ar dates on muscovite from pegmatites that cut metamorphic rocks of the complex yield an age of 2,370—2,420 m.y. (Fedorovsky and Leytes, 1968) in the bald mountains of Burpala on the Kalakan River, and 2,510—2,540 m.y. (Mironyuk et al., 1971) in areas to the north, on the Nalstak River. Similar muscovite pegmatites among Archean gneisses on the Khani River gave an age of about 2,900 m.y. by the same method on mica (data of M.Z. Glukhovsky). The Archean pegmatites of Eastern Siberia do not contain any muscovite (with the exception of rare retrograde occurrences), but various muscovite-bearing pegmatites are abundant in the volcano-sedimentary complex and in some younger rocks. Formerly all of these pegmatites in the Olekma—Vitim mountain-land were attributed to the Kuanda granitoid complex which is younger than the Mesoprotozoic Udokan Group (Salop, 1964—1967). The Kuanda granitoid, however, is 1,900 m.y. old, and the other granitoids are somewhat older. It is probable that they form a separate complex which is given the name "Charodokan" by Mironyuk et al. (1971). Pegmatite with allanite and muscovite, cutting Archean iron formation in the southernmost part of the Chara block close to exposures of volcanic and sedimentary rocks, also belong to the complex. The Pb-isotope age of these pegmatites (allanite) is determined as 2,600—2,700 m.y. (Salop, 1964—1967).

The age of 2,670 m.y. obtained by K—Ar dating of amphibole from garnet—amphibole orthoschist of the Borsala Group, developed on the Oryus—Miele River, is important in establishing the age of metamorphism of the volcano-sedimentary assemblage of the Olekma—Vitim mountain-land (Mironyuk et al., 1971). It appears that the Olondo and its analogues, including the lower part of the Muya Group, were formed before 2,600—2,700 m.y., but later than 3,500 m.y. (the metamorphic age of the Archean gneiss), and consequently, all of them belong to the Paleoprotozoic.

It has been pointed out (Salop, 1964—1967) that the Chuya sequence in the North Baikal Highlands, the Sarma (Muya) Group of the Western Baikal region and the Garga Group of the Upper Vitim basin all correspond to the Muya Group.

In the *western part of the Aldan shield* the Subgan Group, which is locally developed in the Subgan graben, corresponds to the Olondo Group. The lower part of this group is composed of metamorphic schist, and the upper part is various schistose metaporphyritic rocks. It resembles rocks of the Olondo graben.

In the *Stanovoy fold belt* the volcano-sedimentary (metamorphic) sequence of the Yankan—Dzheltula complex, developed in the central part of the Stanovoy Ridge (Sudovikov et al., 1965), and the Kavykta, Chasovinsk and Sobolinsk Formations as a whole (the Amazar—Nikitkin Group) in the southwestern part of this area (Kirilyuk, 1966) are close analogues of the Muya Group. Their similarity with the Muya Group includes the presence of identical intrusions of gabbroid, diorite and plagiogranite. The K—Ar age (2,000—2,100 m.y.) of metamorphic rocks of the Yankan—Dzheltula complex is probably "rejuvenated" by superimposed thermal processes. The possibility that the metamorphic complex of the Stanovoy Ridge includes both Paleoprotozoic and Mesoprotozoic rocks as does the Muya Group, cannot be ruled out. The orogeny that closed formation of this complex could be assigned to the Karelian cycle.

On the southern margin of the Siberian platform, west of the Baikal mountain area, analogues of the Muya Group are reported from the *Sayan region*, where they are represented by volcano-sedimentary rocks in the Urik—Iya and Onot grabens (Buzikov et al., 1964; Dodin et al., 1968). The presence of banded iron formation is a peculiar feature of these rocks (in the Sosnovy Bayts and Urik Formations). As stated above, the same strata are developed in the Muya Group of the Northwestern Baikal region. Carbonate rocks are present in the lowermost Paleoprotozoic of the Sayan region (the Ondot and Kamchadal Formations), and in the stratotype section for these rocks in the Baikal mountain area (the Bulunda Formation). New data of A.Z. Konikov indicate that in the Urik—Iya graben, an essentially terrigenous unit, the Daldarma (Belorechensk) Formation unconformably overlies sedimentary and volcanic rocks, whereas in the Onot graben, its analogue (the Iret Formation) overlies a volcanic sequence without any break. Both of these formations are considered correlative with the Zama Subgroup of the Muya Group which is Mesoprotozoic. Thus, in the Urik—Iya graben, the orogeny that concluded the Paleoprotozoic Era, manifested itself not only by a change in rock type (formation of thick clastic sequences), but also by formation of an angular unconformity.

Paleoprotozoic strata are probably present in *Khamar-Daban*. The Precambrian stratigraphy of this area is still poorly known and the contacts between different metamorphic units there are mostly tectonic. Correlation of these rocks with stratotypes is obviously suspect. The thick Khangarul sequence ("group") of metamorphosed terrigenous rocks with subordinate marbles and amphibolites (basic metavolcanics?) is possibly Paleoprotozoic. The relationships of these strata to Archean rocks are not certain. Many workers have suggested that it is an upper unit of the Archean complex, but its lower grade of metamorphism in comparison with the Archean rocks, and the abrupt change in degree of metamorphism in some places suggest that it is younger (Paleoprotozoic). The Khangarul sequence is very similar to the Garga Group which is developed in an area far removed from the zone of deep faults and is therefore poor in volcanogenic rocks.

In Khamar-Daban the Irkut Formation and its analogues (the Naryn and Ingasun Formations) which consist of marble interbedded with highly metamorphosed terrigenous and probably, volcanic rocks (Dodin et al., 1968), is considered to be Paleoprotozoic, but in a higher stratigraphic position than the units discussed above. This formation has no close analogues in the stratotype sections of the Baikal mountain area, but it has some resemblance to part of the Paleoprotozoic section in the Urik—Iya graben of the Sayan region.

In the *East Sayan fold belt* and adjacent areas of the Altai—Sayan fold zone, for instance in Tuva, Paleoprotozoic supracrustal rocks are closely related to conformably overlying Mesoprotozoic strata. In the Uda—Derba Group (Alygdzher and Khana Formations), and in the Irkutsk—Oka zone, the corresponding part of the Kitoy Group (Arkhut and Irkut Formations). These thick groups include both Paleoprotozoic and Mesoprotozoic rocks, because their lowermost units (Alygdzher and Arkhut Formations) are very similar to the Khangarul and Garga successions in composition, structure and degree of metamorphism. They also resemble the Teskhem and Mugur Formations of Tuva. The Mugur Formation includes jaspilite, which is characteristic of Paleoprotozoic assemblages. The Arkhut Formation according to Mankovsky (see Dodin et al., 1968) transgressively overlies Archean rocks. The upper formations of both groups (Khana and Irkut Formations) are composed essentially of carbonate or terrigenous—carbonate rocks similar to those of the Irkut Formation of Khamar-Daban and the Urik Formation of the Sayan region.

Oncolites are reported in the uppermost part of the Derba Group (Zhaima Formation). They are rare in the Paleoprotozoic, but are common in younger subdivisions of the Protozoic. The Zhaima Formation is unconformably (locally conformably) overlain by the Kuvay Group which contains oncolites typical of the Neoprotozoic. Analogues of the Zhaima Formation in Tuva (Bilinsk and Chartyssk Formations) are conformably overlain by the Aylyg and Naryn Formations with the same oncolites. Both groups are made up of two incomplete sedimentation macrorhythms, each beginning with terrigenous strata, and finishing up with carbonates. It seems that only the strata of the lower macrorhythm are Paleoprotozoic. The terrigenous or carbonate—terrigenous rocks that initiate the upper macrorhythms (Sigach and Urtagol Formations) represent the beginning of the Mesoprotozoic stage of sedimentation. Such a two-cycled subdivision of the sections in the Derba and Kitoy Groups is probably due to oscillatory tectonics caused by Paleoprotozoic orogenic movements.

In the *Yenisei Ridge area* the Yenisei Group consists of highly altered terrigenous and volcanic rocks. These rocks are similar, both in composition and stratigraphic sequence, to the Khangarul and Garga Groups of the Sayan and Upper Vitim areas, and partly to the Alygdzher and Arkhut Formations of Eastern Sayan. All these units are of eugeosynclinal aspect but have a limited amount of volcanics. The diastrophism at the end of the Paleoprotozoic Era was strong in this area, because in the Yenisei Ridge area Paleoprotozoic granite is widely developed, and the Yenisei Group makes up a separate structural stage, quite distinct from the overlying Mesoprotozoic one.

In the *Taymyr fold belt* Paleoprotozoic strata are subdivided rather arbitrarily because there are few dates that yield "original" ages. The composition of rocks there (orthoamphibolite, amphibole and biotite—garnet gneiss) suggests an eugeosynclinal terrigenous—volcanic assemblage and it is probable that they may be correlated with the Yenisei, Khangarul, Garga and other groups of similar composition.

The relationships of amphibolitic gneisses in the northwestern part of Taymyr (designated the Paleoprotozoic Trevozhnaya Group by Zabiyaka, 1974) to overlying terrigenous rocks of the Mesoprotozoic Khariton Laptev Coast Group, are obscured by reworking during the strong Mesoprotozoic orogeny which resulted in the formation of mantled gneiss domes. In the eastern part of Taymyr an angular unconformity is reported between the Paleoprotozoic and Mesoprotozoic complexes (Ravich, 1963).

In the *Olenek uplift of the Siberian platform* the basement rocks are tentatively assigned to the Paleoprotozoic. These include the Eyekit Group of metamorphosed terrigenous rocks which are cut by basic and acid intrusions (Krasil'shchikov and Vinogradov, 1960; Biterman and Gorshkova, 1969). K—Ar dating of mica from granite and from metamorphic rocks of the group gave dates in the range 1,950—2,080 m.y. These correspond to the Karelian orogenic cycle, but it is also possible that they date the time of Mesoprotozoic "rejuvenation" processes, and the Eyekit Group, and cross-cutting intrusive rocks are older.

This argument is supported by the presence of a great unconformity between the Eyekit Group which forms the basement of the platform, and the horizontally overlying non-metamorphosed (Neoprotozoic) rocks (Solooli Group) which are the oldest rocks of the platform cover. If the Eyekit Group and the abyssal intrusive rocks cutting it were Mesoprotozoic, then it would hardly be possible for them to be so lithologically different. Finally, there must have been a long period of erosion for the metamorphic and intrusive rocks to be exposed and for the folded area to be peneplaned. Jaspilite pebbles are reported in the basal conglomerate of the Epiprotozoic Maastakh Group which unconformably overlies the Eyekit Group (jaspilite is not known from outcrops of the Eyekit Group). In Siberia jaspilite mostly occurs in Paleoprotozoic sequences.

Paleoprotozoic supracrustal rocks are probably also present in Precambrian sections of the Central Asiatic fold system, but they have not been radiometrically dated as yet. They are subdivided into the Kazakhstan and Tien-Shan successions.

In Western *Kazakhstan* the volcano-sedimentary rocks in Ulutau, which form the Aralbay (up to 6,500 m) Group, and the conformably overlying Karsakpay (up to 4,000 m) Group are considered to be Paleoprotozoic. Both groups comprise various altered volcanics of liparite—dacite, keratophyre, basalt and spilite composition, interbedded with tuffaceous rocks and metasedimentary rocks such as schist, phyllite, marble and metasandstone. The presence of jaspilite units at several horizons is typical. The Aralbay Group unconformably overlies gneisses of the Archean Bekturgan Group, and both of these are unconformably overlain by the Maytyube Group which is either Lower Neoprotozoic or Mesoprotozoic. The Zhiydinsk Group (volcanics) was defined by Zaytsev and Filatova (1971) as overlying the Karsakpay Group and underlying the Maytyube Group. This group is probably equivalent to the upper part of the Karsakpay Group. Folding and intrusion of granitoids probably preceded deposition of the Maytyube Group.

In Northern Kazakhstan (Kokchetav Massif) the sedimentary and volcanic rocks of the Borovsk Group (5,000 m) which includes the Yefimovsk and Imanburluk Formations are considered to be Paleoprotozoic. These rocks overlie gneisses of the Archean Zerenda Group, but are transgressively overlain by the Lower Neoprotozoic (?) Kuuspek Group. Some beds of hematitic schist are present in the Yefimovsk Formation.

In *Central Asia* the Paleoprotozoic is only tentatively subdivided. This has been done only in Northern Tien-Shan, where the Kirgiz (Makbal) Group is possibly Paleoprotozoic. It is younger than the Archean gneiss complex developed there. The upper boundary is not certain. The overlying Precambrian rocks are not dated, but are probably Neoprotozoic. Thus, the Kirgiz Group may in fact be Mesoprotozoic. It is composed of various schists, quartzites, marbles and amphibolites under epidote—amphibolite facies of metamorphism. According to Ognev and Bel'kova (see Shatalov, 1968) the Kirgiz (Makbal) Group is similar to the metamorphic complex in the eastern part of the Central Asiatic system. It is developed in Kuruktag (China, Sinkiang province) where it unconformably overlies Archean gneisses.

Rocks of this erathem are probably widely distributed in a number of areas in *China*, but a Paleoprotozoic age is known only for the Anshun Group of the Liaoning province. This group is made up largely of highly altered basic lavas, represented by amphibole, chlorite, actinolite orthoschist and amphibolite interbanded with phyllites, and iron formations (jaspilites) of great economic value. It is unconformably overlain by the Mesoprotozoic Liao Ho Group. According to Polevaya, Li P'u (1965), and Tugarinov and Voytkevich (1970) muscovite from pegmatites cutting the Anshun rocks gave an age of 2,250—2,330 m.y. (by K—Ar analysis) but these values are slightly "rejuvenated" because allanite from the same pegmatite, dated by Pb-isotope analysis gave an age of 2,560 m.y. Thus, the upper age limit (the age of the granite and associated metamorphism) is about 2,600 m.y. That is, it corresponds to the Kenoran orogeny which ended the Paleoprotozoic Era.

The age of the Wutai Group ("system") which forms the basement of the older rocks in Shansi province is even less certain. The studies by Ma Hsing-yuan (1962) show the group to be formed of amphibole, chlorite, mica—quartz and other metamorphic schists, and also calcareous phyllite, among which bands of altered spilite, tuff and iron formation (jaspilite of the "Anshun type") are reported. Most of the group is made up of greenschists which were probably formed from basic volcanics. Because of this the Wutai Group is sometimes called "the system of greenschist". Conglomerates of the Mesoprotozoic Huto Group unconformably overlie the Wutai rocks. Micas (muscovite and biotite) from pegmatite cutting the Wutai amphibolite, dated by the K—Ar method, gave an age of 1,860 m.y. (Tugarinov and Voytkevich, 1970). This means that they are of the same age as the Karelian orogeny which ended the Mesoprotozoic Era. This indicates that the Wutai Group should be attributed to the Mesoprotozoic. However, the Wutai rocks suffered, at least, two orogenic cycles, the most intensive migmatization and granitization being at the time of the Liuliang movements (Ma Hsing-yuan, 1962, p.156) which occurred after deposition of the Huto Group. Thus, the pegmatite age values given above are probably related to the second orogenic cycle (Liuliang or Karelian), and the Wutai Group may be much older.

In *North Korea* the Machkholen Group is probably correlative with the Anshun Group. It is made up of various schists, gneisses and amphibolite among which bands of dolomitic marble and jaspilite occur. The latter are closely associated with orthoamphibolites (metavolcanics). This metamorphic group is unconformably overlain by Neoprotozoic sedimentary rocks of the Sanvon Group. Biotites from schists of the Machkholen Group fall into the 1,600—1,700 m.y. range, but these values are probably strongly "rejuvenated", because the Machkholen rocks were metamorphosed and folded at least twice.

Metamorphic rocks developed in the *Soviet Far East* are also very similar to the Anshun Group. These include the sedimentary and volcanic groups of the Khankai Massif (including the Turgenevo, Nakhimov and Tatjana Formations), of the Burein anticlinorium (where the Ambardakh, Saganar and Lepikan Formations occur), and of the Khingan—Burein Massif (including the Amur Group, which consists of the Tulovchikhino, Dichun and Uril Formations). All these sequences are made up of metamorphosed rocks under amphibolite, and less commonly, under greenschist-facies conditions. They include various schists, including orthoschists, gneisses, amphibolites, quartzites and marbles. In the Khankai massif they overlie the Archean gneiss—granulite complex (the Iman Group). In other places they are the oldest Precambrian rocks. They are everywhere unconformably overlain by Mesoprotozoic rocks but their age has not yet been determined by isotopic methods.

Paleoprotozoic strata are widespread in Southern Asia in the *Indian plat-*

form. The Dharwar Group (or Supergroup) of Southern India is the stratotype. Indian geologists rank it as "a system". The succession varies from place to place. The most complete section of the group is in the so-called "Shimoga—Dharwar schist belt", where the succession of rocks (from top to base) is as follows:

Upper part of the Dharwar "system"	ferruginous quartzite limestone argillite and carbonate schist quartzite and conglomerate
------------ Unconformity	
Middle part of the Dharwar "system"	banded iron formation limestone and dolomite phyllite, chlorite schist, etc. conglomerate
------------ Discontinuity (break)	
Lower part of the Dharwar "system"	lava of different composition usually metamorphosed, banded iron formation greenstones, amphibolite and hornblende gneiss older gneiss and volcanics

based on Rama Rao (1940, 1962), and Krishnan (1960a, b).

This sequence cannot be considered certain because the stratigraphy of the Lower Precambrian of India is poorly known. The Dharwar Supergroup is considered to be made up of three parts, or groups, separated by stratigraphic breaks. The most widely distributed is the lower sedimentary—volcanogenic group which is represented in all the areas of Southern India. This group is probably the only one that should be included in the Dharwars. It is probable that the upper part of the supergroup is a discrete unit of younger (Mesoprotozoic) age, which has been included with the underlying volcano-sedimentary assemblages which are more highly metamorphosed and deformed. In the following discussion the Dharwar Group is used to mean only the lower and the middle parts of the supergroup. Having had the opportunity to observe these units during a field trip of the 22nd session of the International Geological Congress in 1964, part of the description given below is based on first-hand observation.

The Dharwar Group is a typical volcano-sedimentary assemblage such as is found in many geosynclinal belts. The volcanics are metamorphosed diabases, spilites (including pillow lavas), and keratophyres. They are largely confined to the lower and middle parts of the group, but lavas are found at all levels. Phyllite, micaceous schist, metasandstone, quartzite and carbonate rocks are more important in the upper part of the section. Finely banded iron formations (jaspilites) are characteristic and usually associated with metavolcanics. The thickest units of banded iron formation are in the central part of the group. Iron ores that were remobilized during metamorphosm and also hypergene secondary deposits are associated with these thick units. In the fold belts that involve the Dharwar succession there is a clearly defined metamorphic zonation. This zonation is to some extent controlled by older structural elements such as deep faults, granite intrusions, etc. Rocks of this group are metamorphosed to about middel greenschist facies or amphibolite facies. In a number of synclinorial structures there appears to be an increase in metamorphic grade towards the base of the group. This is probably due to the remobilization of the Archean basement during formation of structures of the gneiss-dome type.

The volcano-sedimentary assemblages are intruded by gabbroic and hyperbasite intrusions, and also large plutonic bodies of late orogenic granite in a number of places. Locally, in narrow zones, synkinematic granitization is reported. The Klosepeth Massif is among the best-known post-Dharwar (Paleoprotozoic) plutons. It is made up of various granites, mainly pink to pink-grey in colour, that are coarse-grained and commonly porphyritic. These granites are associated with pegmatite and lamprophyre.

The Dharwars are characterized by long linear folds but dome-shaped structures are also reported. The geological maps of India clearly show that the Dharwar Group (and its analogues) occur in definite belts. They probably correspond to Paleoprotozoic geosynclinal systems which surrounded the older platforms composed of Archean rocks. This is suggested by the regular distribution of the Paleoprotozoic fold belts and also by the presence of regular changes in facies, folding and metamorphism, across the strike of the system.

The Dharwar sequence is not the oldest in India, as is commonly suggested by Indian geologists. These sequences unconformably overlie an Archean gneiss—granulite complex. Locally (in the Kolar area) the former have conglomerates at their base (Salop, 1966). The upper limit of the group is given by the unconformably overlying sedimentary Neoprotozoic Cuddapah Supergroup. If, in fact,the Mesoprotozoic age of the upper part of the group is correct, then Mesoprotozoic strata also unconformably overlie the Dharwar (s.str.) strata (see Chapter 6). New data confirming these views on the age and complex structure of the Dharwar Group were published by Srinivasan and Sreenivan (1972).

In most cases the K—Ar ages of the Dharwar metamorphic rocks and granites fall between 1,900 and 2,200 m.y. These values are certainly strongly "rejuvenated" due to later thermal processes. Some "relict" K—Ar dates from amphibole in amphibole schists yield an age of 2,700 (2,630) m.y., and a single date on muscovite from micaceous schist gave 2,800 (2,700) m.y. (Pichamuthu, 1971). A Rb—Sr isochron plotted for eight samples of Dharwar metavolcanics gave an age of 2,345±60 m.y., and an isochron representing ten samples of the post-Dharwar Klosepeth granite gave 2,380±30 m.y. Other samples gave an isochron of about 2,700±400 m.y., which Crawford (1969) considers to be the probable age of these granites. The Pb-model age of Dharwar amphibolite from the Kolar area is 2,900 m.y. (Vinogradov and

Tugarinov, 1964), but this dating is not certain. It is likely that the 2,600—2,700 m.y. interval was the time of Paleoprotozoic orogeny.

Thus, it is concluded that the Dharwar Group of India is similar in age, lithology and stratigraphic sequence to the other Paleoprotozoic eugeosynclinal complexes described above.

In the northeastern part of the Indian Peninsula, in the state of Orissa, rocks very similar to those of the Dharwar Supergroup are present. These comprise the iron-rich rocks of the Orissa Supergroup. Many geologists have studied this supergroup and proposed various stratigraphic subdivisions (Krishnan, 1960a). The latest studies by Prasada Rao et al. (1964) suggest that the supergroup is made up of six groups separated by unconformities and consisting of metavolcanics, chlorite, amphibole, micaceous schist, phyllite, quartzite, metasandstone, conglomerates and iron formation (jaspilite) at several different horizons. Manganese-bearing schist is reported in one of these units. The groups are separated by periods of intrusion of gabbro and ultrabasic rocks and some granites. Some Indian geologists consider this stratigraphic scheme to be unnecessarily complicated. Thus, the unconformities between the groups are not certain. The gabbro and ultrabasic intrusions may not separate the groups, but perhaps accompanied various periods of extrusive lava formation. This may also be the case for some subvolcanic sodic granites (granophyres) which were perhaps comagmatic with acid volcanics.

The possibility remains, however, that the Orissa Supergroup, like the Dharwar Supergroup, includes strata that should really be assigned to different Precambrian eras. Its lower part, which contains abundant volcanics and is jaspilite-bearing, may be Paleoprotozoic, whereas the upper part, which is separated from these rocks by an unconformity and emplacement of granite, and is largely composed of conglomerate, quartzite, schist, greywacke and slightly altered lavas (upper part), is Mesoprotozoic. The Orissa Supergroup unconformably overlies the Archean gneiss complex of the Eastern Ghats (which includes charnockite and granulite). The Paleoprotozoic age of the lower part of the supergroup is indicated by the intrusive Singbhum granite. The age of the granite is 2,700 m.y. by various methods. Lava and diabases that occur at the top of the supergroup have an age of 1,665—1,745 (1,600—1,700) m.y. by K—Ar analysis (Pichamuthu, 1971). This evidence favours a Mesoprotozoic age for the upper part of the section.

North America

Paleoprotozoic supracrustal and plutonic rocks are widely distributed in the Canadian—Greenland shield and in many parts of the North American Cordillera. In a large region forming the southeastern part of the shield they may never have been deposited, or possibly they were only locally developed in narrow zones of tectonic subsidence. This region includes the ancient

Ungava block, the adjacent Nain province, the northern part of the Grenville province, Baffin Island and Boothia Peninsula in Arctic Canada, and also the northern part of Greenland. It probably represents a large uplifted cratonic massif that is here called the Baffin platform (protoplatform).

Paleoprotozoic supracrustal assemblages are everywhere represented by volcanic and sedimentary eugeosynclinal rocks metamorphosed under greenschist facies and in some places under amphibolite-facies metamorphism. Typical miogeosynclinal formations are either absent or still to be discovered, but some groups which contain few volcanics and which contain widely developed terrigenous rocks almost have a miogeosynclinal aspect. Strata attributed to facies transitional between mio- and eugeosynclines are poorly known. Some of these may represent parts of thick eugeosynclinal sections that are characterized by predominance of sedimentary rocks.

Paleoprotozoic volcano-sedimentary assemblages are very similar in composition in many areas of the *Canadian—Greenland shield*. Altered basic volcanics are predominant, mostly pillow lavas, together with greywackes and various tuffs. Intermediate and acid volcanics and arkoses are less abundant. Still less common are quartzites and carbonate rocks. Conglomerates and tuff-conglomerates are reported at several stratigraphic levels, in some cases forming thick horizons. Basal conglomerate or arkose is locally observed (Chapter 4). In many groups finely banded iron—silica (in some cases iron—silica—carbonate) rocks (jaspilites) are present. These are generally spatially related to volcanic rocks.

The internal stratigraphy of the volcano-sedimentary assemblages and their degree of metamorphism vary from place to place. It is partly for this reason that groups may be given different names although they occur quite close to each other (Table III, inset plate). In the Superior tectonic province the volcano-sedimentary assemblages commonly show a threefold stratigraphy. The top and the bottom are largely composed of clastic rocks with subordinate amounts of volcanics. The middle part is characterized by abundant volcanics. However, this sequence is not universal.

Different names are used for the terrigenous strata in different areas (Couchiching, English River, Pontiac, etc.). The thickest part of the section, the volcanogenic one is given a general name — Keewatin-type. This name is used in many cases for the lower and middle parts of the section, or even for the whole group. Some workers suggest that the Lower Pontiac terrigenous rocks are essentially a facies equivalent of the Abitibi volcanics. The upper terrigenous or volcanic—terrigenous rocks have no common name. They are mostly known by local names such as the Timiskaming, Dickinson, Knife Lake, etc. These strata commonly overlie volcanics with an erosional break, and are therefore considered as a separate group.

Formerly the Canadians considered the break between the Keewatin (and its analogues), and the Timiskaming (and its analogues) to be due to a significant orogeny during which folding and granitic intrusions took place

(Laurentian orogeny). However, later work showed that, in a number of cases both sequences are conformable, and conglomerates of the Timiskaming (and its correlatives) contain clasts of older pre-Keewatin granite as well as of granitoids (often derived from sub-volcanic complexes) that are considered to be genetically related to the Keewatin volcanics (Bass, 1961; Boutcher et al., 1966; Holubek, 1968; Morey et al., 1970). The existence of a separate Laurentian orogeny is now denied by most Canadian Precambrian workers. In the well-known tectonic scheme of Stockwell (1964) there is only one post-Timiskaming (Kenoran) orogeny corresponding to the Algoman orogeny of previous classifications.

The geological structure of the Saganaga Lake area, near the Canada—U.S.A. border, is one example where there seems to be evidence of the Laurentian orogeny and associated granites. New studies (Hanson et al., 1971) have confirmed the previous suggestions (Gruner, 1941; Grout et al., 1951) that the Saganaga granite cuts the Keewatin-type volcanics and that the Knife Lake conglomerates unconformably overlie both. These studies also showed that the Saganaga granite and later Ikarus and Gold Island granitoids cutting the Knife Lake conglomerate, do not differ greatly in age. The Rb—Sr isochron age of the former is 2,710±560 m.y. (with a Pb-isotope age of 2,750 m.y.). The age of the latter is 2,690±480 m.y. The Saganaga granite or quartz diorite (tonalite to be more exact) and the so-called Northern Light "gneisses" probably represent intrusive and subvolcanic units accompanying outflow of acid and intermediate volcanics. Many American workers are of this opinion; in particular, Morey et al. (1970) considered the hypabyssal intrusions of granitoid (mainly of granodiorite and diorite composition) to have formed throughout the whole period of Keewatin lava accumulation. Thus, they cannot be attributed to the Laurentian orogeny.

It is difficult to choose a Paleoprotozoic stratotype from among the many well-studied sections in the Superior tectonic province. Probably the section in the Noranda—Malartic—Val d'Or region, near the Ontario—Quebec boundary is most appropriate (Gunning and Ambrose, 1940; MacLaren, 1952; Latulippe, 1966). In this area the lower part of the Paleoprotozoic section is the Pontiac metasedimentary sequence which is commonly a facies variant of volcanic rocks of the generally overlying Malartic succession. According to Holubek (1968) the conglomerates in the lower part of the Pontiac succession include clasts of Archean granite and locally (Kinojewis Lake) lie directly on the granite basement. The Malartic Group is largely composed of basic volcanics, but its upper part includes acid lavas and pyroclastic rocks. Overlying these volcanics (in some cases as a facies equivalent) are metasedimentary and partly volcanic rocks known as the Kawagama, Black River and Cadillac sequences. All of these volcano-sedimentary groups are cut by granites which are dated at 2,700 m.y. by different methods. Paleoprotozoic rocks (including granites) are overlain unconformably by the Mesoprotozoic Huronian succession.

In the Steep Rock area in the Superior province Jolliffe (1966) established an interesting and peculiar section in the Lower Paleoprotozoic where the contact with the underlying Archean granite can be seen. The geology of this deposit was studied both in natural outcrops and in open-pit mines. The Steep Rock Group unconformably overlies gneissic granites and diorite (locally gneissic) and comprises (in upward sequence) basal conglomerate, dolomite, ore band and pyroclastic rocks with pillow lavas (Ashrock Formation). Its general thickness is from 300 to 1,000 m. Thick volcanogenic strata of Keewatin type are present higher in the section (Fig.16).

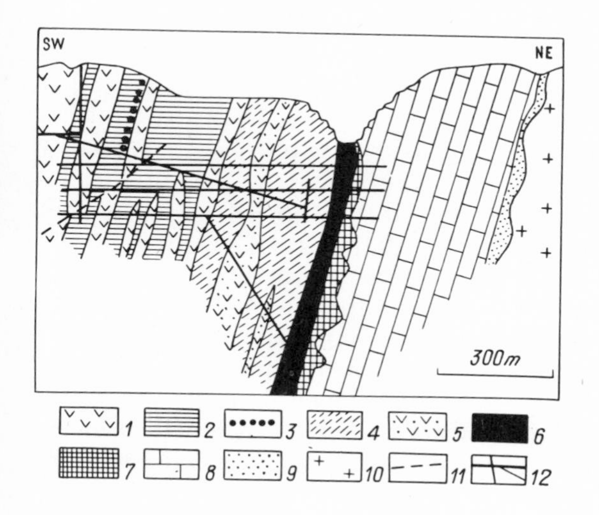

Fig.16. Geological section in the area of Errington Mine, Steep Rock Lake. According to Jolliffe (1966).
1—3 = *volcanic complex* (Keewatin-type): *1* = basic lava; *2* = tuff; *3* = conglomerate; *4—9* = *Steep Rock Group*: *4* = ash tuff; *5* = pillow lava; *6* = goethite zone; *8* = dolomite; *9* = basal conglomerate; *10* = granite; *11* = faults; *12* = mine workings.

The basal conglomerate of the Steep Rock Group lies in depressions in the basement rocks and includes pebbles and boulders of the underlying granitoids. It locally interfingers with arkoses and granite conglomerates. Its maximum thickness is 200 m. In some places the conglomerate is missing, and the dolomite lies directly on granite, in which case it contains rare quartz grains and small pebbles. The dolomite is fine to cryptocrystalline, and is up to 500 m in thickness. Stromatolites are present as also are peculiar concentric radial structures called *Atikokania* by Ch.D. Walcott (1912), but they are probably inorganic (Hofmann, 1971). An erosional unconformity is present between the ore band and the dolomite. The ore rocks lie on a karsted and strongly eroded dolomite surface, but locally they lie directly on the granite. In the lower part of the ore band there are siliceous rocks with manganese minerals, goethite, hematite, kaolinite and gibbsite. Jolliffe compared these formations with recent bauxite of the "terra rosa" type. Up-section there are rocks that are largely composed of goethite and hematite with subordinate amounts of quartz and kaolinite including irregular masses of ferruginous pisolitic bauxite. In the uppermost part of the ore band there is abundant colloform pyrite. Pyroclastic rock and pillow lavas conformably overlie the ore band, and in turn are conformably overlain by thick (some kilometres) volcanics.

The occurrence of dolomite with organic remains, at the base of the volcanic complex in the Steep Rock area confused many geologists, who attributed the Steep Rock strata to the Archean. The pre-Huronian age of these strata is beyond doubt both on the basis of geological observations, and also because of radiometric age determinations which yielded more than 2,500 m.y. (Jolliffe, 1966). K—Ar dating of phyllite from the Steep Rock Group gave the age of metamorphism at 2,800 (2,700) m.y. (Lep and Goldich, 1964).

The upper age limit of the Paleoprotozoic in the Superior province is almost everywhere reliably established. Isotopic dating (by different methods) of metamorphic rocks and Kenoran granites, in most cases falls in the range 2,550—2,750 m.y. The most reliable dates are in the interval between 2,650 and 2,750 m.y. The major stage of Kenoran folding magmatism and metamorphism occurred approximately 2,700 m.y. ago and the younger values are "rejuvenated" (Hanson et al., 1971). The upper limit of the Paleoprotozoic is given by the fact that in many areas the Keewatin—Timiskaming sequences are unconformably overlain by Mesoprotozoic formations (Table III, inset plate).

In the Churchill province many volcano-sedimentary and metamorphic groups are given different names in different areas. These include the Tazin, Kaminak, Sandfly, Wasekwan, Amisk Groups, etc.). In a general way these correspond to those described above. They are analogous in lithology and in the threefold subdivision of the section. Sometimes they are referred to as greenstone strata of Keewatin-type or simply Keewatin. As in the Superior province many units contain iron formations associated with these volcanics. Their K—Ar and Rb—Sr metamorphic age is about 2,600—2,800 (2,520—2,700) m.y., but in some areas the ages are strongly "rejuvenated" by later metamorphic events.

In the Slave and Bear provinces the Yellowknife Group is closely analogous to the Keewatin—Timiskaming complex. The succession in this group varies locally, but generally the lower part is made up of basic (frequently pillow lavas), rare acid metavolcanics, tuff and agglomerate. The upper part either conformably, or in some cases with a stratigraphic break, overlies the lower part. It is composed of metasedimentary rocks, predominantly altered greywacke, arkosic quartzite, arkose, conglomerate and tuffaceous rocks. Terrigenous rocks are not present at the base of the group so that the basic volcanics directly overlie gneisses and granites of the basement. Comagmatic intrusions of gabbro—diorite, emplaced during deposition of the group, are locally associated with metavolcanics according to J.B. Henderson (1972). The Yellowknife rocks are also cut by younger granite, with a Rb—Sr isochron age of 2,650 m.y. (Green et al., 1968). The Pb-isotope age of this granite is 2,610—2,690 m.y. (Thorpe, 1971). The K—Ar method usually yields somewhat "rejuvenated" values (2,400—2,500 m.y.).

The Wilson Island Group is an analogue of the Yellowknife Group in the southeastern part of the Slave province in the Great Slave Lake area (adjacent to Churchill province). It is composed of acid and rare basic metavolcanics, and at the base, a thick unit of quartzite and subarkose. It is probable that the schists and metavolcanics are in the upper part of this sequence, but they are separated from the other units by a fault. The K—Ar metamorphic age of the group and of the granites cutting it, is 2,500 m.y.

The Paleoprotozoic strata in the Slave province and other tectonic provinces of the Canadian (Canadian—Greenland) shield are unconformably overlain by various Mesoprotozoic groups (Table III, inset plate).

In Arctic Canada stratigraphic equivalents of the Keewatin—Timiskaming complex are the sedimentary—volcanogenic Prince Albert Group (Melville Peninsula), Mary River Group and possibly the Hoare Bay (Baffin Island) Group which locally includes large iron-ore deposits of jaspilite type. At the base of the first two groups there are metaconglomerates which lie unconformably on the Archean gneiss—granulite complex (Jackson and Taylor, 1972; Campbell, 1974). The older granite gneisses surrounding rocks of this group are locally strongly remobilized. In these cases the original stratigraphic relationships are masked by the intrusion of newly formed granites.

Highly metamorphosed sedimentary—volcanogenic rocks known as the Clare River complex or Hermon Group may also be Paleoprotozoic. These rocks occur in the southern part of the Grenville tectonic province (M.E. Wilson, 1965), where they unconformably overlie the Archean Grenville complex without an obvious break. The Flinton Group (tentatively considered to be Mesoprotozoic) unconformably overlies these rocks. Metavolcanics under amphibolite facies are most characteristic of the Hermon Group, but usually it contains some original textures and structures such as amygdules and pillows. Basic metavolcanite commonly pass upwards into acid ones. In addition to volcanics, tuffs and greywackes are widely distributed, and metaconglomerates are also present. Isotope dating of the Hermon Group rocks yields "rejuvenated" values (Sethuraman and Moore, 1973). This complex is tentatively assigned to the Paleoprotozoic because of its position in the Precambrian section, because the lithological makeup of the rocks is similar to that of other Paleoprotozoic groups of other provinces, and because of the fact that the Paleoprotozoic greenstone belts of the adjacent Superior province strike in an easterly direction into the Grenville province. Possibly some areas of metavolcanics and metagreywackes of the Keewatin-type, together with gneissic granite of the Kenoran cycle, are developed in the Grenville province near the Grenville Front. In the Grenville province greenstones were possibly twice deformed, and recrystallized (in the Mesoprotozoic and

Neoprotozoic), unlike those of the Superior province which suffered only slight deformation and metamorphism.

Sedimentary and volcanic rocks metamorphosed to the greenschist and amphibolite facies in *Greenland* (the Umanak, Egedesmine, Malene and Tartoq Groups) are considered to be Paleoprotozoic. These rocks are present in Southwest Greenland. The stratigraphic position of the first two groups, which are very similar in composition, is not certain because the base of the Umanak Group is not known, and the Egedesmine Group is in fault contact with an Archean gneiss—granulite complex. Isotopic dates are not available for these groups. They are considered to be Paleoprotozoic because they are at a lower grade of metamorphism than the Archean gneiss—granulite complex. The Umanak Group is overlain by the Karrat Group which is similar to the Mesoprotozoic Vallen Group.

They differ from rocks of similar age in Canada by the presence of carbonate interbeds (in some cases they form a significant part of the section). In the Umanak Group these rocks occur mainly in the middle part of the section (Marmorlik Formation, Table III, column 48). The Malene Group is mainly metavolcanics (locally pillowed). These rocks unconformably overlie the Amitsoq gneiss, with an age of 3,750 m.y. The Malene Group is cut by the Nuk granitoids which are probably genetically related to the volcanics. An early rare phase is composed of gabbroid and diorite. The later phase is the more widely distributed gneissic Nuk tonalite. The Rb—Sr isochron age of the latter is 3,040 m.y. (McGregor, 1968; Bridgwater, 1973). The late tectonic Qorqut is dated by the same method as 2,580 m.y. old, and by K—Ar analysis on biotite and amphibole at 2,700 m.y. old (Pankhurst et al., 1973).

The Tartoq Group which is developed in the southernmost part of the island in the vicinity of Ivigtut village is also very similar to Paleoprotozoic strata of the Keewatin-type in Canada. It is made up of basic metavolcanics (under epidote—amphibolite facies) with subordinate beds of metasandstone, talc schist, conglomerate and agglomerate. This group (Higgins and Bondesen, 1966) unconformably overlies the Archean (pre-Ketilidian) gneiss—migmatite complex, and is itself unconformably overlain by Ketilidian rocks (Vallen and Sortis Groups) which are Mesoprotozoic.

The iron-rich volcano-sedimentary Isua complex is of great interest. It is developed in a small area of Western Greenland about 150 km northeast of the town of Godthaab close to the margin of the ice-sheet. These strata occur among granite gneisses. They comprise quartzites and metagreywackes in the lower part. These rocks are overlain by garnet—chlorite and carbonate schist with banded iron formations (jaspilite), and by greenstone metabasite with ultramafic lenses (Moorbath et al., 1973). These rocks include some schistose conglomerates with clasts (up to 15—20 cm) of gneissic potassic granite and possibly of quartz porphyry (Bridgwater, 1974). The total thickness of these strata is from 2 to 3 km. The rocks show an alternation of greenschist and amphibolite facies. Pb-isochron (Pb—Pb) analysis on samples of ferruginous rocks yielded an isochron of 3,760 m.y. A similar age is reported for granite gneiss (3,700±140 m.y.) by the same method and also by the Rb—Sr isochron method. This granite gneiss is possibly of the same generation as the Amitsoq granite gneiss of the Godthaab area.

Geological relations between the granite gneiss and the Isua complex are not certain, because strong deformation has resulted in structurally conformable contacts (Moorbath et al., 1973). However, the presence at the base of this sequence of clastic rocks, the presence of conglomerates with granite clasts, and, finally, a rather low grade of metamorphism of these rocks, suggest that the supracrustals are younger than the gneiss—granite complex. Very old Pb—Pb ages obtained from iron formations of the Isua strata may be questionable. These results were the first attempt at dating ferruginous rocks by the Pb—Pb method. Possibly there is a peculiar geochemical environment for accumulation and migration of lead and uranium in ferruginous sediments. Perhaps the date does not indicate the time of metamorphism, but rather the time of deposition which could have been shortly after formation of the gneiss complex. Some specialists in the Pb—Pb method suggest the possibility of dating pre-metamorphic events by this technique. Almost all analysed samples of ferruginous rocks have a very similar lead isotopic composition. This may influence isochron precision to some extent.

If we take 3,760 m.y. as the age of metamorphism of the Isua strata then there must have been still older granites than those of the Amitsoq gneiss complex, because the conglomerate contains abundant granite boulders. These older units must have an age of the order of 4,500 m.y. because of the very long intervals between the tectono-plutonic cycles of the Early Precambrian. Such an assumption seems highly improbable, or at least there are no reliable geological data to support it.

Because of the strong similarity between the Isua rocks and other volcano-sedimentary ferruginous rocks of the Canadian—Greenland shield (in particular, the association of jaspilites and volcanics) the Isua complex is here tentatively considered to be Paleoprotozoic although this argument needs further documentation.

In the basement of the *North American plate* Paleoprotozoic strata are not certainly identified, but the Vishnu complex may be of this age. It constitutes the Precambrian basement in the Grand Canyon of the Colorado River. Available isotopic dates suggest a Mesoprotozoic age (see Chapter 6).

The Sherry Creek Group is undoubtedly Paleoprotozoic. It occurs in the fold zone of the *North American Cordillera* in Southwestern Montana. This group is composed of metasedimentary rocks — marble, quartzite, micaceous and sillimanite schist interfingered with amphibolite formed as a result of metamorphism of basic lavas. Jaspilite-type iron formations are present in rocks of this group (associated with amphibolites). The Sherry Creek Group unconformably overlies the Pony gneiss—granulite complex (Pony Creek complex) which is dated at 3,300 m.y., and the group itself was cut by granites and metamorphosed 2,730 m.y. ago (Rb—Sr isochron analyses). The rocks of this group subsequently suffered Mesoprotozoic metamorphism (Brookins, 1968), accompanied by granite intrusion (about 1,850—1,900 m.y. ago, by both Rb—Sr isochron and K—Ar methods).

General Characteristics

Paleoprotozoic supracrustal complexes are typified by abundant volcano-sedimentary rocks. They are similar to eugeosynclinal rocks of the Phanerozoic. True miogeosynclinal deposits are quite rare and in many cases it is not certain that they are really Paleoprotozoic. Some eugeosynclinal volcanogenic complexes have abundant sedimentary rocks including arkoses and quartzites that, in some respects, are similar to miogeosynclinal deposits. However, these rocks are generally much less extensive than typical miogeosynclinal deposits. It is not certain that platform deposits existed, but since it appears that in the Paleoprotozoic vast cratons already existed, such formations may have been developed, but were destroyed by later denudation.

In Eastern Siberia, in the Aldan shield there is a very peculiar type of sedimentary—volcanogenic sequence. These rocks formed in narrow grabens cutting the southern extremity of the older Aldan—Okhotsk platform close to and at a high angle to, the Baikal—Stanovoy geosynclinal belt. These regions of subsidences appear to be related to faulting of the platform basement. They resemble younger Protozoic and Phanerozoic aulacogens, but the strata developed in them differ from those of typical aulacogens in containing abundant volcanics, being strongly folded, metamorphosed and in the presence of intrusive gabbroids and granites. Fedorovsky and Leytes (1968) called these regions "geosynclinal troughs" but this is a poor name because it is usually applied to deep linear subsidences in geosynclinal systems, and these structures are quite different. The name taphrogens is proposed for these structures; the taphrogenic strata that fill them are older equivalents of the aulacogen type of strata.

Eugeosynclinal strata in different regions of the Northern Hemisphere are similar in composition, degree and character of metamorphism, and partly in stratigraphic sequence. A threefold sequence is characteristic of most of these successions. The lower part is largely sedimentary rocks with subordinate volcanics, in the middle part there is a strong predominance of volcanics and pyroclastic rocks, and the upper part is largely clastic and tuffaceous rocks with some volcanics. This is a generalized sequence, for in fact, the sequence may be much more complex. In many areas the lower sedimentary sequence is either thin or missing. Volcanics commonly lie directly on eroded Archean rocks, with only local development of a basal conglomerate.

In many areas there is a basal conglomerate containing clasts of underlying Archean rocks. Sandstones, schists and tuffs are predominant in the lower section, with thick-to-thin units of marble and dolomite. Locally a fossil soil is developed, as at Steep Rock, Canada.

The middle, largely volcanogenic, complex is everywhere the most complete and of greatest thickness. It is the most characteristic type of supracrustal rock in the Paleoprotozoic. There are various kinds of volcanics of basic, intermediate and acid composition, together with tuffs and subordinate sedimentary rocks. The volcanics include basalt, diabase and spilite (pillow lavas), dacite, keratophyre, albitophyre, and rarely, porphyries. Basic lavas predominate, especially in the lower part of the sections. Acid lavas and pyroclastics are common in the upper parts of the section, but are not present in all cases.

Basic and acid volcanics may be interlayered. In many areas subvolcanic intrusions of comagmatic rocks are present. These include albite granite porphyry, granophyre, plagiogranite (trondhjemite), granodiorite, diorite, gabbro diabase, gabbro and hyperbasite that were formed at the same time as extrusion of the lavas or at the final stage of volcanism. The sedimentary rocks include sandstone, mostly greywacke with subordinate quartzites, schist, tuff, siliceous rocks and iron formation; carbonate rocks are rare.

The volcano-sedimentary assemblages of Paleoprotozoic sequences (with the exception of the presence of quartzites) are similar to those of the spilite or spilite—keratophyre association of geosynclinal areas. However, in Phanerozoic fold belts this association is typical only of an initial rather short stage in the development of the geosynclinal system, and is confined to the internal (eugeosynclinal) belt. In the Paleoprotozoic it is present practically throughout the whole tectonic cycle and covers practically the whole area of the geosynclinal system. The presence among some of the Paleoprotozoic volcano-sedimentary rocks, of beds of such mature rocks as quartzites (with relict clastic texture) again differentiates these rocks from typical eugeosynclinal assemblages of the Phanerozoic. This may be related to their particular geological setting — geosynclinal volcanic belts developed close to vast peneplaned platforms.

Comparison of the chemical composition of Paleoprotozoic volcanics with younger volcanic sequences is of great interest. Such an analysis has not yet been done for the northern continents as a whole. The available data on analyses of this kind are not satisfactory. Recently some interesting papers were published by Anhaeusser (1971), Glikson (1970, 1972) and M.J. Viljoen and R.P. Viljoen (1969) in which the chemistry of Paleoprotozoic volcanics of Africa and Australia (and partly of Canada) is discussed. The authors argue that the ancient greenstones have a chemical composition similar to oceanic tholeiites, but in comparison to similar rocks of younger age have slightly higher Fe/Mg ratios and lower K/Rb ratios. They are also characterized by a lower content of Al and Ti and higher values for K, Ba, Sr, Zr and Y. High FeO/Fe_2O_3 ratios are typical and may indicate the low degree of oxidation of the rocks. In the lower part of many Paleoprotozoic volcanic sequences, ultrabasic volcanics with an exceptionally high MgO content (up to 40%, and on average, 30.6%) are present. These rocks are called komatiites. Recently komatiite was reported from Keewatin-type sequences in Canada (Brooks and Hart, 1972; Pyke et al., 1973). Preliminary data by Soviet geochemists indicate that many of the chemical characteristics of the older volcanics of the southern continents, are also typical of the same type of rocks in Karelia, the Kola Peninsula and Eastern Siberia.

The upper part of the Paleoprotozoic, where it survived later erosion, is largely represented by clastic rocks such as polymictic conglomerates, sandstones, siltstones and schist interbedded with volcanics and tuffs in various proportions. This part of the section is particularly characterized by poorly sorted greywacke sandstones formed by erosion of the underlying volcanics. Study of the Timiskaming Group of Canada (Pettijohn, 1943) indicated that the major common feature of Archean sedimentation was poor sorting of the clastic material. However, according to the classification used here, the Timiskaming is not Archean, but Upper Paleoprotozoic. The characteristics described by Pettijohn are found only in Timiskaming-type sequences and not in the Lower Precambrian as a whole.

The Upper Paleoprotozoic units commonly lie with a stratigraphic break, but without angular unconformity, on the underlying volcanics, and in some places on associated intrusive rocks. As mentioned above, these relations, and also the presence of various granitoid clasts in conglomerates (both older granitoids and some that were comagmatic with the acid lavas) led to the erroneous conclusions of the existence of an orogenic cycle in the interval between formation of the volcanic and clastic rocks (Laurentian orogeny).

The presence of clastic rocks in the upper part of many sections indicates that these sections are rather complete. This indicates that not only the "roots" of the Paleoprotozoic orogens are preserved, as some workers have argued.

Rather poor development of carbonate strata is typical of Paleoprotozoic rocks. Carbonate rocks are abundant only among sequences in Eastern Sayan and Tuva on the northern margin of the vast Central Asiatic geosynclinal region.

Organic remains are rare in Paleoprotozoic sequences. There are some poorly studied stromatolites. In the Northern Hemisphere they are known only from two places in Canada: the Steep Rock Group (Walcott, 1912) and in the Yellowknife Group. Similar remains are known from Paleoprotozoic strata in some areas of Africa. In the siliceous rocks of the Soudan iron formation, which is associated with volcanics of the Ely Group, microscopic remains of blue-green algae are reported (Gruner, 1923; Cloud et al., 1965).

The most characteristic Paleoprotozoic rocks are banded iron formations associated with volcanics. Unlike Archean banded iron formations which are usually interbedded with metabasites (basic volcanics), the Paleoprotozoic ores are less intimately associated with volcanic rocks. The iron-rich units commonly occur in a sedimentary association enclosed in volcanics or in sedimentary and tuffaceous rocks which pass laterally into lavas. However the iron formations are invariably associated with volcanics. Banded iron formation of this kind, associated with the Keewatin-type volcanics is given a special name by Canadian geologists (Goodwin and Schklanka, 1967) — the Michipicoten- or Algoma-type. The latter name may be applied to all similar ores of the Paleoprotozoic.

This type of iron formation is subdivided into several subtypes, which commonly represent facies. Goodwin (1966) and Goodwin and Schklanka (1967) established three major facies in the Keewatin-type volcano-sedimentary assemblages of Canada: oxide, carbonate and sulphide. The oxide and partly carbonate facies are widely distributed in other parts of the world. These facies are made up of finely banded silica—magnetite, magnetite—siderite—silica and magnetite—siderite rocks (jaspilites) in which ore bands, non-ore and low-grade ore bands are intimately interfingered. Pyrite and pyrrhotite are common. Hematite or hematite mixed with goethite occurs in some places among the ore minerals,but in most cases these minerals are secondary and formed as a result of oxidation and hydration of magnetite, carbonate and carbonate—sulphide ores. The sulphide facies consists of banded carbonate—sulphide ores which pass along strike into poorly banded and massive sulphide ores. The common sulphides are pyrite and/or pyrrhotite.

In the Michipicoten area (Ontario, Canada) Goodwin (1966) stated that iron formations of all three subtypes are situated at the contact between the lower acid volcanics of rhyolite—dacite composition, and the upper basic andesite—basalt volcanics, and that one subtype may pass laterally into another. The oxide facies iron formation (magnetite—silica) generally occurs far from the volcanic centres in a sedimentary association of schistose greywackes which are a facies equivalent of the volcanics. Goodwin argued that iron formation of the Michipicoten or Algoma type is a direct product of volcanic activity. He showed that the highest concentration of iron (in pyrite and siderite) coincided with the thickest volcanic accumulations, and that the maximum iron accumulation was confined to the vent funnel facies of volcanic accumulations. During volcanic eruption sulphur and carbon dioxide were precipitated together with iron. Silica gel was considered to be the result of leaching of silica from the acid volcanics. These ideas are not yet proved.

The iron minerals in Algoma-type ores are generally in a low oxidation state (mostly protoxide and sulphurous iron) and may possibly indicate the deficiency of oxygen in the Paleoprotozoic atmosphere and hydrosphere. The syngenetic ores, including jaspilitic iron formations, of the younger Precambrian erathems differ greatly from those of the Paleoprotozoic, both in composition and in associated rock types. The Algoma-type banded iron formations may serve as a "marker horizon" for age determination and correlation of Paleoprotozoic rocks.

Conspicuous facies variation, mainly involving lateral equivalence of volcanic and sedimentary rocks, is typical of many Paleoprotozoic volcano-sedimentary sequences. Abundant sedimentary (largely clastic) rocks are developed in the supracrustal complexes close to older platforms or median massifs (blocks), and volcanics predominate in fault zones. A facies zonation of this kind is definitely established for the Paleoprotozoic rocks of the Baikal fold belt (Salop, 1964—1967). In Canada there are several volcano-sedimentary groups with abundant clastic rocks. These include the unnamed group in the Baker and Schultz Lake areas (Keewatin district, N.W.T.), the Prince Albert and Mary River Groups (Canadian Arctic) and others near the margin of the Baffin protoplatform, and the pre-Assean Group of the Thompson area near Mystery Lake (Manitoba) on the margin of the Pikwitonei block.

The thickness of these sequences is highly variable. In some geosynclinal belts or zones it is as high as 10,000—12,000 m, whereas in others it does not exceed 2,000 m (this decrease in thickness is original, and not due to later erosion). In the eastern part of the Baltic shield (Soviet Karelia) in the upper part of a rather thin volcano-sedimentary sequence (Gimola Group — 1,500 m) rocks of the upper clastic part of the section are present. The thickness of these strata varies greatly, but never exceeds 4,000 m. Bogdanov (1971b) showed that thickness and compositional variations were related to the tectonic development of the region — minimum thicknesses are developed on older uplifts, and greater thicknesses are present on the margins of these uplifts and in areas of tectonic subsidence. Strata of similar age and composition in the western parts of the shield, in the central part of a Paleoprotozoic geosyncline, are up to 12,000 m thick.

Metamorphism of Paleoprotozoic rocks is distinctly zonal, commonly linear zonal. In some cases zones of higher metamorphism are associated with deep fault zones (Salop, 1964—1967). Rocks under greenschist and epidote-amphibolite facies conditions with many relict structures and textures are common. Amphibolite-facies rocks are less common and are usually associated with large plutonic granitoid masses. Rocks regionally metamorphosed under granulite-facies conditions are virtually absent, but some examples of this facies and rocks of pyroxene hornfels type are reported. Metasomatic granitization, anatexis and migmatization occur locally in the cores of mantled gneiss domes.

There are abundant plutonic rocks of the Kenoran cycle. Compound gabbro plagiogranite (quartz diorite) and granodiorite intrusive complexes are the most characteristic. They are associated with volcanogenic strata and were possibly derived from deep magmatic chambers which earlier supplied the lavas. Ultramafic intrusions are generally related to deep faults.

Unlike the equidimensional or amoeboid fold systems (gneiss ovals) of the Archean strata, Paleoprotozoic structures are characterized by development of linear or arch-like grouping of the folds, the orientation of which is commonly determined by deep faults around cratonic blocks (protoplatforms).

The greenstone belts are commonly folded into gneiss domes, the cores of which are occupied by remobilized Archean granitic gneisses and Kenoran granites. The greenstone belts commonly form synclinal keels between the domes. They may be of irregular shape (Figs.11 and 17).

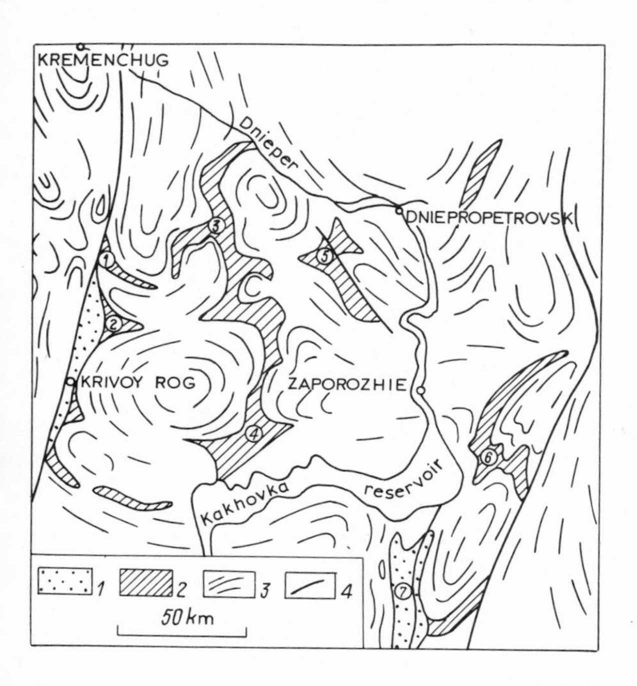

Fig.17. Tectonics of the Dnieper region (Ukrainian shield). Revised after Kalyaev (1965). *1* = Mesoprotozoic ferruginous rocks; *2* = Paleoprotozoic greenstone ferruginous rocks; *3* = Archean granite gneiss and gneiss remobilized at the end of the Paleoprotozoic and partly at the end of the Mesoprotozoic; *4* = large-scale faults. Circled figures in the diagram show *synclinoria*: *1* = Zheltorechenskaya; *2* = Krivoy Rog (Annovsk); *3* = Verkhovtsevo; *4* = Chertomlyk; *5* = Sura; *6* = Konka.

This complex combination of linear and dome-shaped structures within generally elongated fold zones is one of the most characteristic features of Paleoprotozoic tectonics. Another common characteristic is the presence of deep fractures that in some places form a system involving faults in two directions. It is established (Salop, 1964—1967) that such a system of fractures (with northeast and northwest strike) determines the shape of the southern margin of the older Angara platform (protoplatform). Possibly the shapes of some other ancient cratons were also fault-controlled. In the Baikal fold belt some deep intrageosynclinal faults, formed at the beginning of the Paleoprotozoic, remained active up to the Cenozoic and controlled the distribution of mobile zones, different facies and the emplacement of magma (Salop, 1964—1967).

Study of the lithology, facies and tectonic structure of Paleoprotozoic sequences leads to the conclusion that, at the very beginning of the Paleoprotozoic Era, well-defined geosynclinal belts and cratonic blocks of variable size (platforms or protoplatforms and median massifs) were formed. The permobile stage in the earth's evolution came to a close and was replaced by a platform—geosynclinal stage of a particular kind (Fig. 18).

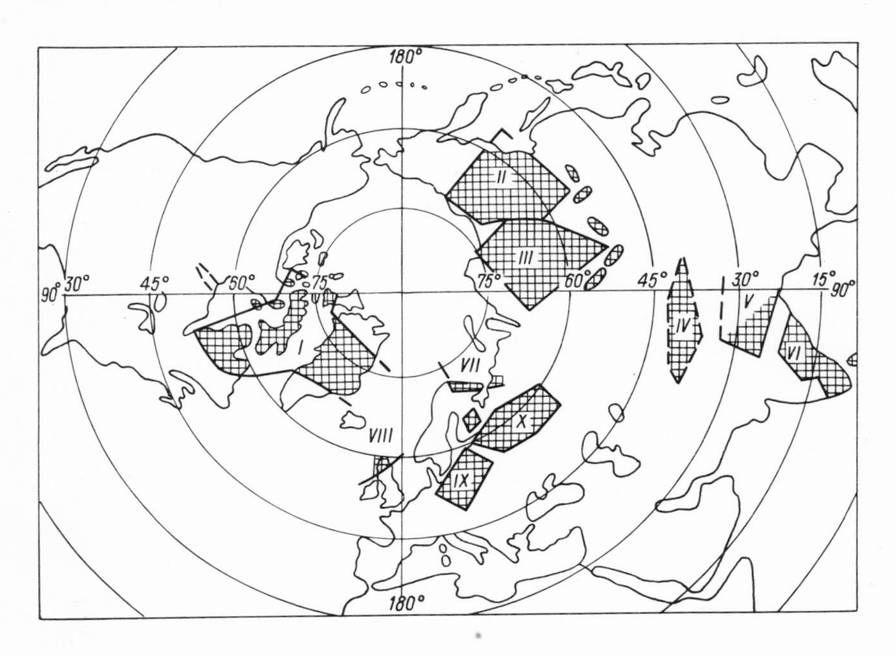

Fig. 18. Paleoprotozoic cratonic blocks—protoplatforms and median massifs (partly presumed).
Protoplatforms: *I* = Baffin; *II* = Aldan—Okhotsk; *III* = Angara; *IV* = Tarim; *V* = Ganges (Bundelkhand); *VI* = Hyderabad—Ceylon; *VII* = Barents; *VIII* = Scourian; *IX* = Middle European (Baltic—Bohemian); *X* = North Russian.

The metallogeny of the Paleoprotozoic is also rather peculiar. Among the primary deposits, banded iron formations of Algoma type are the most characteristic. Small occurrences of these ores, or large low-grade deposits, are present in all regions of the Northern Hemisphere. A number of important economic deposits of this type are also present. These ores have generally undergone secondary enrichment. In Europe economic deposits of this kind include the Kostamuksha and Gimola of Karelia, the Olenegorsk in the Kola Peninsula (if in fact it is in the Paleoprotozoic Tundra Group and not in the Archean Kola Group as some workers believe), the Bjornevatn in Northern Norway, the Grigensberg and some others in Sweden, the Verkhovtsevo, Chartomlyk and others in the Ukraine. Similar deposits in Asia include the

Sosnovy Bayts of the Sayan region, the Karsakpay in Kazakhstan, Anshun and some others in China, Musan in Korea, Orissa (Singbhum) and some others in India. In North America similar rocks are the Michipicoten of Canada, the Soudan in the U.S.A. and others. Iron—manganese deposits of sheet type reported in India and Canada are of lesser importance. The Steep Rock iron—manganese deposit in the south of Canada is rather unique; it is associated with a fossil soil at the base of the Paleoprotozoic.

In sedimentary rocks associated with volcanics or at the base of the sedimentary—volcanogenic groups there are high-alumina sillimanite and kyanite schists which are porential sources of raw materials for alumina and refractories (the Khangarul Group of Eastern Siberia). In some areas (Sayan region, North Korea) large magnesite deposits are present in dolomite rocks occurring among volcanics.

Gold, and in some cases gold—polymetallic deposits in greenstone sequences, possibly related to hydrothermal processes which accompanied lava outflow, are of great economic importance. Large gold deposits of this type occur in Canada, where they are associated with the Keewatin volcanics. Recently, this type of mineralization was found in Paleoprotozoic greenstones of the Ukraine. In Karelia, Southern Finland and in Canada pyrite ores are locally present in the volcanics.

The major economic deposits of the Paleoprotozoic are those of iron, manganese, gold, polymetallic ores and pyritic ores. They are genetically and spatially related to volcanic rocks.

The age limits of the Paleoprotozoic Era are well defined by different isotopic methods in most areas. The lower boundary is placed at the end of the Saamian orogeny, about 3,500 m.y. ago; the upper boundary is the end of the Kenoran orogeny, about 2,600 m.y. ago, so that the era involves about 900 m.y.

CHAPTER 6

THE MESOPROTOZOIC

Strata formed after the Kenoran orogeny, and up to the end of the Karelian orogeny are included in the Mesoprotozoic Erathem, that is the time interval of 2,600 (or 2,800—2,600) to 1,900 (or 2,000—1,900) m.y. In most places the boundaries of the erathem are significant unconformities formed as a result of the above-mentioned orogenies, but in several areas of Central Asia Mesoprotozoic strata apparently conformably overlie the Paleoprotozoic.

Mesoprotozoic supracrustal (and plutonic) rocks are widely developed on all continents and include mio- and eugeosynclinal rocks and deposits of platforms and subplatforms (aulacogens, etc.). In most areas they are well dated isotopically: stromatolites and microphytolites are reported from some sequences.

Regional Review and Principal Rock Sequences

Europe

The best Mesoprotozoic sections in Europe are those of the East European platform and the adjacent fold belt of the Urals. The Karelian complex (supergroup), widely distributed in the eastern part of the *Baltic shield*, especially in Soviet Karelia and adjacent areas of Finland, may serve as a European and world stratotype of Mesoprotozoic platform and miogeosynclinal strata. The following groups (in ascending order) are recognized in this complex: the Tunguda—Nadvoitsa, Sariolian, Segozero, Onega, Bessovets (Ladoga), and Vepsian. The Segozero and Onega Groups are confined to the Jatulian subcomplex (V.Z. Negrutza and T.F. Negrutza, 1968; Salop, 1971a; Bogdanov et al., 1971). The sequence of rock types in the Karelian complex is shown in Table 2.

The platform deposits of the Karelian complex are not very thick and are only slightly metamorphosed. There are numerous breaks in the succession. Quartzites and basic intrusives (diabase porphyry) and subaerial and subaqueous (shallow water) lavas are widespread. These formed under quiet tectonic conditions on small platforms. At the margins of these platforms they pass into miogeosynclinal-type strata of great thickness. These show fewer stratigraphic breaks, higher metamorphism and are of a somewhat different composition (Salop, 1971a). Miogeosynclinal regions in the younger Mesoprotozoic contain rhythmically bedded flysch-like deposits (e.g. the Ladoga Group). Karelian units of the Savo—Ladoga region and of the Northern Lapland—White Sea area have distinct geosynclinal characteristics and may be taken as a Mesoprotozoic miogeosynclinal stratotype (Salop, 1971a). In these regions the same groups as those of the platform are recognized, but

TABLE 2

The Karelian complex: groups, formations and rock types

Group (thickness, m)	Formation (thickness, m)	Principal rock types
Vepsian (2,250)	Shoksha (1,650)	red and grey sandstone (including quartzite) and slate
	Petrozavodsk (600)	dark quartzite or quartzite—sandstone and slate
Bessovets (1,000)	—	fine-grained sandstone, siltstone, shale interbeds
Onega (1,100—2,000)	volcanogenic (300—500)	basic volcanics
	sedimentary (800—1,500)	dolomite and siltstone with interbedded hematite ores; stromatolites; siltstone; arkose and oligomict sandstone; quartz conglomerate, locally with granite pebbles
Segozero (900—1,400)	sedimentary—volcanogenic (200—400)	tuff-sandstone, schist, metadiabase
	sedimentary (700—1,000)	dolomitic sandstone, slate, dolomite interbeds; phyllite, siltstone; quartzite, quartz and oligomict conglomerate, arkose
Sariolian (0—1,200)	volcanogenic (0—400)	basic volcanics, tuff, tuffite
	sedimentary (0—800)	boulder—pebble polymict conglomerate (partly tillite?), gritstone, arkose, quartz conglomerate
Tunguda—Nadvoitsa (0—1,900)	volcanogenic (up to 1,800)	acid volcanics, tuff; basic volcanics, tuff
	sedimentary (0—120)	quartzite, pyritized quartz conglomerate; arkose, conglomerate with granite pebble

some of them are given other names (see Table I).

Table 2 shows that the Karelian complex is of composit structure. The various Karelian groups are either conformable or slightly unconformable. In certain areas some groups are missing or reduced in thickness by contemporaneous erosion. Local development is most characteristic of the two lower (pre-Jatulian) groups. The Segozero and Onega (Jatulian) Groups are of wider distribution and are generally used to define the general character of the complex.

In the author's opinion the Tiksheozero and the Pebozero Groups, in addition to the groups enumerated in Table 2, belong to the lowermost part of the Karelian complex. They occur only in a small area of Northern Karelia within the limits of the Lapland—White Sea zone. Usually these groups are considered to be pre-Karelian (Paleoprotozoic) on the basis of doubtful comparison with the Paleoprotozoic Parandova Group (Bogdanov, 1971b; Bogdanov et al., 1971; etc.). However, the Pebozero Group is slightly unconformable on the Tiksheozero Group and is overlain in turn by the Tunguda—Nadvoitsa (Karelian) Group without marked unconformity (at least there is no angular discordance) (Bogdanov and Voinov, 1968). All of these rocks display a similar metamorphic and tectonic history. The succession in the Tiksheozero and Pebozero Groups resembles that of other groups of the Karelian complex and differs markedly from that of the Paleoprotozoic greenstone complexes. The Tiksheozero Group (400—1,000 m thick) is composed of three cycles. Each begins with coarse terrigenous rocks, locally altered into gneisses, and is overlain by two-mica, high-alumina staurolite—kyanite schist and gneiss. At the base of the first cycle there is a meta-arkose and quartzite unit with lenses of polymictic conglomerate. It unconformably overlies gneiss of the Archean Belomorskaya Group. The upper part of the third rhythm also contains amphibolite. The lower part of the Pebozero Group (up to 3,000 m thick) is comprised of a two-mica, garnet-bearing gneiss with subordinate schist and quartzite. The upper part consists of amphibolite, metadiabase and tuff with subordinate biotite gneiss, schist and quartzite.

In all the successions assigned to the Karelian, with the exception of the upper one (the Vepsian), the lower part of the succession is composed of sedimentary (largely terrigenous) rocks and the upper part is dominantly volcanic. Volcanics are most abundant in the locally developed Lower Tunguda—Nadvoitsa and Pebozero Groups. In the Onega Group, and locally in the Segozero Group, there are carbonate units containing large stromatolitic bioherms, and in some cases, oncolites. Some of these stromatolites are unusual and poorly studied forms, but others resemble known columnar—branching forms, characteristic of the first (Lower Riphean) phytolite complex (e.g. *Omachtenia*, *Kussiella*). Some of the forms are somewhat similar to the Upper Riphean *Jurusania* and *Gymnosolen* (Krylov, 1966). K.N. Konyushkov and M.E. Raaben (personal communication, 1972) suggest that the stromatolites described by Butin (1966) as *Sundia*, but which Krylov (1966) combined with the gymnosolenides, are probably representatives of the kussiellid group, which is typical of the first phytolite complex. Oncolites from the Vepsian Group closely resemble those of "Riphean" complexes of the Urals and other areas (Garbar and Mil'shtein, 1970).

The succession, stratigraphic position and geochronological limits of the Karelian sequence are very reliably established. Thus, it may be recommended not only as a regional rock stratigraphic standard for the Mesoprotozoic, but even as a world stratotype of this erathem.

The Karelian complex unconformably overlies Paleoprotozoic greenstone-belt assemblages and intrusive granites dated at 2,600—2,800 m.y. old. The Karelian complex (including the uppermost Vepsian Group) is cut by pretectonic intrusions of basic rocks (the gabbro—diabase dikes, for example), which give a Pb-isotope and K—Ar age of 2,000—2,150 m.y. (Gerling et al., 1966; Sakko, 1971; etc.). Metamorphic minerals in Karelian rocks and syntectonic granites that intrude them in miogeosynclinal zones yield an age of 1,900 m.y. by different isotopic methods (Salop, 1971a). In stable platform areas syntectonic granites are absent and the Karelian rocks are cut only by the intrusions of rapakivi granite dated at 1,600—1,650 m.y. old. Carbonate rocks of the Onega Group have been dated at 2,300±120 m.y. by the Pb-isochron method (V.S.E.G.E.I. Laboratory, analyst A.D. Iskanderova, collection of L.J. Salop). The age of basic lavas of the Segozero Group is 2,500 m.y. (K—Ar method, Svetov, 1972). Mineralization of the Shokscha sandstone (Vepsian) is dated by the $^{207}Pb/^{206}Pb$ method at 1,800 m.y. (Tugarinov and Voytkevich, 1970). These datings of carbonate rocks and lavas are likely to reflect the time of sedimentation and volcanism.

The subplatform or miogeosynclinal rocks in the Kola Peninsula, in Eastern and Northern Finland, in Sweden and Norway, are closely comparable to the Karelian complex. The proposed correlation is shown on Table I (inset plate). The data on which this is based are given elsewhere (Salop, 1971a). New data confirm these correlations. It is now established that both the Keyvy Group and equivalent rocks in adjacent areas of the Kola Peninsula unconformably overlie the sedimentary and volcanic rocks of the Paleoprotozoic (Lebyazh'ya and Voronya—Kolmozero successions) which were metamorphosed 2,600—2,800 m.y. ago (Garifulin, 1971; Golovenok, 1971; Mirskaya, 1971). Formerly the Keyvy Group was erroneously combined with these. The Keyvy Group conformably underlies the Imandra—Varzuga complex which is very similar to the Karelian complex, and is developed to the south of the main outcrop belt of the Keyvy Group.

According to Kozlov and Latyshev (1974) the Imandra—Varzuga Group is cut by the Moncha—Tundra layered gabbro—norite pluton. Formerly this pluton yielded anomalously old K—Ar dates, but recent age determinations by Rb—Sr isochron analysis (both rocks and minerals) yielded 2,020 m.y. (Birck and Allègre, 1973).

The Keyvy and Chartvurt Groups are probably the oldest Mesoprotozoic units in the Baltic shield. They were formed by redeposition of material from soils developed on a peneplaned surface of Archean and Paleoprotozoic rocks. Both of these groups may be correlated with the Tiksheozero Group of Northern Karelia and also with high-alumina rocks developed south of the Korva and Salny tundra on the Kola Peninsula (the Korva beds), where they are overthrust (from the north) by an Archean granulite complex. V.Z. Negrutza (personal communication, 1970) considered the metamorphosed terrigenous and carbonate rocks of the Pestsovaya Tundra Formation to transgressively overlie the Keyvy Group. He describes a basal conglomerate at the base of the unit. Formerly this formation was also attributed to the Keyvy Group. On the basis of lithological similarity and because of the presence of stromatolites, the Pestsovaya Tundra Formation is considered to be correlated with the Onega Formation of the Karelian complex and with the corresponding part of the Imandra—Varzuga Group. K—Ar analysis on mica from schists of the Keyvy Group yielded an age of 1,900—2,000 m.y. so that the metamorphism of this group of rocks took place during the Karelian orogeny.

Within the limits of the *Russian plate* the red quartzites (Konosha) of the Kresttsy graben (near Kresttsy village) and of Yulovo—Ishim (Pachelma aulacogen) may be compared with units of the Karelian complex. These rocks are assigned to the Mesoprotozoic for they occur at the base of the platform cover on older crystalline rocks and are unconformably overlain by Neoprotozoic formations. Glauconite from the Konosha sandstone yielded an age of 1,700 m.y. This is one of the oldest glauconite ages in the world, but is probably still "rejuvenated". The Kresttsy sandstone is cut by diabase dated at about 1,560 m.y. All of these rocks are probably correlative with the Vepsian Group which is the uppermost group of the Karelian sequence.

In the *Ukrainian shield* the Mesoprotozoic is represented by rocks of miogeosynclinal aspect. The Krivoy Rog and Frunze Mine Groups of the Krivoy Rog, and also the Belozerka and Pereversevsk Groups of the Belozerka iron-ore area are examples.

The Krivoy Rog Group (sensu stricto) unconformably overlies Paleoprotozoic metabasite and Paleoprotozoic granite gneisses. It comprises two formations; the lower one, called the Skelevatka (500 m) is mostly arkose and phyllites, the upper one, known as the Zhelesorudnaya (1,400 m), is characterized by interfingering of metamorphic schists and finely banded siliceous iron formation (jaspilite). Some workers consider the Krivoy Rog Group to include the underlying metabasite (the "metabasite formation") and the overlying Frunze Mine Group (the "Upper Krivoy Rog Formation"). However, the presence of stratigraphic breaks and marked difference in rock types in the Krivoy Rog area support subdivision into three separate groups. These are called the metabasite (Verkhovtsevo), the Krivoy Rog, and the Frunze Mine Group. The lower one is here assigned to the Paleoprotozoic but it is possible that the Krivoy Rog Group includes Mesoprotozoic basic metavolcanics, younger than those of the underlying Verkhovtsevo Group. The Krivoy Rog sequence generally lacks volcanic rocks, but rare bands of talc schist in the lower parts may represent metamorphosed ultramafic volcanics. The lowest formation is mainly composed of quartz arenites and arkoses.

The Frunze Mine (Rodionovo, Ingulets) Group unconformably overlies the Krivoy Rog Group. It is mostly known from drill holes and mine workings in a complex tectonic setting, so that it is not completely understood. It is probable (Kalayev, 1965; Semenenko, 1965) that an iron-rich unit (about 10 m thick) overlies stratigraphic units of the Zhelesorudnaya Formation of the Krivoy Rog Group. Above the iron-rich unit, with evidence of a hiatus, are quartzites (20—60 m), slaty sandstones (20—60 m), slaty carbonates (300—900 m), a thick (1,000—1,500 m) sandstone—conglomerate unit (possibly with a stratigraphic break) and finally, a slaty unit (300—1,000 m thick). In the lower part of the formation, bands of massive and schistose ferruginous rock and ferruginous sandstone are reported. These formed by resedimentation of the Krivoy Rog jaspilite. The slaty carbonate formation contains dark grey "coaly" and "coaly-carbonate" slate (phyllite) and thick bands of dolomite and dolomitic limestone. Organic (?) remains similar to *Corycium* Sederholm are locally present (Belokrys and Mordovets, 1968). *Corycium* was originally described from the Bothnian sequence of Finland which is Mesoprotozoic. The conglomerate in the overlying unit is in the form of thick (up to 500 m) wedges which thin out in a short distance. The conglomerate is composed of pebbles of rocks of the Krivoy Rog Group and granite (mostly Archean plagiogranite).

The most probable correlatives of the quartzite unit are rocks of the Segozero Group (or the lower part of the Onega Group), the slaty-carbonate unit with the Onega Group, and the sandstone, conglomerate and shale units with the Ladoga Group of the Karelian complex. This correlation is based on the similarity of sequence, of rock types and on the occurrence of organic remains in the carbonates. If this correlation is accepted, then the Krivoy Rog Group would be equivalent to the lower groups of the Karelian complex, the Tunguda—Nadvoitsa and possibly in part the Sariolian. The fact that beds of siliceous iron formation are present in metamorphosed strata of the Finnish Lappi sequence, which are close analogues of the Tunguda—Nadvoitsa Group, also favours this correlation. From the point of view of this correlation the find of isolated large sandstone boulders of dropstone type which are probably of marine glacial origin in the Zhelesorudnaya Formation of the Krivoy Rog Group (Gershoig et al., 1974) is of great importance. Rocks of this origin are also recorded in the Sariolian Group (Escola, 1965) and in the Pechenga complex of Kola Peninsula (the Karelian complex correlative) (Zagorodny, 1962).

The Belozerka Group which occurs on the left bank of the Dnieper River, in the area of the Belozerka iron-ore deposit, is an analogue of the Krivoy Rog Group. It also consists of two formations; a sandstone slate overlain by ferruginous strata. It differs from the Krivoy Rog Group in containing porphyritic acid volcanics which are locally abundant. The Belozerka Group unconformably overlies Archean amphibolite (3,500 m.y. old) which encloses granite and granite—gneiss bodies. It is unconformably overlain by terrigenous rocks of the Pereversevsk Group. The Pereversevsk Group is subdivided here for the first time on the basis of study of drill-core material by the author, and on data provided by geologists working on the deposit (M.V. Mitkeev, E.M. Lapitsky and G.F. Gusenko, personal communication, 1971). The group is composed of two formations: a lower one made up of quartzite, slate and sandstone (300 m) and an overlying unit of sandstone and conglomerate (more than 600 m). The lower formation has a composition similar to that of the lower formation of the Frunze Mine Group, but it is thicker and corresponds to the three lower formations of the Krivoy Rog area. If this is the case then the carbonate-rich strata are missing (eroded?). The sandy, conglomeratic unit that unconformably overlies the lower formation is almost identical to the sandstone—conglomerate formation of the Frunze Mine Group.

The Pb-isotope age of uranium-bearing sulphides from the matrix of basal conglomerates of the Skelevatka Formation of the Krivoy Rog Group is 2,600—2,700 m.y. This could represent the time of sedimentation, as suggested by Tugarinov (Tugarinov and Voytkevich, 1970). Clastic zircon and monazite from the lower formation of the Krivoy Rog Group gave a Pb-isotope age of 2,800 m.y. (Shcherbak et al., 1969). Thus, deposition of the basal beds of the Krivoy Rog Group began soon after metamorphism of the sedimentary and volcanic Paleoprotozoic rocks.

Metamorphic minerals (mostly micas) from the overlying Frunze Mine and Pereversevsk Groups yielded K—Ar ages in the range of 1,700—2,000 (2,100?) m.y. These dates correspond to the time of the Karelian orogeny. The same age (1,800—1,900 m.y.) was produced by whole-rock K—Ar dating of phyllite from the Frunze Mine and Pereversevsk Groups (V.S.E.G.E.I. Laboratory). K—Ar ages of porphyries from the Belozerka Group are 1,895 and 2,215 m.y. (whole-rock samples, V.S.E.G.E.I. Laboratory). The first value is from strongly schistose porphyry and possibly reflects the time of metamorphism. The second is from a more massive and relatively unaltered rock and may be the approximate age of the rock.

The Kursk and Oskol Groups of the *Kursk magnetic anomaly* (the Voronezh Massif) have much in common with the groups discussed above. The Kursk Group and the Krivoy Rog Group are almost identical in rock types, stratigraphic position and structure. The Kursk Group unconformably overlies the Paleoprotozoic Mikhaylovsk Group and Archean gneisses, and is itself overlain unconformably by the Oskol Group which is correlative with the Frunze Mine Group of the Ukraine. Like the Krivoy Rog Group, it consists of two formations, a sandstone-slate unit and an overlying iron-rich formation. The conglomerate matrix of the lower formation is reported to contain syngenetic uranium-bearing pyrite which yielded a Pb-isotope age of 2,730 m.y. This date possibly indicates the time of accumulation of the basal beds of the Kursk Group as was the case in the lower formation of the Krivoy Rog Group (Tugarinov and Voytkevich, 1970).

The Oskol Group (locally known as Tim, Kurbakin, Yakovlevo, Voronezh, etc.) is mainly composed of slates with beds of dolomite, limestone and schungite (which is characteristic of the Onega Group of the Karelian complex). In some areas the sedimentary rocks are replaced by acid and basic lavas and tuffaceous rocks.

In the lower part of the Oskol Group (as was the case in the Ukraine) there are some beds that formed as a result of resedimentation of iron formations. The K—Ar age on micas from the rocks of the Kursk and Oskol Groups is in the range of 1,750—2,000 (2,100?) m.y. The Kursk (and Oskol?) Group is cut by pre-tectonic basic intrusions of the Trosnyan—Mamon complex with a K—Ar age of 1,950 m.y. (Chernyshev, 1972).

Mesoprotozoic miogeosynclinal strata are widely distributed in the *Southern Urals* in the Bashkir anticlinorium where they include the Burzyan Group — stratotype of the so-called Lower Riphean. Probably this group should also be regarded as a para-stratotype of the miogeosynclinal Mesoprotozoic of East Europe, for it contains abundant stromatolites and phytolites of the first ("Lower Riphean") complex. Unfortunately, its age limits are not well defined by isotopic dating.

The Burzyan Group includes three formations: the Ayya, Satka and Bakal (in ascending sequence). The Ayya Formation is a complex stratigraphic sequence and some have suggested that it should be given subgroup rank and subdivided into three formations; the Navysh (450 m), composed of tuff sandstones, conglomerates and basic volcanics; the Lipovsk (600 m), made up of conglomerate and arkose; and the Verkhneay (1,400 m), composed of conglomerates, sandstones, quartzites, slates and phyllites. Locally, stratigraphic breaks are present between the formations. The overlying Satka Formation (2,400 m) is largely composed of dolomite and limestone with slaty interlayers, and the Bakal Formation at the top of the section is made up of phyllitic slate with interbeds of dolomite and limestone. In the Satka and Bakal Formations the following stromatolites have been identified: *Kussiella kussiensis* Kryl., *Collenia frequens* Walc., *C. undosa* Walc., *Conophyton cylindricus* Masl., and others; and also oncolites: *Osagia pulla* Z. Zhur.; and catagraphs: *Vesicularites rotundus* Z. Zhur.

The Burzyan Group unconformably overlies gneiss and amphibolite of the Taratash complex (>2,700 m.y. old), and is unconformably overlain by the Kuvash (Mashak) Group (Lower Neoprotozoic?). It is cut by the Berdyaush pluton (rapakivi granite) with an age of 1,560—1,600 m.y. (Salop and

Murina, 1970) by the Rb—Sr isochron and Pb-isotope methods (on zircon which gave similar values based on three major isotope ratios). However, this granite does not accurately define the upper boundary of the group for it was intruded during a post-tectonic phase in the development of the Urals geosyncline, much later than the folding that followed deposition of the Burzyan Group.

A K—Ar age of 1,900—2,000 m.y., reported for some small granite massifs situated in gneisses of the Taratash complex, is regarded by some as the lower age boundary of the Burzyan Group. However, these values may be the age of the post-Burzyan (Karelian) reactivation of the basement to the Burzyan Group. The fact that rocks of the Taratash complex occur in the core of dome-like structures which are surrounded by Burzyan Group rocks, favours this opinion. These structures are similar to mantled gneiss domes formed by metamorphic and tectonic remobilization of the crystalline basement rocks. A similar age is reported for a gabbro stock that cuts the Yamantau Group, which is stratigraphically equivalent to the Burzyan Group. Biotite from this gabbro gave an age of 1,870 m.y. by the K—Ar method (V.S.E.G.E.I. Laboratory, collection of Yu.R. Bekker). Thus, Lower Riphean strata of the Urals are probably older than is usually thought to be the case.

The rocks of the Burzyan Group resemble those of the Karelian complex in general stratigraphic sequence and lithology. However, detailed (formational) correlation is difficult. The broad miogeosynclinal belt of the Urals has a thicker Mesoprotozoic section that may include equivalents of some groups of the Karelian complex. Purely on the basis of lithological similarity, the Navysh Formation may be correlated with the Tunguda—Nadvoitsa Group, the Lipovsk Formation with the Sariolian Group, the Verkhneay Formation with the Segozero Group and the Satka and Bakal Formations with the Onega and Bessovets(?) Groups. Correlation of the upper units is facilitated by the presence of stromatolites and sheet deposits of hematite in carbonates of both the Urals and Karelia. Recently a Pb-isochron age of 2,470±500 m.y. was obtained from stromatolitic dolomite of the Satka Formation (V.S.E.G.E.I. Laboratory, analyst A.D. Iskanderova). The analysed rocks were of very low metamorphic grade so that this figure may be the age of sedimentation. Thus, the so-called Lower Riphean may in fact be Mesoprotozoic. More isotopic data are required to resolve this problem.

The Kyrpinsk Group (platform-type), known from drill holes in the Kama region adjacent to the Urals, is more easily correlated with the Burzyan Group than with rocks of the Karelian complex (Jacobson, 1968, 1971a). The most complete sections were examined in drill holes known as the Arlan-36 and Or'yebash-82. The Arlan hole penetrated about 500 m of flat-lying grey-to-black argillite and mudstone with interbeds of light dolomite and pink sandstone (the Arlan Formation). Unlike other argillites of the younger sedimentary cover, these rocks are highly consolidated, silicified, and in places, phyllitized. Glauconite from these rocks yielded an age of 1,535 m.y.

(Kazakov et al., 1967), but this may be a "rejuvenated" value. Microphytolites of the first complex are present in carbonate interbeds.

The Arlan Formation passes gradationally into the Kaltasinsk Formation which is made up of light dolomite interbedded, at its base, with argillite of the Arlan type. A more complete section is present in the Or'yebash-82 drill hole, where the formation reaches 1,700 m in thickness. The part of the Kyrpinsk Group that is known from bore-hole data is very similar to the middle part of the Burzyan Group, both in lithology and sequence. The Arlan Formation is similar to the Verkhneay Formation exposed along the Ay River, upstream from the town of Kusa. This formation is made up of grey-to-black, phyllitized argillite interbedded with dolomite and sandstone. The Kaltasinsk Formation is similar in every aspect to the Satka Formation of the Burzyan Group (Jacobson, 1968). Strata similar to the Kyrpinsk Group are not known elsewhere in the sedimentary cover of the Russian plate. Microphytolites of the third complex, which usually occur in the upper part of the Neoprotozoic, are present in the Kaltasinsk Formation. However, a Neoprotozoic age cannot be accepted, because it evidently contradicts the geological and radiometric data.

Recently the Kaltasin unmetamorphosed dolomites were dated (V.S.E.G.E.I. Laboratory, analyst D.I. Iskanderova). The Pb-isochron age obtained was 2,150±60 m.y.; it probably reflects the time of sedimentation.

Thus, data from drill holes have shown that, in the area around Kama, Mesoprotozoic platform formations are present in a region of Neoprotozoic strata. They occur in a deeply submerged zone of the Kaltasinsk aulacogen where the depth to basement is about 8,000 m on the basis of geophysical data (Yarosh, 1970). The Kyrpinsk Group is exposed only to a depth of 5,000 m, so that the remaining 3,000 m may either be made up of the lower part of the group, or of older (Mesoprotozoic) groups. Recent drilling in the vicinity of the town of Ocher (100 km west of Perm) revealed light pink quartzitic sandstones under the Arlan or Kaltasinsk dolomite.

Metamorphic strata of the Northern Urals may be of Mesoprotozoic age and miogeosynclinal facies. They are subdivided, in ascending sequence, into: the Man'khobeyu (quartzite), Shchoku'rya (marble, phyllite, mica schist), Oshiz (mica quartzite), and Puyva (quartz—mica and graphitic schist, metadiabase) Formations (Belyakova, 1972). Algae of *Murandavia magna* Vol., and *Nelcanella* type are reported in carbonate rocks of the Shchoku'rya Formation. According to K.B. Korde (personal communication, 1970), similar forms occur in the Gonam Formation (lowermost Neoprotozoic) of the Uchur Group in the Aldan shield. *Murandavia magna* Vol. is also known from the Onega Formation of the Karelian complex, and the tubular algae *Nelcanella*, from the Satka and Bakal Formations of the Burzyan Group (V.E. Zabrodin, personal communication, 1970). L.T. Belyakova (1972) believes that the upper two formations (Oshyz and Puyva) are part of a separate group that may be correlated with Middle Riphean strata of the

Southern Urals (which we attribute to the Middle Neoprotozoic). K—Ar dating from feldspar from granite that cuts the Puyva Group (1,370 m.y.) suggests that this correlation is invalid. This value is thought to be too low because of the method used, the material dated, and because the rocks in question were subjected to Hercynian folding.

The metasedimentary rocks of the Uraltau region (Southern Urals) are probably also Mesoprotozoic. They form part of the Maksyutovo complex which consists mainly of quartzite, micaceous quartzite, muscovite—quartz, graphitic and other types of schists. Basic volcanics are present in one formation of the complex (the Kayrakli Formation). This formation and the overlying Karamali Formation contain lenses and thin beds of marble with possible organic remains (oncolites?). The rocks of this complex differ from the Mesoprotozoic of the Bashkir anticlinorium in lithology, degree of metamorphism and in tectonic style. The tectonic style in the Uraltau region is characterized by gneiss-dome structures. These rocks resemble the corresponding strata of Karelia. There are no reliable isotopic dates for rocks of the Maksyutovo complex; an α-Pb age of 1,000—1,600 m.y. was obtained from zircons in the schist and quartzite but is not considered significant.

In Mugodzhary, the Verkhnekumak (Uzunkayrak) Group is conditionally assigned to the Mesoprotozoic. It is made up of micaceous and graphitic quartzite, mica—quartz schist, phyllite, tremolite—actinolite rocks and porphyritic igneous rocks. Ferruginous quartzite (jaspilite?) occurs locally in this group, but is, as yet, poorly studied. This group unconformably overlies the Taldyk Group which is probably Paleoprotozoic.

The Bothnian rocks (Tampere Group) of Western Finland are an example of eugeosynclinal facies of Mesoprotozoic age. The Elveberg, Vargfors and Larsbo—Melar Groups of Southern Sweden are other similar examples from the Svecofennian region of the *Baltic shield*. All of these groups are characterized by thick successions of metamorphosed greywacke schists with local intercalations of basic and intermediate metavolcanics. In Sweden the rocks of eugeosynclinal aspect unconformably overlie leptites and Early Svecofennian granites that intrude them. The relationships between the Bothnian Group and the underlying(?) metamorphic and crystalline rocks is not certain. A basal polymictic metaconglomerate (Suodeniemi—Lavia area) and quartzite, and metaconglomerate (in Southern Finland, Tiirismaa area) may indicate that the Bothnian Group unconformably overlies a Paleoprotozoic schist—amphibolite complex (Salop, 1971a).

Laitakari (1969) reported the presence of quartzite xenoliths in numerous diabase dikes that cut the Bothnian Group. Such quartzites are not known in the host rocks. The presence of these quartzite fragments indirectly supports the idea that basal quartzites of Jatulian type underlie the volcanic and sedimentary units of the Bothnian Group. It has been suggested (Salop, 1971a) that the basal part of this group may be correlative with the Sariolian and Segozero Groups of the Karelian complex, and that the major (upper) part may be equivalent to the Onega and Ladoga (Kalevian) Groups.

Many Finnish and Scandinavian geologists consider the Mesoprotozoic strata of the Svecofennian region and older supracrustals, to form a single complex. They argue that the differences in the rocks are mainly due to different metamorphic grade. This idea was criticized in an earlier publication (Salop, 1971a). New studies in the Western Ladoga region near the boundary between the Karelides and Svecofennides showed that the Onega (= Sortavala or Pitkyaranta) and Ladoga (= Bessovets) Groups of the Karelian complex are continuous into the Svecofennian area. These groups show some sedimentary facies changes, and the metamorphic grade increases (up to the amphibolite facies) in a westerly direction. Locally both groups are transformed into migmatitic gneiss. It has also been established that the Onega Group, and in some places, the Ladoga Group, unconformably overlie Paleoprotozoic amphibolite and plagiogranite, or Archean gneiss and granulite which are exposed in the cores of mantled gneiss domes. The unconformity is commonly obscured by intensive post-Ladoga deformation and granitization.

The Svecofennian formations are cut by granites with a K—Ar age of about 1,800 m.y., and the Pb-isotope age (on zircon) is 1,900 m.y. (Kouvo and Tilton, 1966). Thus, the Svecofennian granites are of the same age as those that cut the Karelian formations in miogeosynclinal zones.

The Osnitsk Group outcrops in the western part of the *Ukrainian shield* and in adjacent areas of the Russian plate (Volyn'—Podolia). It represents the eugeosynclinal facies of the Mesoprotozoic, and is composed of metamorphosed greywacke and arkose, basic and acid volcanics and tuff. It unconformably overlies Archean or Paleoprotozoic gneiss, and is cut by the Osnitsk granite dated in the 1,300—1,800 m.y. range by the K—Ar method on biotite (Khatuntseva, 1972). The granite dates are almost certainly too young because the Osnitsk Group and the Osnitsk granite are older than the Pugachev and Ovruch Groups which are about 1,800 m.y. old. Also, the groups in question are older than the rapakivi granites of the Korostensk complex which are reliably dated at about 1,700 m.y. Most probably the Osnitsk granite was contemporaneous with the granite of the Kirovograd—Zhitomir complex (1,800—1,900 m.y.).

In *Eastern Byelorussia* possible correlatives of the Osnitsk Group are the miogeosynclinal Zhitkovichi Formation (Group?) which consists of metavolcanics (porphyries and schists formed from keratophyre, diabase and andesite) and metasedimentary rocks (phyllite and micaceous quartzite). This formation (as seen in drill cores) unconformably overlies Archean gneiss. It is cut by granites in the age range 1,730—1,900 m.y. according to K—Ar analysis (on micas), and a Pb-isotope age (on zircon) of 1,860 m.y. (Pap, 1967, 1971).

In the fold belts of Western Europe Mesoprotozoic supracrustals are not definitely recognized, though in some areas, for instance in *Scotland*, the Karelian (Inverian) orogeny was intense and retrograde metamorphism of

Archean rocks took place, together with granitoid intrusions of about 1,900—2,100 m.y. It is possible that supracrustals of the so-called sub-Moine complex, in the Northern Highlands of Scotland, belong to the Mesoprotozoic. These rocks unconformably overlie Archean gneiss of the Lewisian complex and were themselves deformed by the Laxfordian events (1,400—1,600 m.y. ago). These rocks are preserved as biotite and amphibole gneisses, amphibolitic, graphitic, micaceous, kyanitic and other schists, and marbles together with irregular bodies of basic and ultrabasic rocks (Pavlovsky, 1958). In the Gairloch area, the Loch Maree Group is an example of this sequence. It is composed of greywacke metamorphosed to greenschist and amphibolite facies, quartzites, and pelites with thick amphibolites and subordinate bands of marble and ferruginous rock. This group is attributed, by British geologists (Bennison and Wright, 1969) to the Laxfordian cycle which culminated in the so-called Laxfordian orogeny. Metamorphism and plutonism associated with this orogeny is considered to have taken place 1,700—1,600 m.y. ago, but new Rb—Sr isochrons provided a date of 1,850±50 m.y. ago, indicating that the Laxfordian metamorphism may be part of the widespread Karelian cycle (Lambert and Holland, 1972).

Mesoprotozoic geosynclinal strata are probably present in the *Spitsbergen Archipelago*. In the west and southwest parts of Western Spitsbergen Island there are two groups which are possibly Mesoprotozoic: the Eimfjellet Group and the unconformably overlying Deilegga Group (Birkenmajer, 1964; Winsnes, 1965; Krasil'shchikov, 1973). There is an angular unconformity between the Isbjörnhamna garnet—mica schist and marble, which are very probably of Paleoprotozoic age, and the Eimfjellet Group. The latter consists of amphibolite, quartzite (mainly in the lower part of the group) and schists with metaconglomerate horizons. The Deilegga Group (3,500 m) is composed of schist with interbeds of dolomite and quartzite; conglomerate occurs at its base. Above it are Neoprotozoic microphytolite-bearing strata.

In the New Frisland Peninsula of Western Spitsbergen the Mossel Group (5,000 m) may be correlated with the above-named groups or only with the Deilegga Group. It appears to lie unconformably between the Archean(?) gneiss of the Atomfjella Group, and the Middle(?)—Upper Neoprotozoic strata of the Veteranen Subgroup, but the contacts are generally tectonic (Krasil'shchikov, 1973). According to Harland (1961) this group is composed of a lower unit of garnet—biotite schist which is overlain by a thick succession of phyllite, quartzite and marble.

Asia

The Mesoprotozoic strata in the older belts that surround the Siberian platform and, especially in the Baikal fold belt, are the most complete and best known. This region is proposed for the miogeosynclinal and eugeosynclinal Mesoprotozoic stratotypes of Siberia and possibly of all Asia.

The miogeosynclinal stratotype is to be the Udokan Group (about 13,000 m thick). It occurs in the northwest part of the *Baikal fold belt* in the Olekma—Vitim Highlands. Three subgroups are recognized, which in turn are subdivided into formations (Salop, 1964—1967). The lower or Kodar Subgroup (up to 6,000 m) includes the Sygykhta, Orturyakh, Boruryakh, Ikabiya and Ayan Formations. It is comprised largely of terrigenous rocks of marine aspid formation: various dark-coloured biotite-and graphite-bearing schists, metamorphosed siltstones and sandstones (quartzitic in part). The latter are usually present as thin and rare interbeds, but approximately in the middle part of the subgroup quartzite occurs as thick bands that interfinger with dark schist (the thick Boruryakh Formation). In the lower part of the subgroup horizons of marble are enclosed by sandstones and schists.

The Chine Subgroup (from 1,800 to 4,500 m), conformably overlies the Kodar Subgroup. It consists of four formations: the Inyr, Chetkanda, Aleksandrov and Butun, of which the lower two are essentially terrigenous, composed largely of orthoquartzite with subordinate bands of phyllite. The two upper formations contain thin and thick horizons of carbonate rocks (marls, limestones and dolomites). North from the Kodar—Udokan area, in the Chenchin Mountains, the Butun Formation is called the Bulbukhta Formation. The three lower formations of the Chine Group formed in coastal-marine environments with extremely shallow water (ripple marks and desiccation cracks are common) and the Butun Formation accumulated in a "lagoonal", highly mineralized basin (indications of evaporites include crystal moulds after halite, albite—scapolite or zeolite-bearing rocks, etc.).

The Kemen Subgroup (up to 4,500 m) is the uppermost in the Udokan Group. It is composed of the very thick Sakukan Formation of cross-bedded orthoquartzite and arkosic sandstone, and the overlying Namingu Formation of siltstone. The older formation is characterized by cross-bedded sandstone (locally ripple-marked) with clastic magnetite (martite). Where slightly metamorphosed, some of these rocks are of a pink-red colour. A horizon of cupriferous sandstone is present in the upper part of the Sakukan Formation. The Sakukan Formation overlies the Butun Formation with local erosional contacts and basal conglomerates are present. Strata of the Kemen Subgroup were deposited in subaqueous and subaerial deltaic environments. The Udokan Group includes formations typical of both transgressive and regressive conditions, and thus provides a variety of sedimentary rock types characteristic of the miogeosynclinal Mesoprotozoic.

The stratigraphic position (age) of the Udokan Group is well established. It is younger than the Archean basement and the Paleoprotozoic greenstone assemblages (Olondo Group and analogues) which were metamorphosed and intruded by granites 2,600—2,800 m.y. ago. All the units of the Udokan Group are intruded by the syntectonic granite of the Kuanda complex, and by the late tectonic granite of the Chuya—Kodar complex which is about 1,900 m.y. old on the basis of Pb-isotope, Rb—Sr and K—Ar methods (some scores of determinations). The Teptorgo Group (Lower Neoprotozoic) was deposited on the eroded surface of these granites.

Stromatolites are reported from dolomites of the Butun Formation. They are mostly *Conophyton garganicus* Korol., *C. meta* Kiric., and *Stratoconophyton* Korol. Rare examples of *Collenia frequens* Walc. and oncolites of *Osagia* type are also present. I.N. Krylov (personal communication, 1970) came to the conclusion that the Butun stromatolites resemble those of the Satka Formation of the Burzyan Group (Lower Riphean) of the Urals. However, we consider the Burzyan Group to be Mesoprotozoic. I.K. Korolyuk (personal communication, 1900) reported stromatolites from the Mongosha (Garga) Formation of Eastern Sayan. Some of the stromatolites of this unit appear to belong to the second (Middle Riphean) complex. In the Chenchin Mountains the Bulbukhta (Butun) Formation, which is cut by 1,860—1,900 m.y. old granite, contains the following stromatolites: *Conophyton garganicus* Korol., *Kussiella* sp., *Baicalia bulbuchtensis* Komar., and oncolites *Osagia libidinosa* Z.Zhur. According to Dol'nik (1969) these forms are typical of the first complex (Lower Riphean). Phytolites are abundant in dolomites of the subplatform Namsala Formation which is almost certainly correlative with the Butun Formation. Oncolites from this formation resemble similar structures of the Satka Formation of the Urals and stromatolites belong to *Stratifera* sp., *Conophyton* sp., and *Omachtenia* Nuzh., which are typical of the "Lower Riphean" of Siberia (Robotnov, 1964).

Thus, it appears that in Eastern Siberia stromatolites and microphytolites characteristic of the first, or Lower Riphean complex, occur in Mesoprotozoic strata older than 1,900 m.y.

The Zama Subgroup of the Muya Group in the interior regions of the Baikal mountain belt is probably the most suitable stratotype for eugeosynclinal facies of the Mesoprotozoic in Siberia. However, it lacks abundant organic remains, and isotopic dating of these units is not very reliable. Nevertheless it can be safely correlated with miogeosynclinal strata of the Kodar—Udokan area so that its boundaries and stratigraphic position in the Precambrian section are known. Its correlation with the Udokan Group is possible because of a transitional relationship (quartzite decreases and metavolcanics increase with increasing distance from the miogeosynclinal zone) (Salop, 1964—1967).

The lower boundary of the Zama Subgroup is determined by the fact that it conformably overlies the Kilyana Subgroup, which is attributed to the Paleoprotozoic. Kilyana metavolcanites with interbeds of metagreywacke and tuff pass upwards into thick orthoquartzites, indicating a change of tectonic regime. In some areas thick horizons of subaerial tuff conglomerate and agglomerate breccia are present among porphyritic volcanics near the base of the Zama Subgroup.

The upper boundary of the Zama Subgroup is well defined, because deposition was followed by strong diastrophism and emplacement of intrusions of the Muya plutonic complex. All the K—Ar age determinations from the rocks of this complex yield "rejuvenated" values. The Akitkan Group (Lower Neoprotozoic) transgressively overlies the Zama Subgroup and the granites that cut it. This group contains porphyry, the age of which is well dated (Rb—Sr isochron) at 1,710 m.y. (Manuylova, 1968).

The Zama Subgroup (up to 6,000 m thick) is locally subdivided into formations. The lower part is mainly composed of sedimentary rocks, the middle units are tuffaceous sandstone, aleurolite, dolomite and limestone with interbedded volcanics, while the upper part is predominantly volcanic rocks with some sedimentary intercalations (see Fig.15).

In the Baikal fold belt the Mesoprotozoic geosynclinal strata proposed as stratotypes, are folded, intruded by various igneous rocks and metamorphosed. The metamorphism has a linear zonation such that rocks of the inner part of the fold zone are in the epidote—amphibolite facies (particularly adjacent to large granitoid intrusions), whereas in the periphery of the fold belt, metamorphism is in the lowest greenschist facies.

The strata of the Olondo and Verkhnekhani grabens in the western part of the *Aldan shield* (left bank of the Olekma River) are only slightly deformed and metamorphosed. These are considered as the stratotype of the Mesoprotozoic platform and subplatform (aulacogen) facies. They are comparable to the Udokan Group, developed in the adjacent miogeosynclinal belt. The lowest unit is the Charodokan Formation (250—500 m) comprised of quartzo-feldspathic sandstone and orthoquartzite with a basal conglomerate that overlies the Archean gneiss. Above this is the Namsala Formation (150—225 m), which consists of grey dolomite with interbeds of sandstone and carbonate siltstone, and is overlain by the Khani Formation (60—600 m) of grey and dark-grey finely bedded siltstone with interbeds of sandstone. The uppermost unit is the Kebetka Formation. It is made up of thick (up to 1,300 m) cross-bedded quartz sandstone (Mironyuk, 1959).

The Charodokan Formation is closely comparable to the Inyr and Chetkanda Formations; the Namsala and Khani Formations with the Aleksandrov and Butun Formations and the Kebetka Formation with the Sakukan (Nuzhnov, 1968; Nuzhnov and Yarmolyuk, 1968; Salop, 1964—1967). As stated above, Lower Riphean stromatolites and oncolites are present in dolomite of the Namsala Formation. These are similar to those reported from the Butun Formation of the Chine Subgroup. Copper mineralization of the type found in the Sakukan Formation of the Kemen Subgroup is confined to sandstones of the Kebetka Formation. Thus, the strata corresponding to the lower subgroup of the Udokan Group and to the uppermost part of the upper subgroup are probably missing in the platform stratotype.

In a monograph on the geology of the Baikal fold belt Salop (1964—1967) stated that the Chuya strata correspond to the Udokan Group of the Olekma—Vitim Highlands. East of the Patom Highlands in the Amalyk, Chencha and Nechera basins the Chuya strata are so similar to the Udokan

Group that the same formation names can be used. In particular, the lower part of the succession is almost identical; analogues of almost every formation of the Kodar Subgroup of the Udokan Group are present. One difference is the presence of green orthoschist in the lower part of the section. This probably indicates a closer relationship between the Mesoprotozoic formations and the Paleoprotozoic volcano-sedimentary strata than is the case in the Kodar—Udokan area, where, however, a similar relationship exists in the adjacent eugeosynclinal belt. East of the Patom Highlands the upper part of the Chuya succession is locally developed. In the Chencha River basin analogues of the Butun Formation (Bulbukhta Formation) of the Chine Subgroup and of the overlying Sakukan Formation (Kemen Subgroup) are definitely recognized.

In the central areas of the Patom Highlands (Tonoda anticlinorium) and in the North Baikal Highlands (Chuya anticlinorium) the Chuya strata are poorly known, but possible analogues of the formations of the Udokan are present. In particular, the Abchada Formation (dolomite) of the Upper Chaya River (with its contained recrystallized stromatolites of the *Conophyton* group) is closely comparable to the Butun Formation.

East of the Baikal mountain area the Chulman Group of the *Stanovoy fold belt* is probably correlative with the Udokan Group. It is developed in a tectonic wedge that includes the Udokan Group miogeosynclinal belt (Salop, 1964—1967). The Chulman strata are composed of highly altered terrigenous rocks which are very similar to the formations of the Kodar Subgroup in the zone of strong metamorphism (Sudovikov et al., 1965). The K—Ar age of these rocks is about 1,900 m.y., which is the same as that of the intensive Karelian diastrophic cycle (deformation, metamorphism and magmatism) in the Stanovoy area.

The analogues of the Udokan Group and of its subplatform stratotype are sedimentary (commonly very slightly metamorphosed) rocks of the Yarogu, Davangro—Khugda and Atugey—Nuyama grabens in the central and southeastern parts of the *Aldan shield* (Gunin, 1968; Kostrikina, 1968; etc.). The Yarogu strata, which are developed in the graben of the same name in the Amedichi River basin are typical. They comprise predominantly metasandstone with phyllites and quartzite. Marble is present in the lower part. The upper age boundary is defined by small granite stocks with K—Ar ages (micas) of 1,850—1,900 m.y. Conglomerates of the Konkulin Formation which are here considered to be Lower Neoprotozoic, overlie these strata and the intruded granites.

The slightly altered subplatform (taphrogenic) terrigenous and volcanogenic strata are rather unusual. These rocks are developed in grabens east of the Aldan shield within the limits of the Ulkan trough (Zabrodin, 1966; Nuzhnov, 1967, 1968; Gamaleya et al., 1969). The lower part (Toporikan Formation) is up to 450 m thick and consists of quartzite sandstones which transgressively overlie the Archean strata. The overlying Ulkachan Formation

(250 m) is subaerial porphyritic basic volcanics and is followed by the thick Elgetey Formation (up to 4,000 m) of subaerial porphyritic acid volcanics with interbeds of tuff and terrigenous detrital rocks and some porphyritic sheets. The K—Ar age of the intrusive quartz porphyry is 1,615 m.y. and the Pb-isotope age of extracted zircons is 1,840 m.y. The Ulkan granite pluton which cuts the Elgetey Formation yielded an age in the 1,538—1,800 m.y. range (several K—Ar analyses), and zircons from granite ($^{207}Pb/^{206}Pb$ ratio) yielded a 1,900 m.y. age (Tugarinov and Voytkevich, 1970). Gamaleya (1968) argued that the Ulkan granite and the porphyry of the Elgetey Formation are comagmatic. These sedimentary and volcanogenic strata are overlain by the variegated or red beds of the Lower Neoprotozoic.

On the southern margins of the Siberian platform, west of the stratotype sections of the Baikal area, Mesoprotozoic geosynclinal strata are reported from the Eastern Sayan and Yenisei Ridge.

In the *East Sayan fold belt* the terrigenous Iret and Daldarma (Belorechensk) Formations of Onot and Urik—Iya grabens of the Sayan area (Dodin et al., 1968; Konikov, 1974) are assigned to the Mesoprotozoic miogeosynclinal facies and are comparable to the stratotype in the Kodar—Udokan region. These formations unconformably overlie the Paleoprotozoic volcano-sedimentary sequence and are transgressively overlain by the Lower Neoprotozoic Kalbazyk Group. The metamorphic age of these rocks is 1,900 m.y. (K—Ar analysis on micas). Possibly the Utulik Formation of Khamar-Daban and the Tumashet strata of the Tumashet graben of the Sayan area (Dodin et al., 1968) are of the same composition and age.

The Urtagol Formation of the Kitoy Group of the Irkut—Oka zone is possibly of eugeosynclinal facies. It overlies the Paleoprotozoic Irkut Formation and is overlain with angular unconformity by the Orlik Group of Neoprotozoic age (Dodin et al., 1968). The volcanic and terrigenous Sigach Formation and the overlying Pezina (largely carbonate) Formation of the Derba Group of the Uda—Derba zone are similar. The Sigach Formation conformably overlies the Paleoprotozoic Khana carbonate Formation but records the start of a new transgressive—regressive macrorhythm. All of these units are mainly composed of highly metamorphosed sedimentary rocks of marine origin, but basic metavolcanics (orthoamphibolite) are also abundant (locally making up as much as 30% of the rock assemblage). These strata are a volcano-sedimentary facies formed far away from the zone of maximum volcanic activity and deep fractures. In this respect these rocks are similar to the underlying Paleoprotozoic strata. In the Uda—Derba zone, the Zhaima Formation, which conformably overlies the Pezina Formation, possibly belongs in part to the Mesoprotozoic. The relationships of the Derba Group and the overlying Neoprotozoic Kuvay Group are different in different localities: in some areas they are apparently conformable but in others a large break, or even an angular unconformity, is recorded.

In the *Yenisei fold belt* the miogeosynclinal Teya Group, the Vyatka

(Abalakovo) eugeosynclinal strata, and the Zyryanovo eugeosynclinal strata belong to the Mesoprotozoic. The lower part of the Teya Group (> 2,500 m thick) is biotite, garnet—biotite, and two-mica schist which commonly contains high-alumina minerals (sillimanite or kyanite) with subordinate quartzite, amphibolite, and crystalline limestone (the Karpinsky Ridge Formation). These rocks are overlain by crystalline limestone and silicate—carbonate rocks with interbedded mica schist and quartzite (the Penchenga Formation). Recently jaspilite was reported among these rocks. The lower boundary of the Teya Group is not certain, but all the workers in the Yenisei Ridge area contend that it is younger than the Yenisei Group which is considered to be Paleoprotozoic. Its upper boundary is fixed by the overlying Neoprotozoic Sukhoy Pit Group. The Vyatka (Abalakovo) strata are correlated by many workers with the lower part of the Teya Group which is developed in the western part of the ridge and has tectonic contacts with the surrounding rocks. It is made up of various mica and amphibole schists, intercalated with amphibolite. The amphibolitic rocks were probably derived by metamorphism of basic volcanics. These strata are intruded by granites of the Garevsk complex which have an age of 1,760, 1,800, and 1,900 m.y. based on Pb-isotope analysis on zircon and allanite (Volubuev et al., 1964). The latter value is closest to the time of the Karelian orogeny that completed the Mesoprotozoic Era. The Zyryanovo volcano-sedimentary strata are usually correlated with the Teya Group but their stratigraphic position is not certain, for neither the underlying nor overlying rocks are exposed.

Two major types of Mesoprotozoic strata are reported in the northern border of the Siberian platform in the *Taymyr fold belt*. The Khariton Laptev Coast Group (up to 12,000 m thick) has a miogeosynclinal character; it is composed of metamorphosed terrigenous rocks, largely sandstone, siltstone, slate and phyllite. It is distributed throughout the western part of Taymyr, on the Khariton Laptev coast (Zabiyaka, 1974). The altered sedimentary—volcanogenic rocks of the Chelyuskin Peninsula Group and the conformably overlying Chukcha Group are eugeosynclinal in character. These groups are developed on the east side of the peninsula (M.G. Ravich and O.G. Shulyatin, personal communication, 1968). Both of these groups unconformably overlie the Paleoprotozoic strata (Trevozhnaya Group and Faddei Island Formation respectively), and are in turn unconformably overlain by red beds of the Neoprotozoic Stanovskaya Formation. In North Zemlaya their stratigraphic analogues are the Telman and Partizan Formations (Egiazarov, 1959), which are largely terrigenous and of miogeosynclinal type. The upper part of the Partizan Formation includes acid lavas.

Further to the south, in Asia, Mesoprotozoic strata are developed in China and the southern regions of the Soviet Far East.

In *China* the Mesoprotozoic is reliably subdivided into the Liao Ho Group in Liaoning province and the Huto Group in Shansi province, and corresponding groups in other regions.

The Liao Ho Group (more than 10,000 m thick) occurs mainly in Liaotung Peninsula; it is characterized by intercalation of phyllite and metagreywacke, and in the middle part it contains a thick dolomitic marble (up to 2,000 m) with subordinate phyllite. Oolites (oncolites?) are reported from the dolomite. Magnesite deposits occur in certain dolomite bands of high MgO content. Rare carbonate-rock interbeds occur in the upper part of the section. This group unconformably overlies Archean and Paleoprotozoic rocks that are cut by 2,600 m.y. old granite. The Pb-isotope age on orthite of pegmatite cutting the group is 1,600 m.y. (Tugarinov and Voytkevich, 1970). K—Ar dates from micas are variable, ranging from 1,400 to 1,900 m.y. The Neoprotozoic, platform-type Sinian complex unconformably overlies the Liao Ho Group and the granite that cuts it.

The Huto Group is very similar to the Liao Ho Group; the most complete sections are in the Wutaishan area where it transgressively overlies the Wutai Group and is unconformably overlain by platform deposits, the lower part of which belongs to the Sinian complex. According to data of Ma Hsing-yuan (1962) the Huto Group is subdivisible into two formations; the Toutsun (4,300 m) and the Dunje (up to 3,000 m). At the base of the Toutsun Formation there is a thick mass of metamorphosed conglomerate containing pebbles of the underlying rocks followed by an arkosic quartzite (Nantai) which is overlain by schist with interbeds of quartzite and siliceous dolomite (the Toutsun "roofing slate"). The Dunje Formation is light-coloured dolomite which passes upwards into siliceous limestone. Stromatolites occur locally in the carbonate rocks. The Huto Group appears to be equivalent to the lower and middle parts of the Liao Ho Group. In Shansi province, rocks equivalent to the upper part may have been eroded down to formations of the Sinian complex. The Huto Group is folded, locally rather strongly metamorphosed, and is cut by granite and pegmatite during the so-called Liuliang (Karelian) folding with an age of 1,860 m.y. (K—Ar analysis on micas from pegmatite).

In the *Soviet Far East* the Mesoprotozoic strata appear to be widely developed in the Khankai Massif and in the Bureya and Khingan—Bureya regions (Shatalov, 1968).

In the Khankai Massif the Spask, Mitrofanovo and Lysogorsk Formations are attributed to the Mesoprotozoic. These comprise various micaceous and graphite-bearing schists with interbeds of amphibolite, schist, limestone, metasandstone and metasiltstone. These rocks unconformably overlie the Paleoprotozoic metamorphic sedimentary and volcanic strata and are transgressively overlain by Neoprotozoic sedimentary deposits.

In the Bureya region, within the limits of the so-called Bureya anticlinorium, there are rocks correlative with the above-mentioned formations of the Khankai Massif. These have been given local names and are commonly called, in ascending order, the Myna, Zlatoust and Sagur Formations. They correspond closely, both in composition and degree of metamorphism with formations of the Khankai Massif. They unconformably overlie Paleoprotozoic strata and are overlain by Neoprotozoic sedimentary deposits.

In Khingan (Khingan—Bureya Massif) mica-schist, quartzite, marble and metasandstone with sheets of economic graphite microgneiss (Soyuznaya Formation) and the conformably overlying Ditur Formation which consists of marble, quartzite, and graphitic phyllite are possibly Mesoprotozoic. This complex occupies a higher stratigraphic position than the Paleoprotozoic Amur Group and is separated by a gap from the overlying Neoprotozoic strata (Igincha Formation, etc.).

Reliable isotopic dates are not available from the above-described strata of the Far East, but their Mesoprotozoic age is established by the fact that they overlie strongly metamorphosed eugeosynclinal sedimentary and volcanic strata of the Paleoprotozoic, and underlie microphytolite-bearing and relatively unmetamorphosed Neoprotozoic sedimentary deposits. The Mesoprotozoic rocks of the Far East are generally of miogeosynclinal facies.

In the *Hindustan platform* the Mesoprotozoic is both extensive and complete, but strata cannot be given a certain age assignation because of a paucity of information on stratigraphy of the older strata and because of a dearth of geochronological data. As stated in Chapter 5, the upper part of the Dharwar Supergroup in the "Shimoga—Dharwar schist belt" (Mysore state) and the Orissa Supergroup of Orissa state may possibly be Mesoprotozoic.

The upper part of the Dharwar Supergroup (Upper Dharwar Group) overlies the lower part of the supergroup with angular unconformity. It contains conglomerate at the base and is composed of sedimentary rocks, largely quartzite, argillite, carbonate-rich slate, limestone and ferruginous rocks. These latter are finely banded jaspilites and hematite-rich quartzite. The upper ferruginous rocks differ from the banded iron formation of the lower (Paleoprotozoic) part, in that they are associated with sedimentary (terrigenous and carbonate) strata. This is typical of the Mesoprotozoic iron ores, whereas the latter are commonly interbedded with metavolcanics. The Upper Dharwars are relatively unmetamorphosed and of miogeosynclinal aspect, in contrast to the highly metamorphosed eugeosynclinal strata of the underlying Paleoprotozoic.

The upper two groups of the Orissa Supergroup are of similar composition. They unconformably overlie rocks of the lower units of the supergroup and the Bonai granite which intrudes them. The lower group is represented by conglomerate and greywacke, overlain by quartzite and quartz schist with ultrabasic and basic sills and dikes. The upper group, which is separated from the lower by a hiatus, is composed, at the base, of conglomerate, sandstone and quartzite, overlain by basic lavas and tuffs intruded by dolerite sills. These are followed by schist, limy schist and silicified limestone. Both groups are cut by granite (Prasada Rao et al., 1964).

A radiometric age is available only from lavas of the upper formation which give a K—Ar age of 1,665—1,745 (1,600—1,700) m.y. (Radhakrishna,

1967), but it is probable that these values are too low, due to the loss of argon.

In the central part of Hindustan the Chilpi—Gat Group is probably Mesoprotozoic. It is present in the Bilaspur—Balaghat tectonic zone (Krishnan, 1960a). This group is largely comprised of phyllite, schist and schistose quartzite with ferruginous quartzite and manganese ores in the lower part. It contains a basal conglomerate and unconformably overlies strongly metamorphosed rocks of Archean and Paleoprotozoic Lower Dharwar. It is probably correlative with the Sakoli Group of the Nagpur—Bandara zone.

A miogeosynclinal complex in the northwestern part of the Hindustan Peninsula, in Rajasthan, is probably also Mesoprotozoic. This complex is composed of three groups which, in ascending sequence are, Aravalli, Raialo and Delhi. The Aravalli Group (up to 3,000 m thick) unconformably overlies the Archean granite gneiss of Bundelkand. It is composed of metamorphosed greywacke, arkose, conglomerate and quartzite. The upper part consists mainly of schists and phyllite, with interbedded quartzites and dolomitic marble and limestone. In some localities the group is capped by quartzite and granule conglomerates which unconformably overlie the underlying rocks (Krishnan, 1960a; Nath et al., 1964; Pichamuthu, 1967). This quartzite is usually attributed to the Aravalli Group, but may in fact be younger.

The Raialo Group which was deposited on top of the Aravalli Group after what is interpreted as a small stratigraphic break, is of local distribution and is rather thin (up to 600 m). Some workers include it in the overlying Delhi Group (Sen, 1970). The group is composed mainly of marble, but at the base there is a thin horizon of quartzite, and in the upper part, local interbeds and bands of garnet—biotite slate. The thick Delhi Group (up to 6,000 m) is situated above the Raialo Group. The two are possibly separated by a break. It is subdivided into two subgroups. The lower one, the Alwar, consists of variably altered arkose, granule and coarser conglomerate and predominant quartzite. The upper and coarser (Ajabhar Group) consists of phyllite (or biotite schist), limy slate, marble and calc-silicate rocks.

Metamorphism of the groups is irregular but generally increases from east (lowest degree of the greenschist facies) to west (the amphibolite facies). In western areas the rocks are commonly transformed into gneiss and, in some cases, into lit-par-lit migmatite. All groups are cut by ultrabasics and various granitoids. They form tight NW-trending folds along the edge of the Gang (Bundelkand) protoplatform.

The upper age limit of the complex is fixed by the unconformably overlying Neoprotozoic Vindhyan Supergroup. K—Ar age of metamorphic minerals from the complex and the granite cutting it does not generally exceed 1,100 m.y. The recent data of Sarkar (1972), however, revealed that these values characterize the time of the latest (superimposed) metamorphism at the end of the Neoprotozoic, when intrusion of the so-called Erinpura granite (the Ajmer and Antala granites) took place. Earlier granites which cut the Aravalli

Group are about 2,000 m.y. old, and correspond to the time of the Karelian orogeny. The Beirat granite and the Kisengarh nepheline syenite which cut the Delhi strata are dated at 1,660 and 1,490 m.y. respectively; possibly these ages are "rejuvenated" to some extent.

The Indian geologists formerly correlated the Aravalli Group and the Dharwar Supergroup. However, the former probably corresponds only to the Upper Dharwar, with which it has many common features. Stromatolites in the carbonate rocks of the upper part of the Aravalli Group (Raja and Iqbaludhin, 1968) also favour a Mesoprotozoic age. Single stromatolites rarely occur in the Paleoprotozoic, but in the Aravalli Group, they form relatively large and numerous bioherms.

Phytolites are not present in the Raialo and Delhi Groups. The age of these groups can only be established by their relationships with the Aravalli Group. According to the earlier studies by Heron (1953) the Delhi strata overlie the Aravalli Group with angular unconformity. This is the basis for the proposal of a diastrophic cycle (including granite intrusions) between these groups. If these relationships are correct then the Raialo and Delhi Groups may be attributed to the Lower Neoprotozoic. The isotopic ages from the intrusive rocks cutting these groups do not contradict this supposition (if the age is not "rejuvenated"). However, new observations made by Sen (1970) reveal that the contact between the Aravalli and Delhi Groups is commonly tectonic and the tectonite of the latter was erroneously interpreted as a basal conglomerate. In some places the groups appear to be conformable. On the available data the supposed unconformity is probably not very important and may only be locally developed. Detailed studies are needed to resolve this problem.

In the Hindustan Peninsula, in addition to the Mesoprotozoic miogeosynclinal strata, platform and subplatform facies are well developed. One such sequence is the Bijawar Group which overlies (almost horizontally) the Archean Bundelkand granite gneiss. The group is thin (about 240 m) and is composed of quartzite sandstone, granule conglomerates and limestones overlain by hematite-bearing sandstone. The sedimentary deposits are associated with basic lavas, and tuff, and are intruded by diabase sills and dikes. In the upper part of the group, unsorted boulder and pebble conglomerate occurs, intercalated with aleurolite and schist. Mathur (1960) considered these to be tillites. The Rb—Sr isochron age of a fine-grained basic lava flow is 2,780±365 m.y. (Crawford and Compston, 1970). The rock is not altered and its age may reflect the time of emplacement of the lava and of sedimentation. Probably this value is slightly too old. The date is rather imprecise (large error) but if the 365 m.y. is subtracted then the age is reasonable. Thus the lava was laid down soon after the Paleoprotozoic diastrophism (2,700—2,800 m.y.). It is probable that the Bijawar Group is a platform analogue of the lowermost part of the Aravalli group, which also includes basic lava. The Lower Neoprotozoic Semri Group (Lower Vindhyan) overlies the Bijawar group with an angular unconformity.

North America

In North America the Mesoprotozoic is widely developed and represented by a variety of rock types. In the *Canadian—Greenland shield* there are platform and subplatform (protoplatform) sedimentary deposits which resemble those of the Karelian of the Baltic shield. These deposits occur in the Superior and Slave tectonic provinces (see Fig.7 and Fig.11) which were stabilized after the Kenoran diastrophism. On the western margin of the Slave province the Mesoprotozoic is represented in the Coronation geosyncline, by rocks of both miogeosynclinal (dominant) and eugeosynclinal aspect. Similar facies are also present in the Penokean fold belt on the southern margin of the Superior province, in the Grenville and Churchill tectonic provinces of Canada, and also in Southeast Greenland.

The Huronian Supergroup is usually considered as a platform stratotype of the Mesoprotozoic. It is developed in the southern part of the Canadian shield (north shore of Lake Huron). Formerly Canadian workers attributed it to the Lower Proterozoic, but it is now considered to be Aphebian, according to Stockwell's scale. Recent data on the correlation of this supergroup and other Precambrian units of the Canadian—Greenland shield reveal, however, that the Huronian strata do not represent the whole Mesoprotozoic Erathem, but only its lower part, which is locally developed in North America. Thus, the Huronian Supergroup is the North American stratotype of the platform-type Lower Mesoprotozoic.

The stratigraphic position and age of the Huronian Supergroup are quite certain. The Huronian strata overlie with sharp unconformity the metamorphosed greenstones and sedimentary rocks of the Keewatin- and Timiskaming-type successions and their cross-cutting Kenoran granites. These Huronian strata are cut by dikes and sills of Nipissing diabase and the gabbro norite of the Sudbury pluton, which according to Rb—Sr isochron analyses have an age of 2,000—2,150 m.y. (Van Schmus, 1965; Gibbins et al., 1972), and by the Murray and Creighton granites, the age of which has been found to be more than 2,000 m.y. by the Rb—Sr method.

The Whitewater Group apparently unconformably overlies the Huronian strata. It belongs to the younger Mesoprotozoic.

The stratigraphy of the Huronian succession is well known (see e.g. Roscoe, 1957, 1969; Young, 1966, 1968; and Young and Church, 1966). It was recently established that slightly metamorphosed volcanics and subordinate clastic sedimentary rocks are part of the basal Huronian, and not Archean or transitional between Archean and Huronian as was previously thought. In some areas these basal strata are missing and the succession starts with conglomerates and sandstones of the Matinenda Formation which were formerly thought to be the oldest Huronian strata (Fig.19).

The Huronian is now subdivided into four groups which, in most cases, are separated by erosional breaks, but do not show angular discordance. The

Elliot Lake Group (base) includes the following formations: (1) Livingstone Creek (15—500 m), subarkose and conglomerate; (2) Thessalon (65—2,300 m), basic volcanics, subarkose, conglomerate; (3) Copper Cliff (320—1,200 m), acid volcanics; (4) Matinenda (35—320 m), coarse-grained subarkose, oligomict quartz-pebble conglomerate; and (5) McKim (65—2,300 m), sandstone, siltstone and argillite. The overlying Hough Lake Group includes the following formations: (1) Ramsay Lake (15—65 m, up to 200 m), polymict conglomerate; (2) Pecors (15—260 m), argillite and siltstone; and (3) Mississagi (250—650 m, locally up to 3,000 m?), coarse-grained subarkose. The overlying Quirke Lake Group includes the following formations: (1) Bruce (25—65 m), polymict conglomerate and greywacke; (2) Espanola (160—250 m), limestone, dolomite and siltstone; and (3) Serpent (65—300 m), arkose and subgreywacke. The Cobalt Group, which is the youngest, lies directly on Paleoprotozoic rocks (in the north part of the Huronian outcrop belt). It comprises the following formations: (1) Gowganda (650—1,300 m), polymict and boulder conglomerate, grey and pink arkose, purple and green argillite; (2) Lorrain (2,400 m), quartzite, quartz-pebble conglomerate, arkose; (3) Gordon Lake (260—1,200 m), multicolored mudstones and sandstones; and (4) Bar River (650—2,000 m), mainly white orthoquartzite.

The Huronian strata formed in continental and coastal-marine environments. The rocks commonly contain many shallow-water structures. The Gowganda Formation includes typical glacial conglomerates (tillites); they lack sorting and contain faceted and striated clasts. Associated varved shales include large isolated dropstones, and, at the base of the Gowganda, a striated pavement is locally present (Young, 1968; Casshyap, 1969; Lindsey, 1969). Conglomerates in the underlying Ramsay Lake and Bruce Formations commonly also have a tillite-like appearance. In the upper part of the Cobalt Group there is some indication of saline conditions as evidenced by the presence of gypsum and anhydrite (in the Gordon Lake Formation), and by occurrences of syngenetic copper mineralization (in quartzites of the Bar River Formation).

One of the most important aspects of the Huronian is the occurrence of uranium—gold conglomerate in its lower part. These occur predominantly in the Matinenda Formation, and to some extent in the Thessalon and Livingstone Creek Formations. These deposits are of economic value and are also stratigraphically significant because their origin may be related to a specific environment characterized by a certain atmospheric composition and a particular tectonic regime.

The Mesoprotozoic Whitewater Group (younger than the Huronian Supergroup) is of rather limited extent in the Sudbury area. It occurs in a simple brachy-syncline (basin) and is separated by the sheet-like Sudbury intrusion (lopolith), from the underlying Huronian and Archean rocks. The Whitewater Group is composed (in ascending sequence) of the Onaping Formation (tuff?), the Onwatin Formation (slate), and the Chelmsford Formation (sand-

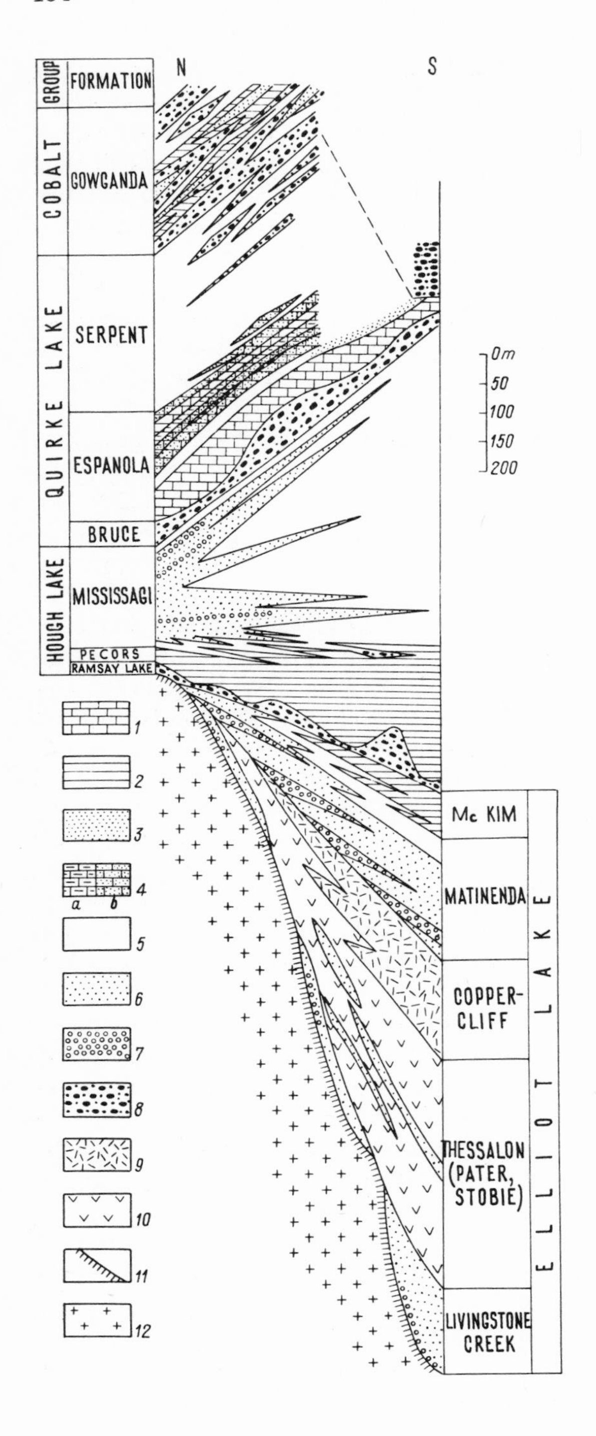

GROUP
FORMATION
N
S
COBALT
GOWGANDA
QUIRKE LAKE
SERPENT
ESPANOLA
BRUCE
HOUGH LAKE
MISSISSAGI
PECORS
RAMSAY LAKE
0m
50
100
150
200
1
2
3
4
a
b
5
6
7
8
9
10
11
12
Mc KIM
MATINENDA
COPPER-CLIFF
THESSALON (PATER, STOBIE)
LIVINGSTONE CREEK
ELLIOT LAKE

stone). The Onaping Formation is considered by some to be fall-back breccias and the products of partial melting resulting from meteorite impact. Minerals that are thought to be the result of shock metamorphism are reported from these rocks, and regularly oriented shatter cones are present around the Sudbury Basin (French, 1967; etc.).

Rb—Sr isochron dating of rocks of the Sudbury area presents some problems. The Onwatin slate and granophyre of the Sudbury irruptive give an age of 1,700 and 1,700—1,720 m.y. respectively (Fairbairn et al., 1968). Granite, cutting the Huronian strata and locally the gabbro norite of the irruptive, gives ages of 2,230—1,780 m.y.; i.e. the geologically younger rocks yield an older age. The data on the Murray granite massif are particularly interesting. This massif is situated on the south side of the Sudbury structure. It intrudes Huronian strata but a dike-like apophysis cuts the gabbro norite. The granite of this massif yields the oldest age values — 2,230 m.y. ($^{87}Sr/^{86}Sr = 0.719$); granite from the apophysis is about 2,050 m.y. old ($^{87}Sr/^{86}Sr = 0.711$), and the granite of small dikes cutting the Murray Massif gives an age of 1,780 m.y. ($^{87}Sr/^{86}Sr = 0.708$) (Gibbins et al., 1972). The high primary strontium isotope ratio in the Murray granites suggests that these rocks were formed as a result of remelting of older crustal material.

Gibbins et al. (1972) suggested that the Murray granite is older than the Sudbury lopolith, and that the granite of the apophysis formed as a result of the Murray granite melting on contact with the basic magma, so that its apparent age is that of the irruptive. However, this is not the only possibility, for the Murray granites may also be younger than the Sudbury gabbro norite, its age appearing to be greater because of the presence of excess radiogenic strontium. The ages given above for the Onwatin Formation and the granophyre are probably "rejuvenated" values.

Strata similar to the Huronian occur in other areas at the edge of the Superior province. For example the rocks near Chibougamau and Mistassini Lakes (Mawdsley and Norman, 1935; Roscoe, 1969) are possible correlatives of the Huronian formations, shown in the correlation chart (see Table III, inset plate, row 29). The presence of glacial units (tillites) in the Chibougamau Formation strengthens its correlation with the Huronian Gowganda Formation (Young, 1973). In the Churchill tectonic province the Montgomery Group (900 m) probably corresponds to the lower part of the Huronian. It is locally developed near the lake of the same name, in the district of Keewatin,

---

Fig.19. Lower part of the succession of the Huronian Supergroup on the north shore of Lake Huron (Quirke Lake area). Compiled according to data of Roscoe (1957, 1969). *1* = limestone; *2* = argillite, shale, greywacke; *3* = greywacke; *4* = intercalation of carbonate rocks with shale (*a*) and sandstone (*b*); *5* = feldspathic quartzite; *6* = gritstone, coarse feldspathic quartzite, arkose; *7* = oligomict conglomerate; *8* = polymictic boulder conglomerate (tillite); *9* = acid volcanics; *10* = basic volcanics; *11* = pre-Huronian fossil soil; *12* = pre-Huronian rocks (volcanics and granite).

west of Hudson Bay. It is made up of grey quartzite, which is locally pebbly and pyritic, and polymictic conglomerate and mudstone. These are underlain unconformably by Paleoprotozoic metavolcanics and granitoid rocks (R.T. Bell, 1971). The group is separated by a poorly defined hiatus from the Hurwitz Group, which belongs to the higher part of the Mesoprotozoic.

On the southwest margin of the Slave tectonic province in the northwest part of the Canadian shield the Wilson Island Group may correspond to the Huronian succession. This group is distributed in the area of Eastern Arm of the Great Slave Lake; it unconformably overlies the older principally Paleoprotozoic rocks and is formed of dark slates and basic lava with tuff, dolomite, and siltstone interbands (Hoffman, 1968). These rocks are unconformably overlain by the Great Slave Supergroup which can be regarded as a stratotype of the upper part of the Mesoprotozoic in the Canadian—Greenland shield.

The northeastern exposures of the Great Slave Supergroup are made up of typical platform deposits. It is composed, almost exclusively, of sedimentary rocks, and is relatively thin (2,500—3,000 m). Deformation and metamorphism are minimal. It increases in thickness towards the southwest (up to 6,500—7,000 m), volcanics are present, and the degree of deformation and metamorphism are higher. Thus the supergroup takes on a geosynclinal aspect in this region. The generalized characteristics of the supergroup (Fig. 20) are described below.

According to Hoffman (1968a, b, 1969, 1973), the Great Slave Supergroup may be divided into four groups which are, in ascending order; Sosan, Kohachella, Pethei and Christie Bay. The Sosan Group overlies the Paleoprotozoic rocks unconformably and is also separated from the Wilson Island and Union Island Groups by a hiatus. It includes the following formations (in ascending sequence): (1) Hornby Channel (250—1,500 m), feldspathic quartzite with interbeds of conglomerate, mudstone and rarely dolomite; (2) Duhamel (up to 300 m), stromatolitic and oncolitic dolomite; and (3) Kluziai (up to 350 m), feldspathic quartzite. The latter lies on the eroded surface of the underlying dolomite and, in the platform area, it lies directly on the Paleoprotozoic rocks. The Kohachella group (450—1,600 m) consists of finely bedded, commonly red, mudstones, argillite, slate, and fine-grained sandstone with rare interbeds of oolitic (?) hematite and stromatolitic dolomite with local cavities after gypsum. In the southwest, the clastics are partly replaced by basic metavolcanite (the Seton Formation).

The Pethei Group (400—600 m) in the northeast is largely composed of dolomite (commonly stromatolitic). To the west it passes into rhythmically interbedded (flysch-like) thin-bedded carbonate argillite, mudstone and greywacke, with local dolomite. The bottom of the group has a characteristic horizon of red clayey limestone. The topmost Christie Bay Group includes the following formations; (1) Stark (up to 750 m), red mudstone and argillite, locally with crystal moulds after NaCl and with lenses of stromatolitic dolomite, and carbonate breccias with a red argillaceous matrix; (2) Tochatwi (600—900 m), cross-bedded red lithic sandstone with conglomerate at the base; and (3) Portage Lake (100—200 m), argillite, overlain or replaced by the Pearson basalt.

Recently, occurrences of sedimentary uranium mineralization of the Huronian type were reported from the quartz-pebble conglomerate of the Hornby Channel Formation of the Sosan Group (Follinsbee, 1972). In the Pethei Group, *Conophyton* and *Jacutophyton* occur among the columnar stromatolites (Hoffman, 1973). The latter form is characteristic of the second and third ("Middle and Upper Riphean") phytolite complexes in the Precambrian sections of Siberia, but in rare cases, it is reported from the first complex (the Lower Riphean—Mesoprotozoic).

To the west the Great Slave Supergroup is cut by granites which have given a K—Ar age of 1,915 (1,845) m.y. The lower Neoprotozoic Et-Then Group overlies it with angular unconformity.

The Goulburn Group is similar to the platform facies of the Great Slave Supergroup. It also has miogeosynclinal analogues in the Epworth and Snare Groups. The stratigraphic sequence of this group (and of its correlatives) is shown in Fig.20. The Goulburn Group unconformably overlies the Paleoprotozoic metavolcanics and granite and is unconformably overlain by a Neoprotozoic sedimentary assemblage (the Tinney Cove Formation, etc.). The K—Ar age of the gabbro diabase, cutting the Goulburn Group, is 1,610 (1,550) m.y. (Tremblay, 1971).

In the Canadian—Greenland shield the Animikie Supergroup is the best-known Mesoprotozoic unit of miogeosynclinal type. It is proposed as the stratotype for North America.

This supergroup is present in the Mesoprotozoic Penokean fold belt, on the south margin of the Superior province. The stratigraphy has been studied in detail by both American and Canadian geologists (Leith et al., 1935; Grout et al., 1951; James, 1955, 1958; Goldich et al., 1961; etc.). The most complete section should be chosen as the stratotype. It occurs in Michigan were the Animikie sequence comprises several formations which are combined into four groups, locally separated by breaks. In upwards sequence these are: (1) Chocolay — composed of tillite-like conglomerate, followed by quartzite, and then slate and dolomite (with local stromatolites); (2) Menominee — formed of quartzite, slate, jasper, conglomerate, rare stromatolitic limestone and volcanics (iron formation is thickest in this group); (3) Baraga— composed of quartzite, slate, greywacke, local volcanics, and interbeds of iron formation; and (4) Paint River — predominantly greywacke, with lenses of siderite. Total thickness of the supergroup is 12,000 m, of which 6,000—9,000 m is in the Baraga Group.

One of the most striking features of the Animikie Supergroup is the occurrence of iron formation, of "Superior type", which differs from the Paleoprotozoic ferruginous rocks in that it is associated with sedimentary rocks

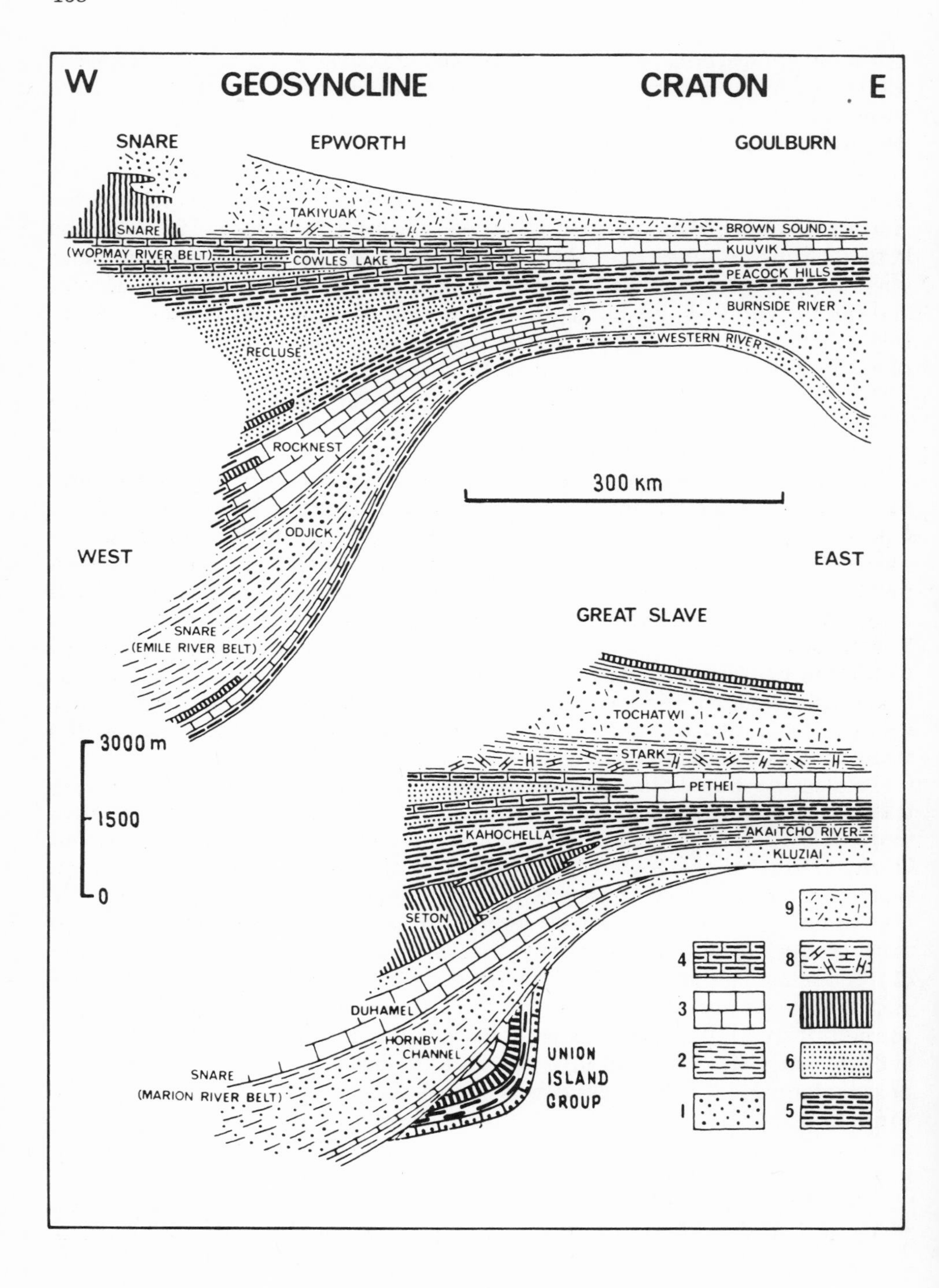
W
GEOSYNCLINE
CRATON
E
SNARE
EPWORTH
GOULBURN
TAKIYUAK
SNARE
(WOPMAY RIVER BELT)
COWLES LAKE
BROWN SOUND
KUUVIK
PEACOCK HILLS
BURNSIDE RIVER
?
WESTERN RIVER
RECLUSE
ROCKNEST
300 KM
ODJICK
WEST
EAST
GREAT SLAVE
SNARE
(EMILE RIVER BELT)
TOCHATWI
3000 m
1500
0
STARK
PETHEI
KAHOCHELLA
AKAITCHO RIVER
KLUZIAI
SETON
DUHAMEL
HORNBY CHANNEL
UNION ISLAND GROUP
SNARE
(MARION RIVER BELT)
1
2
3
4
5
6
7
8
9

rather than volcanics. Microscopic algal remains are common in siliceous layers of the iron formation. These remains are very abundant and well preserved in the Gunflint Formation on the north shore of Lake Superior. This iron formation may be correlated with part of the Menominee group in Michigan. It contains stromatolites which have been identified (Hofmann, 1969) as forms which Soviet workers call *Stratifera* Korol., *Tungussia* Semikh., *Gymnosolen* Steinm., *Kussiela* Kryl. etc. These forms are typical of the Lower, Middle and Upper Riphean of the Urals. Semikhatov (Cloud and Semikhatov, 1969) described a new form (*Gruneria biwabikia*) from this formation. Oncolites of the Osagia group also occur in this formation. M.E. Raaben (personal communication, 1972) after looking through the collection of H. Hofmann, considers Lower Riphean forms of stromatolites to be predominant in the Animikie Supergroup.

The Animikie Supergroup unconformably overlies Paleoprotozoic volcano-sedimentary rocks and granites, and is itself cut by granite which by the K—Ar method and Pb-isotope analyses on zircon yielded an age of 1,860—1,890 m.y. (Banks and Cain, 1969). Aldrich and some other workers (1965) consider 1,900 m.y. as the age of this granite. Earlier intrusive sheets and diabase dikes, that intrude the Animikie Supergroup are 2,000 m.y. old (Hanson and Himmelberg, 1967). The Neoprotozoic Lower Keweenawan Group overlies the Animikie succession with sharp unconformity.

The relationships between the Animikie and Huronian Supergroups are debatable. At present all Canadian workers agree that they are both Aphebian, but formational correlations pose problems. According to one concept, the lower part of the Chocolay Group corresponds to the three lower groups of the Huronian. Stromatolitic carbonates in the upper part of the Chocolay Group (in the Randville or Bad River Formation) would thus correlate with the Espanola Formation of the Huronian. However, no stromatolites are known from the Espanola Formation. The lower part of the Menominee Group would correspond to the Cobalt Group. According to another concept, the Animikie Supergroup lacks strata equivalent to the lower part of the Huronian; its basal units are considered correlative with the Cobalt Group (Young and Church, 1966). Such a correlation is rationalized by the presence in the lowermost part of the Chocolay Group, of tillite-like conglomerate

---

Fig. 20. Succession of the Mesoprotozoic Great Slave Supergroup, Epworth and Goulburn Groups of the Slave and Bear tectonic provinces. After Hoffman et al. (1970), with supplemental data by Hoffman (1969).

1 = cross-bedded quartzite and subarkose, partly conglomerate; *2* = intercalation of siltstone, slate and sandstone; *3* = stromatolitic and oolitic (oncolitic) dolomite, limestone and marl; *4* = intercalated limestones and slates; *5* = slates, slate with concretions and siltstone; *6* = greywacke, rhythmically interbedded with slate; *7* = volcanics, volcanic breccia, tuff; *8* = carbonate metabreccia with red argillite matrix; *9* = "immature" cross-bedded lithoclastic sandstone.

(the Fern Creek Formation) which is compared with the tillite of the Gowganda Formation of the Cobalt Group, and also by the similarity of the quartzites of the Lorrain Formation and the Sturgeon quartzite which overlies tillite in the Chocolay Group. This correlation is also supported by the presence of ferruginous quartzite in the uppermost part of the Huronian.

The second point of view is accepted in the correlation chart used here (Table III, inset plate). The thickest part of the section of the Animikie Supergroup appears to be younger than the Huronian. The Huronian thus represents only the lower part of the Mesoprotozoic, and therefore cannot be used as a world stratotype of the Mesoprotozoic, Aphebian or Lower Protozoic as has been suggested (Semikhatov, 1964).

The Animikie Supergroup can be correlated with the Kaniapiskau Supergroup of the Labrador trough, with the Povungnituk Group of the Cape Smith fold belt, with the Belcher Group of Belcher Islands and the Manitounuk Group of the east shore of Hudson Bay (Table III, inset plate). The stratigraphy of all these groups has been summarized by Dimroth et al. (1970). Correlation is based on occurrence of similar rock sequences, the fact that they have similar structural situations (miogeosynclinal zone surrounding the Superior tectonic province), and similar isotopic results. Iron formations and stromatolitic limestone and dolomite provide marker horizons.

All these correlatives of the Animikie Supergroup unconformably overlie the Archean gneiss—granulite complex (for example in Ungava) and are in turn unconformably overlain by Lower Neoprotozoic rocks. The age of metamorphism and of granites that intrude these rocks is in the 1,665—1,820 (1,600—1,756) m.y. range (K—Ar analysis). Similar K—Ar ages are reported from the Mesoprotozoic rocks of the Penokean fold belt, but comparison with Pb-isotope ages reveals that the former are "rejuvenated". New determinations (Fryer, 1972) of the age of metamorphic schists of the Kaniapiskau Supergroup yielded values of about 1,900 m.y. (Rb—Sr isochrons).

Stromatolites from the Kaniapiskau Supergroup (the Denault Formation) were described in detail by Donaldson (1963). Krylov (1969) considered them to be similar to the Lower Riphean forms of the Burzyan Group of the Urals.

A number of groups belong to the Mesoprotozoic miogeosynclinal strata in the Churchill province: Nonacho, Waugh Lake, Hurwitz, Great Island, Sickle, Missi, Assean, etc. (Table III, inset plate). These groups are largely composed of metasedimentary rocks which are predominantly terrigenous (meta-arkose, quartzite, metagreywacke, etc.), and partly of carbonate rocks. Iron formations are present but are mainly massive hematite and siderite bodies. Similar ores also occur in the Kaniapiskau Supergroup (the Larch Formation) in the northern part of the Labrador trough.

In the Churchill province the most complete Mesoprotozoic section is represented by the Hurwitz Group in the Ennadai Lake and in the Kognak River basin areas (Eade, 1970, 1971; R.T. Bell, 1968, 1970, 1971). At the base of the group grey and greenish-grey boulder and pebble conglomerate and sandstone of the Padlei Formation, (500 m) unconformably overlies Paleoprotozoic metavolcanics and granites or the Lower Mesoprotozoic clastic rocks of the Montgomery group. Young (1973) considered the Padlei Formation (Lower Hurwitz) to be of glacial origin. Above the Padlei Formation are coarse-bedded light quartzites and granule conglomerates, passing upwards into finely bedded quartzites with abundant ripple marks (the Kinga Formation, up to 1,000 m). This is overlain by grey slate, mudstone, and greywacke with interbedded red slate and dolomite which locally are overlain by basic lava and tuff (the Ameto Formation, up to 700 m thick). This unit is followed by greywacke, mudstone, argillite, and dolomite of the "post-Ameto complex" which has a thickness of about 700—1,600 m. The section is crowned by unnamed brown quartzo-feldspathic (locally lithoclastic and quartzose) sandstone, and slate (330—670 m). The carbonates in the middle part of the Hurwitz Group contain numerous stromatolites.

All Mesoprotozoic groups of the Churchill province unconformably overlie Paleoprotozoic strata and many are unconformably overlain by Lower Neoprotozoic rocks. The K—Ar age of micas from metamorphic rocks and granites that cut these rocks is 1,770—1,890 (1,700—1,820) m.y. The K—Ar age of granite that cuts the Waugh Lake group in the area of Lake Athabasca, is 1,970 (1,900) m.y. (Koster and Baadsgaard, 1970). Many of the K—Ar dates are probably "rejuvenated" due to argon loss. The Lower Neoprotozoic strata (Lower Dubawnt Group), unconformably overlying the Hurwitz Group, include porphyritic lavas with a K—Ar age of 1,830 (1,770) m.y., and a Rb—Sr isochron age of 1,730 m.y. (Donaldson, 1965, 1967). Thus, the most probable age for the Mesoprotozoic diastrophism in the Churchill province is 1,900—1,970 m.y.

Mesoprotozoic strata of miogeosynclinal facies are well developed in the northwestern part of the Canadian shield in the Slave and Bear tectonic provinces. Mesoprotozoic strata occur at several places on the periphery of the Slave province. Mesoprotozoic units on its western margin, near the Bear province include the strata of the Snare Group. This group is largely composed of metasedimentary rocks: slate, sandstone, argillite, dolomite, and stromatolitic limestones. Locally altered basic lavas are present. Its stratigraphy is still poorly understood but there are strong similarities to the Goulburn Group, with which it is usually correlated (Fig. 20). The Snare Group is cut by pre-tectonic gabbro—diabase intrusions with a K—Ar age of 1,970—2,175 (1,900—2,100) m.y., and by syntectonic granites with a K—Ar (mica) age of 1,920 (1,850) m.y. (Leech, 1966; Green et al., 1968; etc.).

In the Bear province the following three groups are considered to be the same age as the Snare group: the Echo Bay volcano-sedimentary rocks, the Cameron Bay which unconformably overlies the Echo Bay sequence in the Great Bear Lake area, and the Epworth Group in the Coppermine River Basin.

The Echo Bay Group nonconformably overlies granites and is made up of altered volcanics (mostly basic) and tuffs interbedded with clastic and subordinate carbonate metasedimentary rocks. The presence of abundant volcanics suggests analogy with Mesoprotozoic eugeosynclinal formations, but on the other hand, common occurrence of quartzites and conglomerates is more typical of miogeosynclinal facies. The Cameron Bay is composed of red and grey sandstone, conglomerate, and argillite. The K—Ar of granite that cuts both groups, is 1,970 (1,900) m.y. The Neoprotozoic Hornby Bay Group unconformably overlies the Cameron Bay Group (Donaldson, 1968; J.B. Henderson, 1972; Kidd, 1973).

The stratigraphy of the Epworth Group is better studied (Fraser, 1965; Fraser and Tremblay, 1969; Hoffman et al., 1970). It is similar to the Great Slave Supergroup (see Fig.20). Both groups contain stromatolitic dolomite. Their relationships with underlying and overlying rocks are the same as is the case for the stromatolitic dolomite in the Hurwitz Group of the Churchill province. Lithic sandstones unconformably overlie the upper part of the Epworth Group and corresponding parts of the Great Slave Supergroup and the Hurwitz Group. These red beds may be correlated with similar rocks of the Cameron Bay Group (Great Bear Lake) which unconformably overlie the Echo Bay Group.

The Epworth Group is probably correlative with the Goulburn Group on the east side of the Slave province. This correlation (see Fig.20 and Table III, inset plate) is based on work by Fraser and Tremblay (1969). Micas from granite that intrudes the Epworth Group yielded 1,725—1,925 (1,660—1,855) m.y. by the K—Ar method (Wanless et al., 1968). The true age of the intrusions is probably nearer the greater values.

On the opposite side of the Canadian—Greenland shield the Flinton Group possibly belongs to the miogeosynclinal Mesoprotozoic. It occurs in the southern part of the Grenville tectonic province. According to recent studies by J.M. Moore and Thompson (1972) this group unconformably overlies metavolcanics of the Hermon Group and granodiorite that cuts them. The base of the group consists of quartzite and quartz-pebble conglomerate of the Bishop Formation. These are overlain by crystalline dolomite, limestone, dolomite conglomerate, and graphitic schist of the Myer Cave Formation. The topmost unit is finely bedded micaceous schist and crystalline limestone of the Fernleigh Formation. Thickness of the group does not exceed 625 m, but allowing for tectonic flattening, it is probable that the original thickness was much greater (of the order of 2—3 km). The rocks are metamorphosed in the greenschist and amphibolite facies.

Formerly Canadian workers (Miller and Knight, 1908, 1914; Ambrose and Burns, 1956; Wilson, 1965) placed the Flinton Group together with some older formations under the name "Hastings group". Moore and Thompson considered the Flinton Group to be part of the Grenville Supergroup but such a grouping of rocks of different age (one group unconformably overlies the underlying rocks) is not justified.

The Hermon metavolcanics which are unconformably overlain by the Flinton Group are 1,310 m.y. old (Pb-isotope analysis on zircon) and metamorphic minerals from rocks of the group yielded K—Ar ages of about 1,000 m.y. These values are taken as the age boundaries of the group by Moore and Thompson. However, the rocks of the Grenville underwent intense metamorphism (and deformation) during the Grenville cycle, and commonly give "rejuvenated" age values. The Flinton Group rocks suffered at least three separate deformations so that it is possible that only the last one was associated with the Grenville orogeny. It is suggested that the Flinton group may be Mesoprotozoic. This idea is supported by the fact that the Flinton Group differs markedly from the Neoprotozoic Letitia Lake and Seal groups in the northern part of the same province. They are in some respects more similar to the Huronian Supergroup, but have a greater proportion of carbonate rocks.

Radiometric dating (Rb—Sr isochrons) of different old rocks in the southern part of the Grenville province reveals metamorphism at about 1,700—1,900 m.y. ago (Krogh and Davis, 1967). This is similar to the age of the Karelian cycle.

The metamorphosed sedimentary rocks in the northeastern part of the same province at the southern extremity of the Labrador trough, are possibly of Mesoprotozoic miogeosynclinal type. These strata resemble rocks in the middle part of the Kaniapiskau Group. One of the formations of this sequence contains iron formation similar to the Sokoman Formation of the Labrador trough. However, these rocks are strongly metamorphosed (uppermost amphibolite facies).

The Penrhyn Group outcrops on the Melville Peninsula (Arctic coast of Canada). It is also considered to be an example of mesoprotozoic miogeosynclinal facies. It consists of quartzite, crystalline limestone, limestone and dolomite, metagreywacke and slate. It probably unconformably overlies highly metamorphosed rocks of the Prince Albert (sedimentary and volcanogenic) Group which are exposed in the cores of mantled domes. The Penrhyn Group is cut by granites which give a K—Ar age of 1,755 (1,690) m.y. (Heywood, 1968).

Metamorphic strata in the central part of Baffin Island (Piling Group; Jackson and Taylor, 1972) are probably correlative with the Penrhyn Group. These rocks are exposed in the margins of gneiss domes with gneiss and granite gneiss in their cores. These gneisses were remobilized at the end of the Mesoprotozoic. The base of the group is made up of quartzite (60—300 m) followed by marble (1—760 m), then graphitic quartz-rich gneiss (> 300 m), and finally metagreywacke with subordinate metabasite sheets (several thousand metres in composite thickness). Rb—Sr isochrons suggest that metamorphism took place 1,970±100 m.y. ago. The Mesoprotozoic metamorphism also caused "rejuvenation" of the isotopic age of the underlying rocks which give ages in the 1,950—2,340 m.y. range.

The Green Head Group may also be a miogeosynclinal Mesoprotozoic sequence. It is developed east of the Grenville province, outside the boundaries of the Canadian shield, in the northern Appalachians (New Brunswick, Canada). It consists of crystalline limestone and dolomite, quartzite and subordinate micaceous and graphitic slate (Alcock, 1938; Weeks, 1957). Stromatolites are present in the carbonate rocks. At the end of the last century they were described as *Archeozoon acadiense* Matt. (Mattew, 1890; J. Dawson, 1897). A Mesoprotozoic age is suggested on lithological grounds. The lower boundary of the group is not well defined, but the upper one is determined by the unconformably overlying Coldbrook Group (volcanic) that is probably Epiprotozoic.

Hofmann (1974) recently studied stromatolites from the Green Head Group. He stated that the stromatolites resemble Precambrian forms, ranging upwards from Aphebian time (Mesoprotozoic). Their morphology and microstructure support this conclusion, but they most closely resemble *Tungussia*, *Baicalia*, and *Jacutophyton* which are most common in the Middle Neoprotozoic (Middle Riphean). However, *Jacutophyton* also occurs in Mesoprotozoic rocks (in Canada it is reported from the Great Slave Supergroup) and *Conophyton cylindricus* of the Green Head Group is quite typical of the Mesoprotozoic, though it is also found in the Middle Neoprotozoic. Thus, the proproblem of the age of this group cannot be resolved on the basis of stromatolites. The facts suggest the possibility here, that the Green Head Group is Middle Neoprotozoic.

The metamorphic rocks of miogeosynclinal aspect in the *East Greenland fold belt* are tentatively assigned to the Mesoprotozoic. These include the pre-Carolinian complex of Kronprins Christians Land and the central metamorphic complex of Queen Louise Land. Both complexes are composed of quartzites, marble, slate, or schists which grade into gneiss. These rocks are cut by granite and are unconformably overlain by Neoprotozoic strata. No isotopic ages have been reported.

Mesoprotozoic rocks of eugeosynclinal aspect are common in North America but less widespread than the rock types discussed above. The best sections are in the Canadian—Greenland shield. The Kaniapiskau Supergroup of the east-central part of the *Labrador trough* is proposed as an eugeosynclinal stratotype for the Mesoprotozoic. This proposal is made, not only because these rocks have been studied in detail, and their age is reliably established, but also because of the possibility of detailed (formational) correlation with the miogeosynclinal facies of the Kaniapiskau Supergroup (Dimroth, 1970; Dimroth et al., 1970).

The stratigraphy of the Kaniapiskau Supergroup is complicated by rapid facies changes along, and particularly across, the strike of the Labrador trough. A schematic section across the middle part of the trough is shown in Fig. 21 which shows the relationships between miogeosynclinal and eugeosynclinal regions. The upper part of the supergroup is the most uniform in

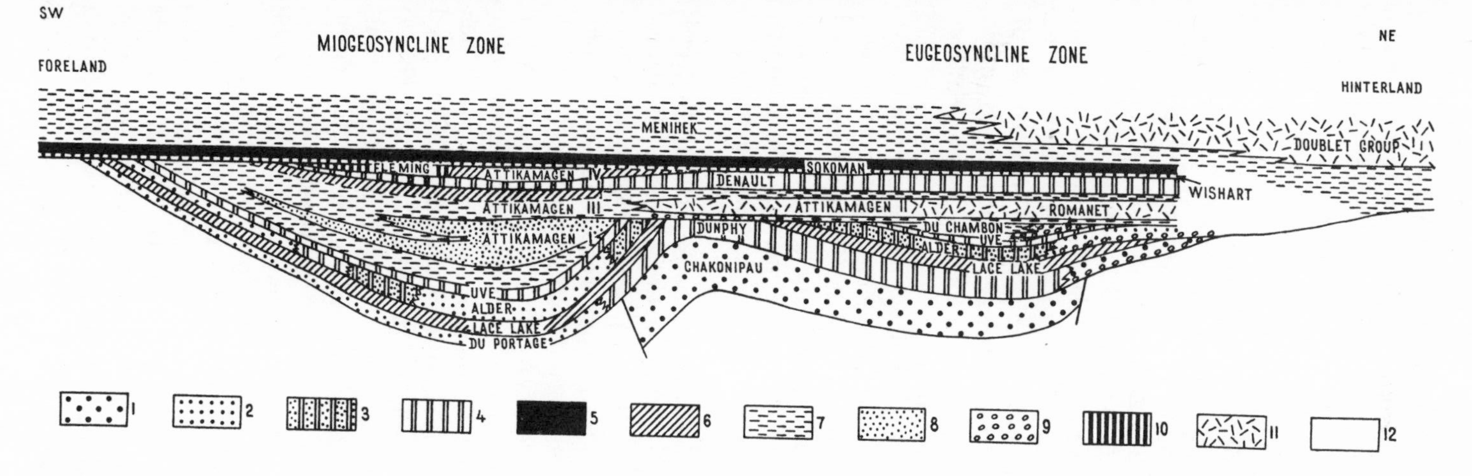

Fig. 21. Transverse profile through the central part of the Labrador trough. After Dimroth (1970).
The upper volcanics of the eugeosynclinal zone are assigned to the Doublet Group and all underlying units of the supergroup to the Knob Lake Group. The Chakonipau and Dunphy Formations are included in the Seward Subgroup, the Lace Lake, Alder and Uve Formations in the Pistolet Subgroup, the Du Chambon and Romanet Formations in the Swampy Bay Subgroup. *1* = arkose, granule conglomerate; *2* = sandstone; *3* = intercalated sandstone and dolomite; *4* = dolomite; *5* = iron formation (jaspilite); *6* = slate with dolomite lenses; *7* = slate and siltstone, partly sandstone; *8* = slate and greywacke; *9* = conglomerate; *10* = breccia; *11* = basic volcanics and tuff; *12* = basement gneiss.

composition and structure. It includes the Sokoman iron formation and the underlying dolomite and limestone of the Denault Formation which contains "Lower Riphean" (?) stromatolites. These formations are good marker horizons in correlation among hetero-facial strata of the Labrador trough. They also provide the possibility of correlating the Kaniapiskau Supergroup with other Mesoprotozoic geosynclinal rocks in other parts of Canada, and in particular with the Animikie Supergroup of the Penokean fold belt (Table III).

The main difference between the eugeosynclinal strata of the Kaniapiskau Supergroup and their miogeosynclinal counterparts is the presence of the submarine volcanics (mostly basic), and the metamorphhic grade.

The metamorphosed sedimentary—volcanogenic Aillik Bay strata are considered to belong to the eugeosynclinal type of the Mesoprotozoic (east of the Labrador trough on the Labrador coast). This group is younger than the Archean gneiss complex, but is probably older than diabase with an age of 2,080 m.y. It is unconformably overlain by the Lower Neoprotozoic Letitia Lake Group. The lower boundary is problematic, so that the possibility that these rocks are Paleoprotozoic cannot be excluded.

A good Mesoprotozoic section of eugeosynclinal type is present in Southern *Greenland*, in the Ivigtut area. It is called the Ketilidian complex and includes two conformable groups, the Vallen and Sortis. These are composed of clastic and carbonate rocks which have undergone greenschist, and in places, amphibolite-facies metamorphism. The upper part of the complex includes thick pillow lavas which are commonly associated with gabbro and gabbro—diabase sills.

According to Higgins (1970) and Higgins and Bondesen (1966) the Ketilidian complex comprises the Vallen Group which includes in upward succession: (1) the Zigzagland Formation (up to 140 m), which begins with conglomerate that unconformably overlies the Paleoprotozoic Tartoq Group, and is followed by quartzite and feldspathic sandstone, then finely bedded, limy and dolomitic slate; (2) the Blåis Formation (50—700 m), which is composed of siltstone, slate, arkose, and greywacke with interbedded graphitic slate and dolomite; and (3) the Graensesø Formation (up to 500 m), which consists of black slate, overlain by dolomite interbedded with slate, followed by locally pyritized slate. Many gabbro sills are present. The Sortis Group includes: (1) The Faselv Formation (100—900 m), which is made up of pillow lava with some interbeds of sedimentary rocks and gabbro sills; (2) the Rendesten Formation (1,100—2,400 m), composed of banded mudstone, slate, chert, and dolomite intercalated with thick beds of pyroclastic rock and diabase sills, with local development of thick pillow lavas; and (3) the Quernertoq Formation (2,200 m) consisting of pillow lava with gabbro and diabase sills. Superior-type iron formation was recently reported from the upper part of the Rendesten Formation (Appel, 1974). Rounded pyrite grains occur in quartz-pebble conglomerate of the Zigzagland Formation. Pyrite of this kind is characteristic of the basal Mesoprotozoic strata of other regions which have economic gold and uranium deposits. Stromatolites are present in dolomites of the Graensesø Formation, and microscopic algal remains are reported from chert intercalated with dolomite. Some of the microfossils are analogous to those of the Gunflint Formation of the Animikie Supergroup (Bondesen et al., 1967; Pedersen and Lam, 1968).

The depositional environment of the Ketilidian rocks is rather peculiar. Higgins (1966) considered the Zigzagland Formation to have accumulated in small closed basins, and the formations of the Vallen Group to have formed in a large marine basin. Eugeosynclinal conditions prevailed only during accumulation of the thick upper part of the Sortis Group.

The K—Ar age of granite cutting the Ketilidian, is 1,815 (1,750) m.y., but this value is probably somewhat "rejuvenated". The Lower Neoprotozoic Qipisarqo Group, which is transected by 1,750 b.y. old rapakivi granite, unconformably overlies the Ketilidian supracrustal complex and granites.

The so-called Nanortalik Ketilidian complex in the southern extremity of Greenland and the Karrat Group in the middle part of the western shore of the Island correspond to the Vallen and Sortis Groups; the Karrat Group overlies the Paleoprotozoic Umanak Group and is cut by granite with a K—Ar age of 1,826 (1,769) m.y. (Berthelsen and Noe-Nygaard, 1965; G. Henderson and Pulvertaft, 1967). The Karrat Group is largely composed of well-preserved metasedimentary rocks. Its lower part (Qeqertarssuaq Formation) is up to 3,000 m thick and is composed of quartzite, locally intercalated with garnet—sillimanite schist, amphibolite and thin beds of marble. In some places relict psammitic texture is preserved in the quartzite. The overlying Nukavsaq Formation is a monotonous flysch-like sequence comprising greywacke, intercalated with subordinate schist and lenses of limy—siliceous rocks. Graded bedding is typical (Pulvertaft, 1973).

In the Eastern Rocky Mountains of North America the Mesoprotozoic of transitional eugeosynclinal—miogeosynclinal type is represented by the Deep Lake Group which is overlain with slight unconformity, by the Libby Creek Group (Houston, 1968). The Deep Lake Group transgressively overlies Archean gneisses and consists of quartzite intercalated with metabasalt that is locally amphibolitized (with relict amygdaloidal structure). These rocks are intruded by metagabbro—diabase sills. The group exceeds 3,500 m in thickness. The Libby Creek Group is subdivided into eight formations which in ascending sequence are: (1) Headquarters Formation (100—900 m), made up of intercalated phyllite and metaconglomerate with rare interbeds of quartzite and dolomite; (2) Heart Formation (up to 1,000 m), composed of quartzite intercalated with phyllite, micaceous schist, and rare metaconglomerate; (3) Medicine Peak Formation (1,900 m), made up of quartzite which is commonly cross-bedded; (4) Lookout Formation (400 m), consisting of quartz—muscovite schist, quartzite and amphibolite; (5) Sugarloaf Formation (thickness not given), made up of ripple-marked quartzite; (6) Nash Fork Formation (thickness not given), comprising dolomitic marble with bands of phyllite and "coaly" slate. Stromatolites are locally present in the marble; (7) Towner Formation (300—500 m), metabasalt which is largely composed of amphibolite and amphibole schist; and (8) French Formation (up to 650 m), of muscovite—chlorite schist and black pyritized phyllite, with quartzite interbeds. According to Young (1974) parts of the Headquarters Formation are glaciogenic.

Both groups are cut by granite with a Rb—Sr isochron age of 1,715 m.y. Schist of the Deep Lake Group gave an age of 1,840 m.y. by the same method (Houston, 1968; Hills et al., 1968). The latter value is probably closer the time of Mesoprotozoic metamorphism and plutonism.

The Vishnu complex, which is at the base of the Precambrian section in the Grand Canyon of the Colorado River, is possibly an example of Mesoprotozoic eugeosynclinal facies. It is composed of metasedimentary rocks, largely micaceous schist, quartzites intercalated with highly altered (in some cases to amphibolite) basic pillow lavas. An upper age limit of the complex provided by dates of 1,600—1,800 m.y. (Rb—Sr isochrons) obtained from metamorphic rocks (Wasserburg and Lanphere, 1965) and of 1,725 m.y. (Pb-isotope analysis) on zircon from granite cutting the complex (Pasteels and Silver, 1966).

In the Mazatzal mountains of central Arizona, in the area adjoining the Cordilleran fold belt, the Ash Creek and the Pinal Groups are analogues of the Libby Creek and the Vishnu Groups. The former is separated by younger strata from other Precambrian rocks. The Ash Creek Group is thick (>7,000 m) and consists of altered basic and acid volcanics and tuffs. The basic lavas are commonly pillowed. The contemporaneous Pinal Group (C.A. Anderson, 1951) is composed of interbedded metavolcanics and metasedimentary rocks (largely schist and meta-arkose). These metamorphic rocks unconformably overlie older (Archean?) migmatitic gneiss and gneissic granodiorite (Blacet, 1966), but are themselves cut by granite. There is a conglomerate at the base of the metasedimentary sequence. The age of the younger granite is about 1,760—1,775 m.y. according to Pb-isotope (on zircon) and Rb—Sr isochron analyses (Lanphere, 1968; Silver, 1969). However, this age may be slightly "rejuvenated", by later polyphase deformation and thermal processes. The Ash Creek Group is combined by some with the younger (Lower Neoprotozoic) Red Rock (Texas Gulch) subaerial volcanogenic group into a single supergroup — the Yavapai. However, this seems unreasonable, considering their different age and lithology.

General Characteristics

The Mesoprotozoic is the first erathem which is characterized by widely developed facies of variable aspect including platform, miogeosynclinal, and eugeosynclinal types. It differs greatly from the Paleoprotozoic, which is composed mostly of rocks of eugeosynclinal facies. This change is related to growth and increase in stability of the Archean cratons.

In many areas platform and miogeosynclinal strata are closely related and show marked similarities. Both types consist of two or more groups, separated by breaks, and have rhythmic transgressive—regressive successions reflecting important epeirogenic movements during sedimentation. The lower groups are usually of local distribution, whereas the overlying groups are more widespread. In the Karelian complex of the Baltic shield and the Huronian of the Canadian—Greenland shield the following sequence is typical: continental and coastal-marine terrigenous strata with abundant volcanics pass upwards into shallow-water, marine, carbonate-rich terrigenous strata with or without subordinate volcanics, followed by coastal-marine and continental red beds lacking volcanics. The same type of succession is typical of many other Mesoprotozoic complexes of both platform and miogeosynclinal type; for example the Great Slave Supergroup, the Epworth and Goulburn Groups in the Canadian—Greenland shield, and many others, but they differ in having fewer volcanics and in being somewhat younger. In the miogeosynclinal Udokan Group of Eastern Siberia marine deposits pass into coastal-marine terrigenous rocks, then a lagoonal carbonate—terrigenous sequence (with evaporites), and finally into red deltaic terrigenous (cupriferous) deposits. Red beds are typical of the upper part of most Mesoprotozoic sections.

Abundant basic volcanics are commonly associated with "mature" sedimentary rocks in Mesoprotozoic sequences. Thus, in the Karelian complex (which is considered the stratotype for the Mesoprotozoic) every group (with the exception of the uppermost) starts with sedimentary rocks and is capped by volcanics. Lava was extruded from fractures, into small seas, continental basins, commonly under subaerial conditions. The volcanic rocks are diabase porphyrite typical of relatively mobile, fractured zones. Volcanics are normally of minor importance in platform deposits. The presence of abundant volcanics (diabase porphyries type) in the Mesoprotozoic platform sequences makes them substantially different from younger platform complexes (especially from the Phanerozoic ones). These successions also differ, in that they have undergone relatively severe metamorphism (usually of greenschist facies), they are relatively strongly deformed and have associated gabbroic intrusions. The Mesoprotozoic platform sequences are thus considered to be unique deposits, reflecting a particular period in the evolution of the earth's crust.

The miogeosynclinal sequences are differentiated from platform deposits by their greater thickness, more continuous sections, higher degree of metamorphism, more intense folding, and also by the presence of granite intrusions. In addition siliceous iron formations are present in miogeosynclinal successions, but are absent from those of platforms. Gold—uranium bearing conglomerates occur in the latter and are lacking in the former.

Cross-bedded, ripple-marked quartzarenites are widespread in both platform and miogeosynclinal sequences of the Mesoprotozoic. In some sequences the quartzites are associated with high-alumina shales, which, during metamorphism, were transformed into pyrophyllite, chloritoid, andalusite, or sillimanite schist. These rocks originated as a result of sedimentation of materials derived from paleosols which were widespread on the ancient platforms. Locally, fossil soils are preserved at the base of Mesoprotozoic successions.

The successions of platform and miogeosynclinal type are represented by arkoses, subarkoses, and lithic sandstones as well as texturally and mineralogically mature sediments. Immature sandstones are more typical of the miogeosynclinal sequences. Lithoclastic sandstones and granule conglomerates are characteristic of red-bed sequences at the top of Mesoprotozoic sections.

Dolomite is the predominant carbonate rock. Stromatolitic (including oncolitic or oolitic) dolomite is common. Stromatolites and oncolites in carbonates suggest relatively shallow-water depositional environments.

In some areas the upper part of the Mesoprotozoic section contains graphitic slates and anthracite-like rocks — the so-called "schungite". Schungite may be regarded as the oldest metamorphosed coal, probably formed by accumulation of remains of blue-green algae.

Some of the altered sedimentary rocks (albitized, scapolitized, and zeolitized slate, dolomite, and quartzite such as those of the regressive sequence of the Udokan Group) were probably formed by metamorphism of evaporites. Evidence of saline and gypsiferous strata is present in the upper parts of some other Mesoprotozoic complexes (e.g. in the Great Slave and Huronian Supergroups of Canada).

Some miogeosynclinal successions include rhythmically bedded flysch-like rocks (for instance, the Ladoga Group of the Karelian complex). These occurrences are similar to the flysch of Phanerozoic geosynclinal regions.

The most important kinds of sedimentary ore deposits of the Mesoprotozoic are those of iron- and gold—uranium bearing conglomerates. The ferruginous rocks are of two types: jaspilitic iron formation and the sheet hematite or siderite type.

The Mesoprotozoic siliceous iron formations differ markedly from older ones in rock association, mineralogy, and in some cases, in structure. In Canada they are called "Superior type". They are characterized by association with other sedimentary rocks (largely shaly miogeosynclinal strata). The Mesoprotozoic jaspilites differ greatly from superficially similar Paleoprotozoic ones which are exclusively developed in "eugeosynclinal" complexes in close association with volcanics. Mesoprotozoic iron formations are seldom found in a eugeosynclinal setting, but where they are they typically occur in a sedimentary, rather than volcanic, sequence. The oxide facies is the most common but carbonate facies and mixed oxide—carbonate subfacies are also present; the sulphide facies occurs infrequently.

The iron ores are generally finely bedded or banded due to the intercalation of chert and ferruginous layers. In the oxide and oxide—carbonate facies the dominant iron mineral is generally hematite, and not magnetite, as is normally the case in Paleoprotozoic jaspilite. In addition to hematite some other minerals, such as magnetite, siderite, greenalite, stilphomelane, and minnesotaite are commonly present. In jaspilites of the Lake Superior area the primary ore minerals are hematite, greenalite, minnesotaite, and siderite; they commonly form very fine oolites or pisolites; the latter are quite peculiar for siderite. Magnetite commonly represents secondary mineral which is the result of ore metamorphism (Laberge, 1964). It is to be noted that oolitic and pisolitic ore structure is missing from the Paleoprotozoic jaspilites. In the same area abundant microscopic remains of blue-green algae and bacteria are reported (Barghoorn and Tyler, 1965). Within the iron formations there is, in some cases, a remarkable continuity of thin layers for distances up to hundreds of kilometres. They are generally confined to definite stratigraphic levels in the middle parts of the Mesoprotozoic miogeosynclinal sequences so that they might possibly be used as marker horizons in regional correlation.

There is no common opinion as to the genesis of "Superior type" iron formations, but the majority of the workers agree that they are chemogenic and formed as a result of precipitation of silica and oxides or carbonates of iron from solutions (or colloids). Some workers consider the material to have been supplied from surface weathering. Others argue that the source was volcanic exhalations. Since most Mesoprotozoic iron formations occur in dominantly sedimentary sequences, it is difficult to establish a relationship with volcanic rocks, even if such a relationship existed. It is difficult to suggest that the genesis of Mesoprotozoic iron formations completely differed from that of Paleoprotozoic ones, which appear to be genetically related to volcanic activity.

There is a second type of iron formation that is characteristic of the Mesoprotozoic. These are massive sheet hematite or siderite ores occurring among carbonate (dolomite) or mixed terrigenous—carbonate miogeosynclinal strata. This type of ore also occurs in younger Precambrian sequences (Neoprotozoic), but is absent or quite rare in the Paleoprotozoic. This type of ore is widely distributed in carbonate rocks of Mesoprotozoic and Neoprotozoic age in the Urals (the Burzyan and Yurma—Tau Groups), and thus, it can be regarded as the "Ural-type".

The uranium- and gold-bearing conglomerates are unique to the Mesoprotozoic Erathem. In many regions associated quartzites are also metalliferous, so that the name gold—uranium bearing conglomerate is not very precise. It has been pointed out (Salop, 1972b) that all larger deposits of this type have a definite stratigraphic position. Such deposits are known from several continents of both the northern and southern hemispheres (Blind River, Canada; Jacobina, Brazil; Witwatersrand, South Africa; Koli-Kaltimo, Finland; etc.). They are confined to strata in the lower part of the Mesoprotozoic and occur before the first extensive carbonate horizons with stromatolite bioherms. The enclosing strata are everywhere of similar composition and facies and appear to have formed in a platform (protoplatform) environment at the margins of cratons or in aulacogens.

There has been much discussion concerning the genesis of gold—uranium mineralization in Precambrian conglomerates. Both hydrothermal and sedimentary ("placer") origin have been strongly supported, but now most workers agree that the primary mineralization was due to accumulation of clastic grains of gold- and uranium-bearing minerals as paleoplacers. However, certain ore minerals and also many structural and textural aspects of ore bodies are best explained as the results of epigenetic processes such as metamorphism, circulation of hydrothermal solutions, and to the effects of intrusive bodies etc.

The following facts support a primary sedimentary (placer) origin for mineralization of the conglomerates: (1) the gold—uranium mineralization is confined to definite stratigraphic horizons (stratigraphic control); (2) the gold—uranium mineralization is confined to certain sedimentary rock types (quartz-pebble conglomerates, quartzites and sandstones); (3) some of the ore-rich zones appear to follow ancient river channels; (4) the ores are commonly layered; (5) the rounded shape of unrecrystallized grains of gold, uraninite, and pyrite in the ores; and (6) gold, uraninite, and pyrite are associated with detrital heavy minerals such as monazite, zircon, cassiterite, chromite, spinel, etc.

The major primary source of the gold (and pyrite) in the conglomerates was the Paleoprotozoic greenstone belts which include hydrothermal deposits of gold—quartz and gold—sulphide types. The source of the uranium mineralization is problematic because no large primary uranium deposits are known in the areas around the uraniferous conglomerates. It has been suggested that uranium-bearing minerals were extracted from pegmatites in gneisses and granites of the basement, but it is probable that the parent formations were volcanic rocks of the greenstone belts which are thought to have supplied the gold and sulphide. This is also suggested by a comparison of the amounts of gold, pyrite, and uranium in conglomerates. The uranium minerals (mainly uraninite) may have been disseminated in the volcanics, and were then concentrated in conglomerates by processes of sedimentary mechanical differentiation of clastic material supplied during erosion of large masses of rock. The facts that uraninite grains are generally small and of limited size distribution (usually about 0.1 mm) and that there is a low, but relatively homogeneous, uranium content (usually 0.01—0.1% U_3O_8) in the rocks also support this idea.

A peculiar feature of the gold—uranium bearing conglomerates is the presence of well-rounded clastic grains of pyrite and uraninite. In modern placers these minerals are not generally preserved as fresh, rounded grains, because of oxidation processes. This fact was used by some as an argument against the placer origin of these ores, but other workers such as Ramdohr (1961) and Roscoe (1969) explained the presence of these grains as being due to a deficiency of oxygen in the early Precambrian atmosphere.

The gold—uranium bearing conglomerates invariably occur in grey-colored (grey and grey-green) rocks and are generally absent from thick red beds which are typical of the upper part of the Mesoprotozoic sequences. In his study of distribution of uranium mineralization in the Huronian of Canada, Roscoe (1969) stated that all horizons of metalliferous conglomerates of economic grade, with typical rounded pyrite and uraninite, are in the grey-colored Elliot Lake Group, whereas the red-colored quartzites and quartz-pebble conglomerates of the overlying Cobalt Group contain hematite instead of pyrite and uranium mineralization is either absent or is very poorly developed (Lorrain Formation).

The lack of gold—uranium bearing conglomerates in the upper part of the Mesoprotozoic is best explained by the presence of sufficient free oxygen in the atmosphere at that time to bring about oxidation of clastic sulphides and uraninite. The appearance of free oxygen was probably related to photosynthetic activity of blue-green algae which flourished at the beginning of the Mesoprotozoic and caused formation of the first significant stromatolitic bioherms (Salop, 1972b).

Thus, the first widespread stromatolitic dolomite of Mesoprotozoic platform and miogeosynclinal sequences generally provides the upper limit for gold—uranium bearing conglomerates. Gold-bearing conglomerates and uraniferous sedimentary units of various kinds occur in younger sequences, but there are no known occurrences of gold- and uranium-rich conglomerates similar to those of the Early Mesoprotozoic. This is probably because the conditions that existed on earth at that time were unique.

Could such conglomerates have formed in strata older than the Mesoprotozoic? Under the proposed genetic model it would be theoretically possible, but it is unlikely that large deposits would have formed. Gold—uranium bearing conglomerates formed not only in response to the geochemical environment, but also to the paleogeographic and tectonic factors. The Paleoprotozoic rocks of many regions are eugeosynclinal sedimentary and volcanic ("greenstone") assemblages. Miogeosynclinal facies and particularly platform facies are rare, and limited in extent. Such deposits were probably largely destroyed by pre-Mesoprotozoic denudation. The Archean strata are typically highly metamorphosed; their tectonic setting and depositional environments were quite different from those of the Mesoprotozoic.The tectonic and paleogeographic setting of Mesoprotozoic times was suitable for deposition of metalliferous conglomerates, for by that time large platforms had developed, there was widespread development of regressive conditions and extensive regions underwent subaerial erosion.

The following factors are necessary for the deposition of uraniferous—auriferous conglomerates; a favourable geochemical environment (scarcity of oxygen in the atmosphere), tectonic conditions favourable to deposition and preservation (stable platforms with associated marginal troughs and aulacogens), a specific paleogeographic setting (large peneplanes with thick soils) and abundant volcanic rocks in the basement complex. The greenstone-belt assemblage is thought to be the source of gold and possibly uranium in the older paleoplacers. Such a combination existed only in the Mesoprotozoic and even then it did not exist everywhere. This concept explains why large deposits of gold—uranium bearing conglomerates are rare and occur only in strata of a certain age. The Mesoprotozoic gold—uranium bearing conglomerates are a particularly striking example of a rock type that was unique in the geological history of the earth.

A remarkable similarity exists in both composition and stratigraphic sequence of Mesoprotozoic subplatform or miogeosynclinal successions of different continents. In some areas the lower part contains gold- and uranium-bearing conglomerate (Elliot Lake Group, Segozero Group etc.). Overlying these are phytolite-rich carbonate rocks (Knob Lake of the Kaniapiskau Supergroup, Sosan of the Great Lake Supergroup, the Onega Group, the Butun Formation of the Udokan Group, etc.). Still higher in the section evaporitic rocks appear (Gordon Lake Formation of the Huronian, Butun Formation of the Udokan Group, etc.), followed by red beds with cupriferous sandstone (Bar River Formation of the Huronian, Onega Group, Sakukan Formation of the Udokan Group).

It is more difficult to find rocks that characterize Mesoprotozoic eugeosynclinal successions. In general, however, volcanics appear to be less abundant than in the corresponding Paleoprotozoic sequences. This is true not only for Mesoprotozoic eugeosynclinal belts such as those of Eastern Siberia, which were formed in the same general region as Paleoprotozoic ones, but also for newly formed geosynclines, such as the Canadian Labrador geosyncline. Possibly this indicates a significant decrease in geosynclinal volcanism with time. Another difference is the relative scarcity of ferruginous rocks in Mesoprotozoic sequences of eugeosynclinal aspect. Jaspilite is reported only from the "eugeosynclinal" facies of the Labrador trough, in a dominantly sedimentary assemblage. The "eugeosynclinal" zone of the Labrador trough is peculiar in some ways and has some "transitional" features similar to those of the adjacent miogeosynclinal zone.

In North America glacial formations are reported from the Huronian Supergroup (Gowganda Formation etc.) and in approximately contemporaneous strata of the Chibougamau Group, Animikie Supergroup, and in the Hurwitz Group, west of Hudson Bay. Similar units are present in the Libby Creek Group (U.S.A.) and in the Bijawar Group of India. Tillites are also known from Mesoprotozoic strata of South Africa (Government Reef Formation of the Witwatersrand Group and Griquatown Formation of the Transvaal Group), and, possibly also from rocks of similar age in Western Australia and South America. Eskola (1965) considered the Sariolian conglomerate of the

Karelian complex to be tillite and compared it with the Gowganda Formation in Canada. At present most workers in Karelia consider these conglomerates to be fluvial deposits. However, the possibility that they are fluvioglacial cannot be ruled out for ice-rafted (?) dropstones are locally present (V.Z. Negrutza, personal communication, 1973). Marine-glacial deposits are also reported in the Pechenga Group of the Kola Peninsula and in the Krivoy Rog Group of the Ukraine. Thus, a significant glaciation (possibly the oldest in the geological record) took place in the Mesoprotozoic.

The Mesoprotozoic also contains much evidence of organic activty. Stromatolitic carbonates are common: these rocks are composed almost entirely of stromatolites and microphytolites. The stromatolitic assemblage is that of the first (Lower Riphean) complex. Typical forms include: *Kussiella kussiensis* Kryl., *Conophyton garganicus* Korol., *C. cylindricus* Masl., *C. lituus* Masl., *Collenia frequens* Walc., various *Omachtenia*, etc. Some of the stromatolites also resemble Middle Riphean, and especially Upper Riphean forms (of *Jurusania* and *Gymnosolen* type) and there are other, as yet, poorly studied forms. Microphytolites are rather rare; they include oncolites, among which the most common is *Osagia libidinosa* Z. Zhur.; rarer forms include *Osagia pulla* Z. Zhur., and *Radiosus tenebricus* Z. Zhur. Katagraphs appear for the first time in the Mesoprotozoic, but are very rare and represented only by the form *Vesicularites rotundus* Z. Zhur.

Microscopic remains such as cells, filaments and more complicated organic forms are present. Some of these forms have consistent morphological characters and can be used for stratigraphic correlation.

The Mesoprotozoic rocks are variably altered, depending on their tectonic position. Certain platform strata are only slightly altered (lowest greenschist facies). A clear metamorphic zonation is present in some geosynclinal sequences. Regionally metamorphosed rocks occur in various sub-facies of the greenschist facies, but in some localities epidote—amphibolite and even amphibolite facies are reported.

The Mesoprotozoic is a period typified by widespread development of platform and miogeosynclinal facies. Paleotectonic maps of Eastern Europe, Northern Asia, and North America show the existence of large cratonic blocks or protoplatforms surrounded by miogeosynclinal zones.

The major structural elements of the Mesoprotozoic in part of Eastern Europe are shown in Fig.22. This region was later consolidated into a single platform. The structural elements of Fennoscandia are depicted as on the tectonic map of the Baltic shield compiled by the author. A full explanation of the subdivisions is given elsewhere (Salop, 1971a). The structure of the covered part of the East European platform basement is based on data from deep drilling and geophysical exploration. Magnetic data compiled by A.N. Berkovsky and others and edited by V.A. Dedeyev (personal communication, 1972) was particularly useful.

An extensive area of Eastern Europe was occupied, in Mesoprotozoic time,

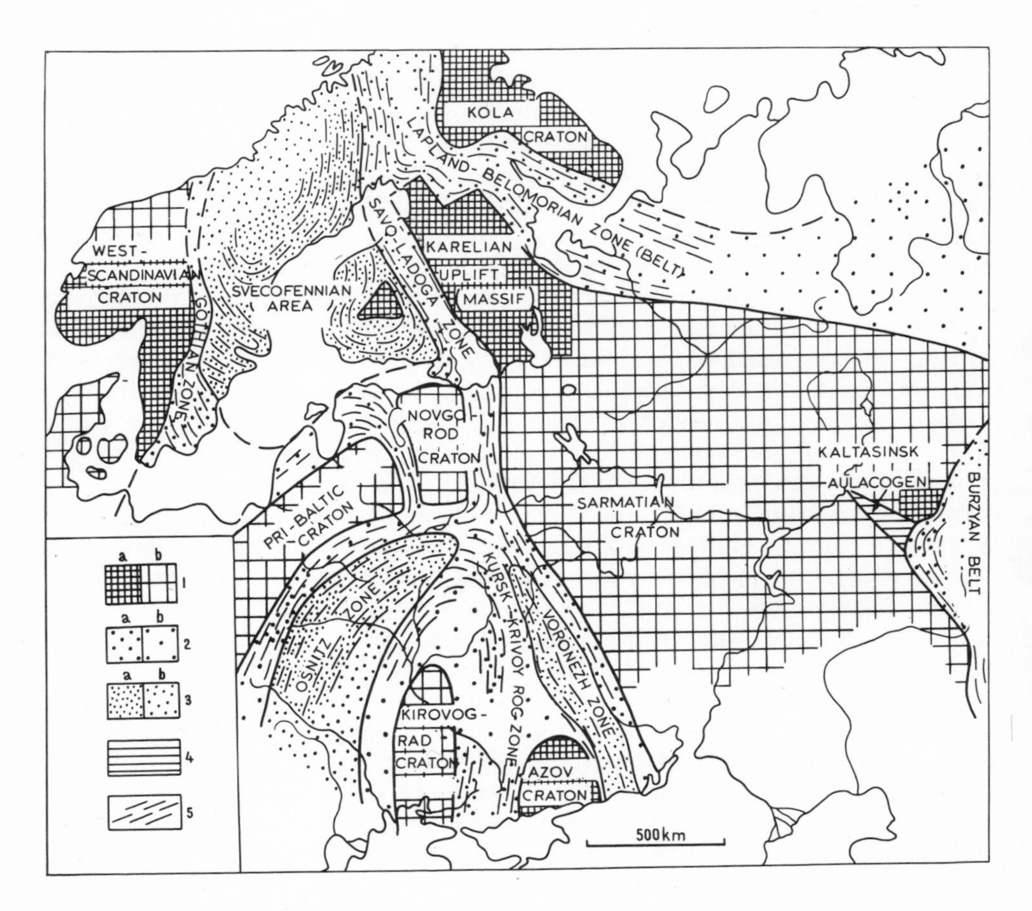

Fig.22. Major structural elements in Eastern Europe during the Mesoprotozoic (paleotectonic map).
1 = cratonic blocks (protoplatforms and median massifs): *a* = exposed, *b* = covered by younger strata; *2* = miogeosynclinal zones (belts): *a* = exposed or known from drillholes, *b* = presumed, covered by younger strata; *3* = eugeosynclinal zones (areas): *a* = exposed or known from drill holes, *b* = presumed, covered by younger strata; *4* = aulacogens; *5* = general strike of Mesoprotozoic fold structures according to geological and magnetic data.

by the Sarmatian craton (protoplatform). At the present time only the northwestern margin is exposed, and is known as "the Karelian Massif" (Kharitonov, 1941, 1957) or the "Jatulian continent" (Väyrynen, 1954). The rocks of the Karelian succession are mainly deposits of platform (protoplatform) facies. Certain younger Precambrian (Neoprotozoic to Eocambrian) or Paleozoic rocks directly overlie the Archean gneiss—granulite complex in the unexposed part of the Sarmatian craton. At some localities (towns of Konosha and Kresttsy), drill holes penetrated red quartzitic sandstones similar to those of the Upper Vepsian of Petrozavodsk and Shoksha of the upper Karelian complex in the Onega region. In the eastern part of the craton (Kama region) there is a northwesterly oriented narrow band of unaltered terrigenous and carbonate rocks known as the Kyrpinsk Group. These rocks formed in the Kaltasinsk aulacogen, that was open to the southeast in the direction of the Burzyan miogeosyncline. Data obtained from drill holes and from geophysics indicates that the thickness of Mesoprotozoic strata in the aulacogen reaches 8,000 m (Yarosh, 1970).

The structure of the basement in the Sarmatian craton is characterized by gneiss fold ovals which are distinctly outlined by magnetic anomalies, particularly in the northern part of the craton (see Fig.14). On the north side of the craton the gneiss ovals are truncated by linear structures which extend into the Lapland—Belomorian miogeosynclinal zone of the Karelides. To the south the craton is covered by thick sedimentary deposits of the Phanerozoic. In this region even geophysical techniques have not permitted definition of the basement structure. The southwestern boundary of the Sarmatian craton is well defined by the Voronezh zone which consists of Mesoprotozoic rocks that are folded about northwest axes. The striking Osnitsk and Kursk—Krivoy Rog zones of linear magnetic anomalies are in this region. Drilling and geophysical studies have shown the presence of folded rocks of the Osnitsk, Kursk, and Krivoy Rog Groups.

The Pri-Baltic and Novgorod cratons are characterized on the map by large oval structures composed of Archean gneiss—granulite rocks. Narrow fold zones of Mesoprotozoic rocks occur between these two cratons and the Sarmatian craton. A characteristic curvature in the strike of the Osnitsk and Kursk—Krivoy Rog fold zones is possibly explained by the presence of the Kirovograd craton (of Archean and Paleoprotozoic rocks) between them. Likewise, branching of the Kursk—Krivoy Rog zones may be due to the presence of the Azov gneiss craton (or several smaller blocks) between these zones.

A peculiar structural curvature is present in the Burzyan miogeosynclinal zone. Its convex side is towards the Kaltasinsk aulacogen. This type of curvature is typical of the boundary between aulacogens and fold zones.

The paleotectonic map of East Europe during the Mesoprotozoic is characterized by the presence of a composite mosaic of large and small cratonic blocks (protoplatforms), surrounded by miogeosynclinal zones. In some cases zones of "eugeosynclinal" aspect occur between them. The picture that emerges differs greatly from that of the Paleoprotozoic, which was characterized by the presence of relatively small stabilized blocks among large eugeosynclinal areas. The general structural setting in the Mesoprotozoic also differed greatly from the younger structures of the East European platform.

The distribution of major Mesoprotozoic structural elements in Siberia has been dealt with in earlier works (Salop, 1958b, 1964—1967; Salop and Scheinmann, 1969). The map shown in Fig.23 is only slightly modified from an earlier one and need not be discussed in detail. The major part of North-

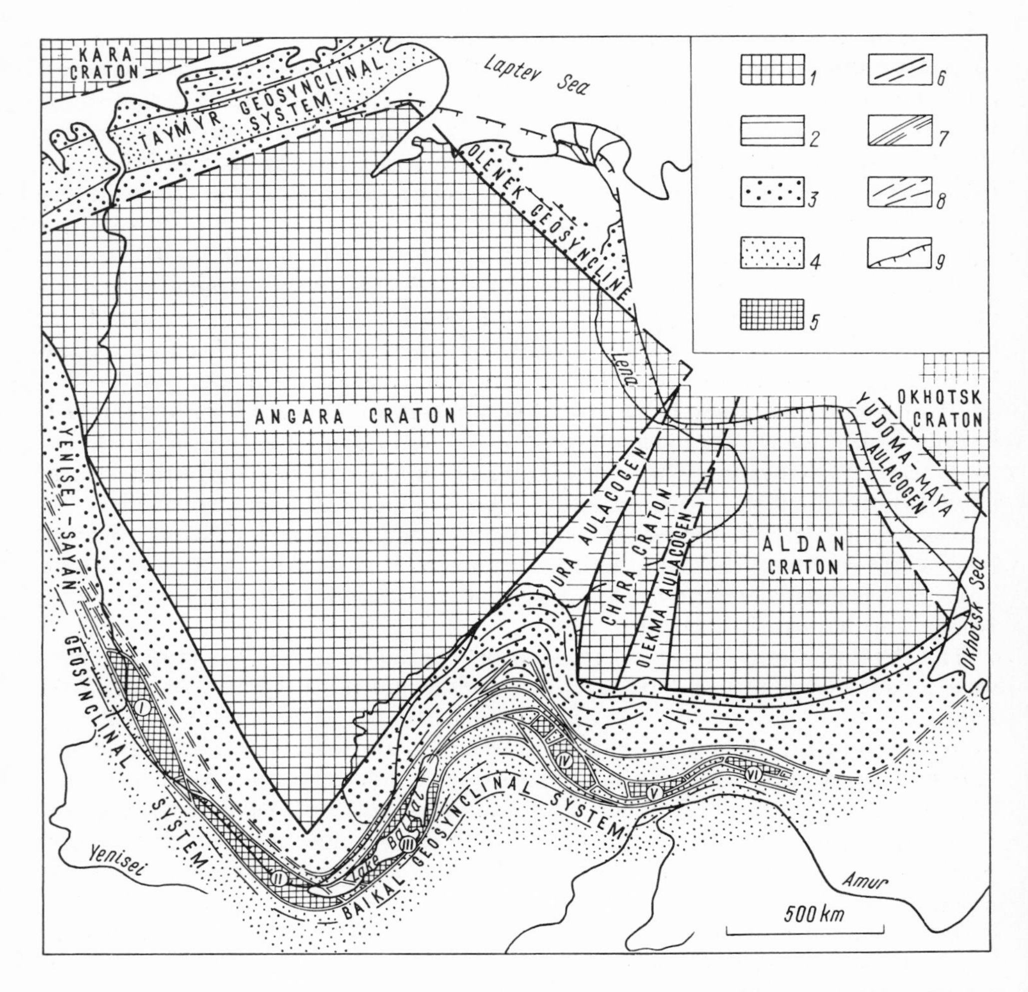

Fig.23. Major structural elements of Northern Asia during the Mesoprotozoic (paleotectonic map).
1 = cratonic blocks (protoplatforms); *2* = aulacogens; *3* = miogeosynclinal zones; *4* = eugeosynclinal zones; *5* = median massifs; *6* = marginal deep faults; *7* = intrageosynclinal deep faults; *8* = strike of Mesoprotozoic folds; *9* = modern boundary of Siberian platform. Circled figures = *median massifs*: *I* = Kansk; *II* = Sayan—Khamar-Daban; *III* = Baikal; *IV* = North and South Muya; *V* = Mogocha; *VI* = Urkan.

ern Asia was occupied by two large cratons (protoplatforms) — the Angara and the Aldan, and a smaller craton (the Chara). Two aulacogens, characterized by relatively shallow subsidence, were developed in the first two cratons. In the easternmost part of the figure, a small craton, the Okhotsk, is shown as being separated from the Aldan craton by the Yudoma—Maya aulacogen.

The aulacogen certainly existed in the Neoprotozoic, but its presence in Mesoprotozoic times is inferred from evidence such as the presence of retrograde metamorphism of the Karelian cycle, affecting adjacent Archean rocks, and from the fact that inherited features are clearly present in other similar structures of Siberia. In northern Asia the Kara craton formerly occupied the region now covered by the Kara Sea. Its existence is evidenced by the location of Mesoprotozoic "miogeosynclinal" and "eugeosynclinal" zones in the Taymyr Peninsula (Zabiyaka, 1974). In Northern Asia, in contrast to East Europe, many major Mesoprotozoic structural elements (protoplatforms and geosynclines) are clearly inherited from those of the Paleoprotozoic. Mesoprotozoic platform deposits have a very limited distribution in the Olekma aulacogen and in grabens in the southern part of the Aldan craton. However, the Precambrian of large parts of Siberia is relatively poorly known (data from a few bore holes) so that the possibility that Mesoprotozoic platform deposits may exist in the inner parts of the Siberian platform, cannot be discounted.

The older cratonic massifs are surrounded by geosynclinal systems with thick external miogeosynclinal zones consisting of Mesoprotozoic terrigenous rocks derived from the cratons (e.g. Udokan Group, Khariton Laptev Coast Group, etc.). The eugeosynclinal belts are situated in the internal parts of the geosynclinal systems. Some Archean blocks may be identified within the younger fold belts. Within these blocks, remnants of the ancient gneiss fold ovals may be recognized (Salop, 1964—1967).

The major Mesoprotozoic tectonic elements of North America (according to the data of Canadian, American and Danish geologists) are shown in Fig. 24. Four large cratons are recognized, the Baffin, Cumberland, Superior and Atlantic, together with some smaller cratonic blocks, the Slave and Baker, and several geosynclinal belts.

The margins of the Superior craton are distinctly outlined by the surrounding Mesoprotozoic fold belts: the Penokean, Grenville, Labrador, Cape-Smith, Belcher and Churchill. In the southeastern part of the craton extensive Mesoprotozoic platform deposits are preserved (Huronian Supergroup and correlatives). This craton probably extended across the continent in a southwesterly direction to the Pacific Ocean, for Archean basement rocks may be present in the Rocky Mountains. They are unconformably overlain by Neoprotozoic rocks, the Mesoprotozoic being missing. This idea is also confirmed by drilling and geophysical data and by the presence of folded Mesoprotozoic strata of geosynclinal aspect, which may be traced as far as California and Arizona. These conform to the general structural grain of the Penokean fold belt on the southeast margin of the craton.

The Baffin craton can only be approximately outlined because much of it is covered by the sea or glacial ice. Mesoprotozoic platform facies are not known from this region. Neoprotozoic strata of shelf facies directly overlie (almost horizontally) the Archean and Paleoprotozoic granitized basement.

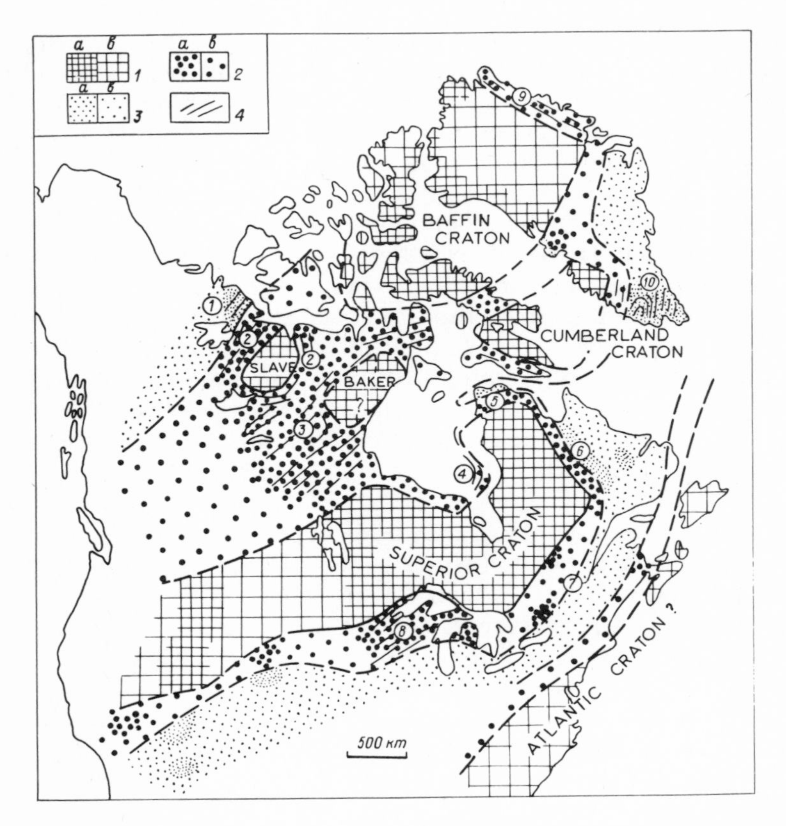

Fig. 24. Major structural elements of North America during the Mesoprotozoic (paleotectonic map).
1 = cratonic blocks (protoplatforms and median massifs); *2* = miogeosynclinal zones; *3* = eugeosynclinal zones; (*1a—3a* = exposed structural elements, *1b—3b* = same elements, covered by younger strata — presumed); *4* = strike of fold structures. Circled figures = *geosynclinal belts*: *1* = Bear Lake; *2* = Coronation; *3* = Churchill; *4* = Belcher; *5* = Cape Smith; *6* = Labrador; *7* = Grenville; *8* = Penokean; *9* = Carolinian; *10* = Ketilidian.

The craton is surrounded by geosynclinal belts: to the southwest is the Churchill belt, to the south the Piling Group of the Fox fold belt and to the east the Carolinian belt. Within these fold belts Mesoprotozoic geosynclinal rocks are folded with axes parallel to the craton margins.

The Cumberland craton is separated from the Baffin craton by the Fox fold belt, from the Superior craton by the Cape-Smith belt, and the Ketilidian belt of Greenland adjoins it on the south and southeast.

The Atlantic craton is tentatively delineated on the basis that Neoprotozoic strata in the Appalachian area lie directly on the Archean (?) gneisses and Mesoprotozoic formations are lacking there. And in New Brunswick the Green Head group of miogeosynclinal aspect is adjacent to the supposed craton.

The Slave province has been established by the work of Hoffman (1969) and J.A. Fraser and L.P. Tremblay (1969). Within this block the Mesoprotozoic is represented by shelf deposits of the Goulburn Group, and equivalents. The block is surrounded by rocks of the Coronation geosyncline (miogeosynclinal deposits of the Epworth Group and Great Slave Supergroup). The Baker block is arbitrarily defined here on the basis of occurrence of only slightly deformed thin Mesoprotozoic strata in the area of Baker and Schultz Lakes, in the dominantly miogeosynclinal Churchill belt. These strata are similar to the miogeosynclinal Hurwitz Group and sometimes are included under the same name. The approximate boundaries of the block are shown on Fig. 24.

The tectonic map of North America clearly shows that the cratons are everywhere bordered by miogeosynclinal zones. This is particularly evident in the case of the Superior protoplatform, which was the source area for clastic material that was carried into the adjoining troughs where various dominantly terrigenous sequences accumulated (Hurwitz, Belcher, Cape-Smith, Kaniapiskau, Animikie, Mojave sequences). The broad Churchill belt (including the Coronation geosyncline) is formed almost entirely of miogeosynclinal-type complexes. Only in its northwest extremity are the eugeosynclinal facies of the Bear province developed. Mesoprotozoic eugeosynclinal complexes are extensive on the east-southeast margin of the Canadian—Greenland shield, extending from the Ketilidian zone of Southern Greenland, through Labrador into the Grenville province and as far west as the Eastern Rocky Mountains and Arizona. It crosses the entire width of North America and includes the sedimentary and volcanic strata of the Ketilidian complex, the Kaniapiskau Supergroup in the internal part of the Labrador trough, the Deep Lake and Libby Creek Groups of Wyoming, the Vishnu complex of Colorado and the Ash Creek and Pinal Groups of Arizona. In the Grenville province the Mesoprotozoic geosynclinal strata may have been largely removed by erosion related to intensive vertical movements at the end of the Neoprotozoic so that these rocks are preserved only in two small areas of the miogeosynclinal zone. One of these localities is east of Chibougamau Lake where the rocks represent the transition zone from platform to miogeosynclinal facies. The other locality, in the southern part of the province, includes the miogeosynclinal rocks of the Flinton Group. The Green Head Group of New Brunswick may belong to the opposite (southeastern) miogeosynclinal zone of the Grenville belt. The facies distribution in the northern part of Labrador suggests that a Mesoprotozoic eugeosynclinal zone may have existed in the internal part of the Grenville province.

Where detailed studies have been carried out in the Northern Hemisphere it appears that the boundaries between cratons and geosynclinal belts are the sites of deep fractures which formed at the same time as the geosynclines. Deep fractures of this kind occur in eugeosynclinal zones at the boundary with regions of miogeosynclinal aspect. Chains of older blocks within younger fold belts also tend to be situated there (see Fig. 23). In southern Siberia such zones have been interpreted as representing island arcs with attendant intensive volcanism (Salop, 1964—1967).

Tectonics of the Mesoprotozoic are characterized by the alignment of fold areas in the form of extensive linear belts. Gneiss domes are much less significant than was the case in the Paleoprotozoic. In some areas Mesoprotozoic domes are the result of reactivation of older domal structures. This can lead to very complex tectonics such as those of the Lapland—Belomorian zone of the Karelides.

In most regions the boundaries of the Mesoprotozoic Erathem are well established, because its strata are separated from the overlying and the underlying rocks by significant unconformities. However, in some areas of Southern Siberia the Mesoprotozoic and the Paleoprotozoic strata appear to be parts of a single structural stage.

The Karelian diastrophism, which terminated the Mesoprotozoic Era, is widespread and intense. The most intense plutonic and tectonic events took place in the interval 2,000—1,900 m.y. ago, as indicated by radiometric dating of syntectonic granites and metamorphic rocks. However, in some areas such as the Ukraine and in the KMA (Kursk magnetic anomaly), there are indications of earlier Karelian events about 2,100 m.y. ago.

Mineral deposits associated with Mesoprotozoic sedimentary and metamorphic rocks are peculiar to that period of geological history. The siliceous iron formations of Superior type and ferruginous carbonates of Urals type include the following: huge jaspilite deposits and associated secondary high-grade ores of the Lake Superior area and of the Labrador trough; the deposits of the Krivoy Rog basin and KMA and several large deposits in India. Important deposits of siderite and hematite of the Satka and Bakal mining areas in the Southern Urals, and relatively small deposits in Karelia and in some regions of Canada and the U.S.A. are of the Urals type. In the Ukraine and KMA hematite deposits of possible economic grade occur among the terrigenous rocks that unconformably overlie siliceous iron formations of the Krivoy Rog, Belozerka and Kursk Groups. These deposits are clastic iron ores formed by erosion of the underlying rocks.

The conglomeratic uranium and gold ores that occur in the lower part of some Mesoprotozoic platform sequences are extremely important economic deposits. The best known is at Elliot Lake (Canada) in the lowermost Huronian. In the Northern Hemisphere other deposits of this type are of moderate size. One example is the Koli—Kaltimo deposit of Eastern Finland. It occurs in clastics of the Segozero Group of the Karelian complex. In Soviet Karelia gold-bearing quartz-pebble conglomerates are present in the same formations. Conglomerates of this very type occur in the Mesoprotozoic strata in aulacogens of the Aldan shield (Vorona et al., 1968).

Magnesite also occurs in rocks of this erathem; it occurs among dolomites, from which it appears to have formed by various epigenetic processes. Large deposits of this kind are known from the Southern Urals. They are confined to carbonate strata that also contain iron ores (Satka and Bakal Formations of the Burzyan Group). An important magnesite deposit occurs in carbonate rocks of the Liao Ho Group in Liaotung Peninsula (China).

A large graphite deposit, associated with graphite-bearing schist and marble, occurs in the Soyuznaya Formation of the Maly Khingan in the Soviet Far East.

The well-known copper deposit of the Outokumpu in Central Finland, is confined to the Kalevian black, carbon-bearing graphitic pyritized slates (correlative with the Ladoga Group). The Mesoprotozoic is also characterized by sedimentary copper deposits associated with red beds. A very large deposit of this type, associated with cupriferous sandstone, occurs in the upper (regressive) part of the Udokan Group of Eastern Siberia. Several small copper deposits are reported from the platform deposits of the Kebetka Formation in the Olekma River basin. Probably the copper mineralization in reddish quartzites of the Bar River Formation (uppermost Huronian of Canada), and also in quartzites of the Onega Group in Northern Karelia (V.Z. Negrutza, personal communication, 1972) are of the same type.

The Mesoprotozoic shelf deposits are also characteristically rich in alumina. These aluminous minerals formed by metamorphism of fossil soils or sediments formed by reworking of such soils. A well-known and large deposit of this kind occurs in the Keyvy Group of the Kola Peninsula; it is made up of kyanite schist and quartzite. In Eastern Finland some small kaolinite deposits (the Pihlajavaara deposit, etc.) and also kyanite—pyrophyllite—kaolinite quartzite deposits (the Hirvivaara deposit, etc.) are located in a paleosol at the base of the Jatulian quartzite (Kajnuu Formation of the Segozero Group). Some geologists (Gladkovsky and Khramtsov, 1967) consider that bauxite in the KMA area at the base of a slate sequence belongs to the Mesoprotozoic Oskol Group. Others (Nikitina, 1971) maintain that the bauxite underlies Paleozoic (Carboniferous) strata and occurs in a fossil soil, developed on ferruginous rocks of the Kursk Group.

Exogenic mineral deposits of the Mesoprotozoic are almost exclusively in sedimentary strata of miogeosynclinal and platform (protoplatform) facies, whereas similar deposits in Paleoprotozoic times occurred in sedimentary—volcanogenic eugeosynclinal complexes.

CHAPTER 7

THE NEOPROTOZOIC

The Neoprotozoic strata can be subdivided and correlated in more detail than was possible for the older sequences, because paleontological methods can be more widely applied, and the presence of syngenetic glauconite makes possible isotopic age determination of many sequences. However, commonly the dates from glauconite give "rejuvenated" values that only provide a minimum age for the time of sedimentation.

As stated already in Chapter 2, the Neoprotozoic Erathem is subdivided into three sub-erathems (Lower, Middle and Upper), by global diastrophic (thermo-tectonic) cycles. The strata of the Middle and Upper Neoprotozoic may also be subdivided on the basis of occurrence of different phytolite complexes. The lower and upper boundaries of the erathem are almost everywhere defined by significant unconformities. These are the unconformities related to the Karelian orogeny (2,000—1,900 m.y.) and the Grenville orogeny (1,100—1,000 m.y. ago). The Lower Neoprotozoic is separated from the Middle Neoprotozoic by a stratigraphic break related to the Vyborgian (Sanerutian) diastrophic cycle (which is considered to be of the second order). It occurred about 1,750—1,650 m.y. ago. The Middle and Upper Neoprotozoic are generally conformable or slightly unconformable with associated local intrusions of basic rocks, related to the Prikamian (Kibaran) orogeny which is also second order (1,400—1,300 m.y.). These relationships clearly define the Lower Neoprotozoic Sub-erathem, but in some cases it is difficult to subdivide the two upper sub-erathems, especially where there are no phytolites or where the phytolites have not been identified according to the Soviet classification. For this reason the Middle and Upper Neoprotozoic are discussed together.

Lower Neoprotozoic: Regional Review and Principal Rock Sequences

Europe

The Lower Neoprotozoic is locally developed in Europe where it is represented by very specific subaerial volcanics and terrigenous—volcanogenic rocks here considered to be of taphrogenic type. The sub-Jotnian Group (sub-Jotnian, the Dala porphyries) of the *Baltic shield*, developed in the Dalarna area of Central Sweden (Magnusson, 1964, 1965) are proposed as a European and world stratotype. This group is subdivided into Upper and Lower Dala Formations which are conformable (Hjelmqvist, 1958). The Lower Dala Formation is composed of porphyrites interbedded with red arkose and quartzitic sandstone, conglomerate and shale. Ripple marks, cross-bedding and rain prints are present in the sandstone. The Upper Dala Formation is largely composed of quartz porphyry, ignimbrite and tuff. The group is several hundred metres thick.

The sub-Jotnian rocks are relatively flat-lying. They overlie various metamorphic and plutonic rocks of the Mesoprotozoic Svecofennian complex, including Late Svecofennian (Karelian) granites about 1,900 m.y. old. The sub-Jotnian rocks are cut by the rapakivi granite which appears to be comagmatic with volcanics dated at 1,650—1,700 m.y. old (K—Ar method). The porphyry gave an age of 1,670—1,690 m.y. (Welin and Lundqvist, 1970) by both K—Ar and Rb—Sr isochron methods but it is probably slightly "rejuvenated". The Middle Neoprotozoic sandstone overlies the sub-Jotnian erosion surface and the rapakivi granite. Thus, the sub-Jotnian Group formed after the Karelian orogeny in the 1,900—1,700 (1,650) m.y. interval.

In Southern Sweden, the Smoland porphyry corresponds to the sub-Jotnian. In Southern Norway similar rocks include the Engerdale porphyry and, in the eastern (Soviet) part of the Gulf of Finland (in Hogland Island), the Hogland porphyry of the Hogland Group is probably correlative.

In the *Ukrainian shield* the volcano-sedimentary complex of the Ovruch Ridge may be correlated with the sub-Jotnian. Drannik (1972) suggested that these strata may be divided into two groups; the Pugachev and the Ovruch.

The rocks in the Belokorovichi graben-syncline belong to the Pugachev Group. These rocks unconformably overlie older rocks including the Mesoprotozoic granite of the Kirovograd—Zhitomir and Osnitsk complexes (about 1,900 m.y. old). The group is composed of two formations, the lower, the Belokorovichi Formation (about 1,200 m), is principally composed of grey and pink-grey lithic sandstone that is underlain by a unit composed of slate and mudstone with subordinate sandstone conglomerate, and locally includes thin diabase sheets. The overlying Ozeryansk Formation (700 m) is made up of green-grey sericite—quartz slate, argillite, and mudstone with a sill of diabase porphyry at the base.

The Ovruch Group is developed in the Slovechan—Ovruch uplift; it also comprises two formations, the Zbran'kovo (340 m), which consists of red porphyry interbedded with amygdaloidal basalt, trachyandesite porphyries, and sandstone; and the Tolkachev Formation (up to 900 m), which is mostly red cross-bedded quartzites. The Tolkachev Formation possibly disconformably overlies the Zbran'kovo Formation. Pyrophyllite-bearing slate beds in the lower part formed from weathered materials derived from a paleosol.

Drannik and Bogatskaya (1967) considered the Pugachev Group to have been intruded by the rapakivi granite of the Korosten pluton and that the Ovruch Group (Zbran'kovo Formation) unconformably overlies the granite. Z.G. Ushakova (personal communication, 1972) concluded that the rapakivi intrusions took place in at least two stages; the earlier intrusion preceded sedimentation of the Zbran'kovo Formation and the later stage intrusion was later than, or contemporaneous with, the porphyry extrusion. Thin-sections of cores from a bore hole (kindly provided by Z.G. Ushakova), revealed a band of cherry-red quartz-porphyry tuff at the base of the Zbran'kovo Formation beneath quartz porphyry and ignimbrite. It is cut by veinlets of granophyre and micropegmatitic granite similar to those present in the underlying rapakivi-granite complex. Extrusion of the acid lava flow probably preceded the granite intrusion, but both were derived from the same magma chamber. Thus, rapakivi granites were intruded both before and after porphyry extrusion. Similar relationships of porphyry and granite are also observed in Central Sweden, in Hogland Island and in some other areas where sedimentary strata of this age occur.

Close genetic and spatial relationships between the sedimentary—volcanogenic strata of the Ovruch Ridge and rapakivi granite are confirmed by isotopic dating. The K—Ar ages of the quartz porphyry and of trachyandesite from the Zbran'kovo Formation are in the range of 1,464—1,530 m.y. (analyses done in the laboratory of V.S.E.G.E.I., on samples of Z.G. Ushakova); these values are possibly too young, because associated diabase yielded an age of 1,690—1,700 m.y. Recently I.M. Gorokhov et al. (1973) determined the age of phyllitized slate (Rb—Sr method) from the Zbran'kovo Formation at 1,370 m.y. ($^{87}Sr/^{86}Sr = 0.733$). This value may give the time of later, low-temperature metamorphism of the rocks. Slightly metamorphosed slates (made up of fine-grained sericite) from the Ozeryansk and Belokorovichi Formations yielded values in the 1,470—1,800 m.y. range (V.S.E.G.E.I. Laboratory, collection of N. Tikhomirova). The wide range of values probably represents both diagenetic and metamorphic processes. The Rb—Sr isochron age of rapakivi granite from the Korosten pluton is 1,720±70 m.y. (I.M. Gorokhov, 1964). These values also indicate that the sedimentary and volcanic strata of the Ovruch Ridge and the sub-Jotnian rocks are contemporaneous.

In the *Russian plate* correlatives of the sub-Jotnian are found in a quartz porphyry present in a bore hole drilled near the town of Glusk (Byelorussia). This porphyry underlies Middle Neoprotozoic red beds. Likewise, amygdaloidal basalt drilled from a drill hole in the Kama region underlies (?) red beds of the Middle Neoprotozoic Tyuryushevo Formation and gave a K—Ar age of 1,650 m.y. Formerly the rocks from this drill hole were thought to be intrusive, but recent studies showed them to be fresh amygdaloidal basalt.

In the western part of the East European platform, intrusions of rapakivi granite and associated basic and alkaline rocks are widespread. These rocks appear everywhere to have formed during a rather short time interval (1,600—1,750 m.y.). They provide a good time-marker for the upper boundary of the Lower Neoprotozoic. The sub-Jotnian and rapakivi intrusions are also cut by diabase dikes which are similar to diabases of the Kresttsy graben with an age of 1,560 m.y. These diabases cut the Mesoprotozoic quartzites but are older than the red beds of the Neoprotozoic Kresttsy Formation.

In the *Southern Urals*, in the Bashkir anticlinorium, the Mashak Group, consisting of subaerial basic and acid volcanics, quartzite (locally red), conglomerate and phyllite (up to 2,000 m thick) is comparable with the sub-Jotnian. This group unconformably overlies the Mesoprotozoic Yamantau complex (correlative with the Burzyan Group), and is in turn unconformably overlain by the Middle Neoprotozoic Yurma—Tau Group. Unfortunately its relationships with the Berdyaush Massif (rapakivi granite and alkaline rocks with an age of 1,570—1,600 m.y., by Rb—Sr and Pb-isotope analyses; Salop and Murina, 1970) are not known. However, since the relationships between the volcanics and rapakivi granites are known, it is possible that the volcanics of the Mashak Group and the granite of the Berdyaush pluton have similar ages.

The *Spitsbergen Archipelago* is another region where the Lower Neoprotozoic is well established. The Kapp Hansten sedimentary—volcanic group (1,700 m), possibly belongs to this sub-erathem. It is exposed in Cape Lapponia, and is made up of red and grey porphyry and tuff, together with agglomerate breccia and beds of phyllitized slate. It unconformably overlies the Archean (?) gneiss—granite complex and is unconformably overlain by the sedimentary Murchison Bay Group which contains Middle and Upper Neoprotozoic phytolites. The porphyry and slate are cut by rapakivi-like granite and show slight contact metamorphic effects. K—Ar dating of this granite yielded ages of 395 and 415 m.y. These values are undoubtedly "rejuvenated" as a result of the Caledonian metamorphism (Krasil'shchikov, 1973).

Asia

Throughout the vast territory of Asia, the Lower Neoprotozoic is reliably defined only in Siberia, but it is probable that strata of this age are present in Kazakhstan, India, and in many other areas of the continent where the Precambrian is poorly known. In Siberia the Lower Neoprotozoic is represented by subaerial sedimentary and volcanic rocks of taphrogenic (sub-Jotnian) type and also by deposits of marine platform and possibly miogeosynclinal type.

The Akitkan Group is the Siberian stratotype of the sedimentary—volcanogenic (porphyry) facies. It occurs in a fault-bounded trough near the western margin of the *Baikal fold belt*. In the lower and upper parts this group is composed of clastics, largely red beds with subordinate acid and basic lavas. The middle part is a thick sequence of red and grey porphyries, ignimbrites and tuffs. These volcanics occur at several levels, separated by beds of sandstone and tuff. The clastic rocks (conglomerate, sandstone, mudstone) accumulated largely in a continental environment. Shallow-marine strata occur only in the uppermost part of the group where they are represented by well-sorted quartzites. The thickness of the group is highly variable, but at a number of localities it reaches 6,000—8,000 m. In the internal zones of the Baikal fold belt the Padra Group is present. This group formed in intermontane grabens. It is correlative with the similar Akitkan Group (Salop, 1964—1967).

The Akitkan Group is cut by granophyric granite porphyry, granite and granosyenite of the Irel complex. These intrusions are thought to be comagmatic with volcanics in the Akitkan Group. The Rb—Sr isochron age of the porphyry is about 1,700 m.y., and of the granite is 1,600—1,620 m.y. (Manuylova, 1968). The dated porphyry is a lava flow in the middle part of the group, so that accumulation began soon after the Mesoprotozoic (Karelian) orogeny. The age of the Irel granite coincides with the age of the rapakivi-type intrusions of Europe. The Irel granitoids are thus similar to the rapakivi intrusions in many respects: high initial $^{87}Sr/^{86}Sr$ ratio, occurrence of fluorite, high iron content of the mafic minerals, presence of an early intrusive phase of gabbroic composition and a final phase represented by diabase dikes.

In the Baikal Range (Western Baikal area) the sedimentary—volcanogenic strata of the Akitkan Group are replaced along strike by the terrigenous, locally red-coloured beds of the Anay Group. These include rare basic lava flows, emplaced in subaerial and shallow-marine environments. In the lower part of this group some beds of high-alumina, chloritoid slate are present among the quartzites. The succession in the Anay Group and its relation to the Akitkan Group have already been discussed in detail (Salop et al., 1974). The most recent dates suggest that rapakivi granite is included in the Primorian complex, which cuts the Anay Group (Eskin et al., 1971). The Middle Neoprotozoic Baikal Group unconformably overlies the Anay and Akitkan Groups.

In the northern part of the Baikal mountain region, in the Patom and North Baikal Highlands, there are rocks that are very similar to the Anay Group. These make up the Teptorgo Group which unconformably overlies the Mesoprotozoic Chuya sequence and Udokan Group and granites intruding them. The rocks of the Teptorgo Group are unconformably overlain by the Middle and Upper Neoprotozoic Patom Group. At the base of the group there is a sequence of quartzite and slate, with interbeds of high-alumina (chloritoid or kyanite) slate or schist, and locally with concretions of diaspore-bearing metabauxite. Higher in the section there are sandstones and slates with diabasic and porphyritic flows overlain by quartzite.

In the area around *Sayan* the Kalbazyk Group (2,700—4,000 m) is Lower Neoprotozoic; it is composed of variegated clastic tuffaceous rocks, and includes both acid and basic volcanics. In lithology it resembles part of the area around the Akitkan, and partly the Anay Groups of Western Baikal (Mats and Taskin, 1971; Taskin, 1971). Many features indicate that the group was mainly formed in a continental environment, as were the above-mentioned groups of the area around Baikal. Phyllites from the upper (Oday) formation of this group gave an age of 1,595 m.y. by whole-rock K—Ar analysis. Volcanics of the lower (Angaul) formation yielded 1,980 m.y. (pyroxene). The first value reflects metamorphism of the rock and is probably somewhat "rejuvenated". The second value is not reliable because of the low potassium content of the mineral used for the analysis. The stratigraphic position of the group is comparable to that of other Lower Neoprotozoic groups discussed above. It overlies with sharp unconformity the Paleoprotozoic—Mesoprotozoic metamorphic rocks of the Urik—Iya graben, and is unconformably overlain by the Middle Neoprotozoic Zuntei Group.

Thus, along the southern margin of the Siberian platform there is an alternation of subaerial sedimentary—volcanogenic and largely sedimentary strata. These rock types replace each other along the strike of the outcrop belt which is situated in the fold belt at the edge of a stable platform.

Another facies type that is common in Lower Neoprotozoic successions is the shallow-marine sedimentary rocks occurring at the base of the platform cover in the northern part of the *Siberian platform*. Surrounding the Anabar shield is the Mukun Group (up to 640 m thick) largely composed of red quartzarenites with shallow-water structures (cross-bedding, ripple marks, sun cracks). This is overlain by the Ust—Ilinsk Formation which consists of a thin (55 m) unit of mudstone with dolomite interbeds. This unit locally disconformably overlies the first formation. The topmost unit is the Kotuykan Formation (up to 450 m) which is made up of variegated (commonly stromatolitic) dolomite. The upper two formations are the lowermost part of the Billyakh Group. Glauconite is a common mineral in the terrigenous rocks. Samples from sandstone of the upper part of the Mukun Group yielded an age of 1,530 m.y. (K—Ar analysis) and from the Ust—Ilinsk Formation several values were obtained in the 1,350—1,480 m.y. range. The scattering of these results suggests that they are all "rejuvenated" by argon migration. Abundant stromatolites are reported in the Kotuykan Formation; these include: *Kussiella kussiensis* Kryl., various *Collonella* and *Conussella*, *Paniscollenia vulgaris* Korol., *Nucleela figurata* Komar, *N. fibrosa* Komar, *Conophyton cylindricus* Masl., etc.; the oncolites: *Osagia pulla* Z. Zhur., *Radiosus kotuikanicus* Milst., etc.; katagraphia: *Glebosites ninae* Korol., *G. gentilis* Z. Zhur., *Vesicularites rotundus* Z. Zhur., etc. These are mostly characteristic of the first phytolite complex, but some forms are more typical of younger complexes, mainly the second one (Tkachenko, 1970).

The Mukun Group overlies a fossil soil on Archean gneiss. The Kotuykan Formation is unconformably overlain by the Yustmastakh Formation of the Billyakh Group. In the lower part of this formation there are stromatolites and microphytolites typical of the second complex (Middle Neoprotozoic).

The Ulakhan—Kurung carbonate terrigenous units of the Udzha Uplift and the Sygynakhtakh (terrigenous) and overlying Kyuyutingde (dolomite) Formations of the Solooli Group in the Olenek uplift, are correlative with the units discussed above. Phytolites of the first complex are reported from the Kyuyutingde Formation (stromatolites: *Collenia frequens* Walc., *Kussiella* Kryl., *Conophyton* Masl., *Stratifera* Korol.; oncolites: *Osagia libidinosa* Z. Zhur., and *Radiosus tenebricus* Z. Zhur.). Glauconite from the Sygynakhtakh Formation is dated at 1,450 m.y. and from the Kyuyutingde Formation at 1,380 m.y. old. These determinations are probably much too young.

On the east side of the *Aldan shield* the Lower Neoprotozoic is represented by two facies types. These are subaerial volcanogenic—sedimentary strata similar to the sub-Jotnian of Akitkan (in the lower part of the section) and shallow-marine strata similar to those north of the platform (in the upper part of the section).

At the base of this sequence the Birinda Formation unconformably overlies acid volcanics of the Elgetey Formation and granites that intrude them. This unit is 180—600 m thick. It is composed of subaerial variegated, commonly cross-bedded, sandstone with conglomerate lenses and amygdaloidal diabasic and trachyandesitic lavas. The Birinda Formation is overlain unconformably by the Konkulin Formation (up to 400 m thick). The latter is made up of coarse-grained polymictic conglomerates interbedded with cross-bedded, commonly red, sandstone. In the upper part of the complex the Uchur Group, which ranges in thickness from 500 to 6,000 m, unconformably overlies the older strata. It includes two formations: the (lower) Gonam Formation, which is largely terrigenous, and the overlying Omakhta Formation, largely carbonate which contains the following stromatolites; *Stratifera* Korol., *Omachtenia omachtensis* Nuzh., *O. utchurica* Nuzh., *Kussiella* sp., *Conophyton garganicus* Korol., *Collenia frequens* Walc., and also oncolites: *Osagia libidinosa* Z. Zhur., and *Radiosus tenebricus* Z. Zhur. (Zhuravleva, 1964; Nuzhnov, 1967). All of these forms are typical of the first phytolite complex. Glauconite from the Gonam Formation gave an age of 1,500—1,550 m.y. (K—Ar analysis), and from the Omakhta Formation, 1,400 m.y., but these values are probably strongly "rejuvenated" due to argon loss. Such argon loss is typical of glauconite dates from old strata which are overlain by thick sequences. The Uchur Group is transgressively overlain by the Maya Group which contains many stromatolites typical of the Middle Neoprotozoic.

The Elgetey volcanics which underlie the Lower Neoprotozoic in the Ulkah trough, are very similar to taphrogenic sequences of sub-Jotnian type. However, they are assigned to the upper part of the Mesoprotozoic because of the isotope age provided by a porphyry and the granite that cuts it (about 1,900 m.y.). It is probable that the taphrogenic stage began earlier in the eastern part of the Aldan shield than in other regions.

The lowest part of the Sukhoy Pit Group in the *Yenisei Ridge* (Korda and Gorbylok Formations) is considered to be an example of Lower Neoprotozoic miogeosynclinal marine strata. The Korda (Maroka) Formation transgressively overlies the Mesoprotozoic Teya Group and is composed of argillite quartzite, phyllite and conglomerate breccia. Its thickness reaches 2,000 m.

The overlying Gorbylok Formation (up to 1,300 m thick), is made up of phyllite with rare limestone interbeds. The Uderey, Pogoryuy and other formations of the Sukhoy Pit Group conformably overlie this section. These are assigned to the Middle Neoprotozoic. Subdivision of the Lower Neoprotozoic within the Sukhoy Pit Group is based on isotopic age determinations on glauconite from the Pogoryuy Formation. Ages up to 1,630 m.y. were obtained. Continuous sedimentation probably took place in the older Yenisei geosyncline during the Early and Middle Neoprotozoic.

In *Kazakhstan* the Maytyube (sedimentary—volcanic) Group in the Ulutau region and the Kuuspek Group (Formation) in the Kokchetav Massif, are probably also Lower Neoprotozoic.

According to Zaytsev and Filatova (1971) the Maytyube Group is very thick (up to 9,000 m), and includes several formations separated by breaks. "This group is composed of porphyry (about 50%) formed from crystalloclastic and lithoclastic tuff of liparite composition and constituting a thick and rather monotonous sequence separated by sedimentary units of about the same thickness. Among the latter there are clastic quartzites, conglomerates, blastopsammitic schist, ferruginous schist with primary clastic quartz, and graphitic schist. Sericite—quartz slate, graphite—quartz slate and phyllite are rhythmically intercalated in the upper part of the group" (Zaytsev and Filatova, 1971, p. 45).

The Maytyube Group unconformably overlies the ferruginous rocks of the Paleoprotozoic Karsakpay Group and is in turn unconformably overlain by the Bozdak and Kokchetav Groups of the Middle and Upper Neoprotozoic. The group is cut by granite with an age of 1,630±150 m.y. on the basis of K—Ar analysis on biotite. Because of its stratigraphic position in the section and the date obtained from the granite, these rocks may be both Mesoprotozoic and Lower Neoprotozoic. The fact that they are so similar to the sub-Jotnian-type successions is an important point in the age assignation accepted here. The liparite volcanism largely took place in a terrestrial environment and the coarse-grained clastic strata interbedded with the lavas were deposited in shallow-water continental basins, by erosion of the newly formed lava piles (Zaytsev and Filatova, 1971). The Maytyube Group is largely composed of typical taphrogenic porphyry (liparite).

The Kuuspek Group (Formation) of the Kokchetav Massif is up to 1,300 m thick. It consists of porphyries similar to those of the Maytyube Group. Usually it is considered to be the upper formation of the Borovsk Group (here attributed to the Paleoprotozoic). These rocks overlie older strata with a considerable break, and in some cases with angular unconformity. The Kuuspek Group is overlain by the Middle—Upper Neoprotozoic Kokchetav Group.

In *India* the Semri Group, of miogeosynclinal (?) type, developed in the northern part of the peninsula, is assigned to the Lower Neoprotozoic. Usually it is regarded as the lowermost unit of the Vindhyan Supergroup (Lower

Vindhyan); however, it differs markedly from the overlying Vindhyan sub-platform groups in that it is strongly folded (Crawford and Compston, 1970). It overlies, with sharp unconformity, the Mesoprotozoic Delhi Group and certain older formations, and is itself overlain unconformably by the Middle Neoprotozoic Kaimur Group. The basal part of the group is a thin unit of quartzite and conglomerate, but by far the greater part (total thickness of 1,000 m) is comprised of slate intercalated with limestone and glauconitic sandstone. In the lower part of the group there is a band of silicified tuff. The group is conformably underlain in places by red slates and includes basal conglomerate and cross-bedded quartzite and sandstone (in Sun River valley, for instance). These strata unconformably overlie the Mesoprotozoic Bijawar Group (Lakshman, 1968).

It was previously suggested (Ahmad, 1954, 1965) that the conglomerate in the lower part of the Semri Group was glaciogenic, but recent work has refuted this idea (Bhattacharya, 1971). Tillite-like conglomerate in fact underlies the Semri Group and forms part of the Bijawar Group (Mathur, 1960).

Glauconite from the upper part of Semri Group sandstone yielded a K—Ar age of 1,100 m.y. An age of 1,400 m.y. was obtained from the rocks in the lower part of the group (Tugarinov et al., 1965a). Tugarinov et al. (1965a) contend that the glauconite yielding 1,400 m.y. was taken from rocks underlying the Semri Group, but Sarkar (1972) considers the sample to have been taken from the lower part of the Semri Group itself. We may conclude that these values are slightly "rejuvenated", for the rocks of the group are strongly folded and metamorphosed. Stromatolites occur in limestones in the middle part of the group. Valdiya (1969) identified *Conophyton cylindricus* (Grab.), *Collenia columnaris* Fent. and Fent. and *Collenia kussiensis* (=*Kussiella kussiensis* Kryl.), forms typical of the first and second phytolite complexes. As already stated, stromatolites of these complexes are fairly common in the Lower Neoprotozoic of Siberia. However, the possibility remains that these identifications are not completely correct.

Altered amygdaloidal basalt of tholeiitic composition underlain by the quartzite and quartz-pebble conglomerate is probably also Lower Neoprotozoic (in addition to the Semri Group). It unconformably overlies the Paleoprotozoic Orissa Iron Ore Supergroup. According to Sarkar (1972) the basalt is dated in the 1,600—1,700 m.y. range. The essentially volcanogenic strata are known as the Dhanjory Group: these are essentially taphrogenic strata.

North America

In North America the Lower Neoprotozoic is commonly represented by sedimentary—volcanogenic strata of taphrogenic type which accumulated in intermontane depressions, grabens, or in troughs formed in a post-geosynclinal stage. As in Europe and Asia, the most characteristic rocks are clastics;

red beds are common, cross-bedded, continental (lacustrine and fluvial) strata are intercalated with subaerial basic and acid lavas (red porphyry and tuff). Sedimentary and volcanic strata of geosynclinal type are rare.

Many different groups could serve as North American stratotypes for the Lower Neoprotozoic, but probably the most suitable is the Lower Dubawnt Group (up to 5,500 m thick), which is widely developed in the northern part of the Churchill province in the *Canadian—Greenland shield*. The strata of this group were described by Donaldson (1965, 1967). It is necessary to subdivide these rocks into two separate groups; the Lower Dubawnt and the Upper Dubawnt. There is an appreciable unconformity between the two sets of strata, but intrusions of alkaline syenite and diabase were probably emplaced at the time of formation of the unconformity. According to Donaldson the Lower Dubawnt includes the following formations (in ascending sequence): (1) the Southern Channel (1,650 m), conglomerate with sandstone interlayers; (2) the Kazan (4,000 m), red cross-bedded feldspathic sandstone and siltstone; (3) the Christopher Island (70—200 m), andesite, latite, trachyte and liparite lava interbedded with red tuffaceous sandstone and agglomerate; and (4) the Pitz (80 m), red and brown porphyry. At the base of the Christopher Island Formation there is evidence of erosion of the underlying rocks. Donaldson placed the Martel syenite between the Christopher Island and Pitz Formations, but it seems more reasonable to exclude these subvolcanic formations from the stratigraphic column for they represent sills, stocks or laccoliths which are likely to be younger or related to the Pitz porphyry. The strata in question subhorizontally overlie the eroded surface of Mesoprotozoic or older rocks and also granites that cut the Mesoprotozoic Hurwitz Group, but are themselves cut by the Martel alkaline syenite, and also by diabase dikes. The K—Ar age of the latter is 1,560 (1,500) m.y. The age of an unaltered porphyry of the Pitz Formation is 1,730—1,770 m.y. by both Rb—Sr isochron and K—Ar analyses. Thus, these rocks may be assigned to the Lower Neoprotozoic.

In the southern part of the same province (Churchill) in the area of Lake Athabasca, the Martin (or Martin Lake) Group is a close analogue of the Lower Dubawnt Group. It is composed of clastic rocks including basic volcanics in the central part (Fraser et al., 1970). The Rb—Sr isochron age of the volcanics is 1,630 m.y. The Martin Group unconformably overlies the Mesoprotozoic Waugh Lake Group and the granite cutting it (1,900—1,970 m.y.) and is intruded by diabase dikes with a K—Ar age of 1,560 (1,500) m.y. The younger Neoprotozoic Athabasca Group unconformably overlies the Martin Group.

The Et-Then Group in the Great Slave Lake Area of the Slave province is probably also Lower Neoprotozoic. It comprises clastic rocks and unconformably overlies the Mesoprotozoic Great Slave Group. The Et-Then Group is cut by diabase with an age of 1,350 (1,300) m.y. Diabases of similar age are also reported in the area of the stratotype, where they also cut the Middle

Neoprotozoic Upper Dubawnt Group. The possibility of a younger age for this (Middle Neoprotozoic) cannot be ruled out.

In the Superior tectonic province the Sioux quartzite is probably Lower Neoprotozoic. It unconformably overlies the Animikie Supergroup and is unconformably overlain by the younger Neoprotozoic Lower Keweenawan Group.

In the Mesoprotozoic fold belts surrounding the Ungava craton the following strata may probably be assigned to the Lower Neoprotozoic: conglomerate, arkose, and basic lavas developed in the Belcher fold belt, the Chukotat Group of the Cape Smith Belt, composed of basalt, quartzite, arkose, and conglomerate, and finally the Sims Formation — quartzite and conglomerate developed in the Labrador fold belt. All these formations overlie Mesoprotozoic rocks with an angular unconformity, but differ from the Lower Neoprotozoic groups in being themselves more intensely folded and metamorphosed. The Chukotat Group is probably formed in a narrow depositional trough (aulacogen?). The upper age boundary of these strata is not established, but their assignment to the Lower Neoprotozoic is based mainly on their close structural relations with underlying Mesoprotozoic rocks and on lithologic similarity to the Letitia Lake Group developed to the east, in the northern part of the Grenville province. The Letitia Lake Group is certainly Lower Neoprotozoic for it unconformably overlies granites formed at the end of the Mesoprotozoic or earlier, but which underwent strong Mesoprotozoic mobilization, judging from the isotopic ages. At the same time this group is cut by alkaline (aegirine) syenite. The age of similar syenite in adjacent areas is 1,765—1,815 (1,700—1,750) m.y. by the K—Ar method. This group is unconformably overlain by the younger Neoprotozoic Seal Group. The Letitia Lake Group is comprised of quartz porphyry, porphyrite, tuff, quartzite, argillite and slate. The rocks are commonly red-coloured; conglomerate occurs at the base of the succession (Brummer and Mann, 1961).

Anorthosite massifs are widely distributed in the outcrop area of the Letitia Lake Group and its probable correlative — the Croteau Lake Group. Rapakivi granites are associated with these massifs. This association is typical in many regions of the world (especially in the Ukrainian shield). The anorthosite—rapakivi intrusions were emplaced shortly after the formation of the Lower Neoprotozoic sedimentary and volcanic sequences. The volcanics are related, possibly in some cases, genetically, to the rapakivi granite. The age of the rapakivi-type intrusions in the Canadian shield is not known, but the anorthosite yields an age of more than 1,540 m.y. This value of 1,540 (1,480) m.y. was obtained by the K—Ar method on biotite from a small granodiorite intrusion that, according to Morse (1969), is rheomorphic, formed in response to emplacement of a large layered olivine gabbro ring-intrusion (Kiglapait). The latter cuts the anorthosite but the genetic relationships are not certain. Thus, this value merely gives the minimum age of the anorthosite (Morse, 1964, 1969). Throughout the world (including Southern

Greenland) the rapakivi granites are dated at 1,600—1,750 m.y. and anorthosites are genetically related to them. Probably the Canadian anorthosites and rapakivi granites are no exception. The Pb-model age of the lead—zinc mineralization in the rocks of the Croteau Lake Group is 1,658±160 m.y. (Cumming et al., 1955).

In Southern Greenland (Ivigtut area) the Qipisarqo Group belongs to the Lower Neoprotozoic. It is composed of terrigenous and volcanic rocks with coarse-grained conglomerate at the base. It overlies Mesoprotozoic Ketilidian strata. The anorthosite, gabbros, and rapakivi granite, belonging to the so-called Sanerutian, are younger. The K—Ar method yielded 1,715 (1,650) m.y. for the rapakivi intrusions but there are also some undoubtedly "rejuvenated" values of about 1,450 m.y. (Bridgwater et al., 1966). Dating by the Rb—Sr isochron method gave an age of 1,758±180 m.y., and Pb-isotope dating on zircon from rapakivi intrusions ($^{207}Pb/^{206}Pb$ ratio) gave 1,724 m.y. and 1,749 m.y. (Van Breemen and Dodson, 1972).

On the east coast of Greenland the Trekant Group is also assigned to the Lower Neoprotozoic. It is composed of red and grey quartzitic sandstones, sandstones, and conglomerates. It overlies the Mesoprotozoic(?) Central Metamorphic Complex with sharp unconformity and is in turn unconformably overlain by the Middle (?) Neoprotozoic Zabra Group (Peacock, 1956).

In the U.S.A. supracrustal rocks of probable Early Neoprotozoic age are exposed only in the Ozark Plateau of the *Midcontinent*, where they are represented by the Hogan and Royal Gorge subaerial porphyry ("rhyolite"). According to one concept this porphyry is overlain by Lower Cambrian strata. It has also been suggested that the porphyry is thrust over these strata (Wheeler, 1965). The Rb—Sr isochron age of the porphyry is 1,420 m.y. so that the porphyry belongs to the middle sub-erathem of the Neoprotozoic, but the rock types are more typical of the lower sub-erathem so that this may be slightly "rejuvenated". It is of importance that rapakivi granite is associated with porphyry in this area.

In the fold zone of the *North American Cordillera* the Lower Neoprotozoic is well established only in the Mazatzal Mountains of North Arizona. The Lower Neoprotozoic is represented by subaerial red porphyries, volcanic breccias, basalt and clastic rocks which in some regions are assigned to the upper part of the Yavapai Supergroup. These sedimentary and volcanic strata (including the Red Rock and Texas Gulch Groups), differ markedly from the older, strongly altered rocks of the Ash Creek and Big Back Groups of the Mesoprotozoic, in having a low degree of metamorphism. The volcanics are unconformably overlain by younger, largely sedimentary deposits of the Middle and Upper Neoprotozoic. Pb-isotope determinations on zircons of the porphyry gave an age of 1,715 m.y. The age of granite and granodiorite, cutting the porphyry, but transgressively overlain by Middle Neoprotozoic rocks, is 1,640—1,655 m.y. according to Rb—Sr isochron, Pb-isotope and K—Ar analyses (Blacet, 1966; Lanphere, 1968; Livingston and Damon, 1968).

In Southeastern Nevada, in the Colorado River basin, near the Mazatzal Mountains, outcrops of rapakivi granite are reported. The granite cuts the Archean gneiss—granulite complex. Age determinations (Rb—Sr method on feldspar and biotite) produced figures in the 830—1,090 m.y. range, but these dates are probably too young. Pegmatite from the basement near the contact with a large rapakivi pluton is 1,630—1,700 m.y. old by whole-rock Rb—Sr isochron analysis (Volborth, 1962). The relationships between the pegmatite and rapakivi are not certain, but the pegmatite is probably related to the older granite. All the age values obtained are probably related to thermal effects of the rapakivi intrusion and give the approximate age of its formation.

General Characteristics

The Lower Neoprotozoic strata were formed during a special stage in the earth's geological evolution. It was at this time that the vast platforms of the present day began to be discernible. The characteristic rocks are subplatform, continental, sedimentary and volcanic strata with local development of continental and shallow-marine sedimentary deposits. Marine platform and geosynclinal units are not very characteristic.

The general features of the subplatform sedimentary and volcanic strata are discussed first. These strata are developed in all continents of the Northern Hemisphere and are similar throughout. Comparable strata are also widely distributed in the continents of the Southern Hemisphere (e.g. lower part of the Carpenterian of Australia and Waterberg of South Africa). Intercalated volcanics and coarse clastics are typical of all these successions. The acid volcanics and sedimentary rocks are commonly red in colour, indicating a terrestrial depositional environment under the action of an oxygenated atmosphere. Lava was mainly extruded from fractures, but eruptions of central type were also widely developed. The sedimentary rocks are fluvial and lacustrine. In certain localities shallow-marine strata pass laterally into continental rocks. The former are commonly confined to the top of the succession. Volcanics predominate in the middle parts of the sections.

Acid volcanics are very abundant. Quartz porphyry is frequently associated with syenite and sub-alkaline rocks, and trachyte and monzonite porphyry are common. These rocks are only slightly altered. The porphyries are usually associated with welded tuff, ignimbrite, volcanic breccia and various tuffs that indicate the presence of explosive-type eruptions. In some areas pyroclastic rocks are more abundant than volcanics. The basic volcanics generally occur among sedimentary rocks but are locally intercalated with acid lavas. More basic types are common in the sedimentary coastal-marine strata. Diabase, basalt and porphyrite (commonly amygdaloidal) are the main components of basic volcanic composition. Transitional types (trachyandesites) are also reported.

Comagmatic hypabyssal or subvolcanic intrusive bodies of acid and basic composition are associated with the volcanics. These intrusions are composed of granite porphyry, granophyre, diorite porphyry and syenite, and also gabbro diabase and gabbro. Holocrystalline granitoid rocks, which commonly have a matrix of granophyre or micropegmatite, are common in the middle parts of thick acid lava flows. Extrusion of lava was interrupted by intrusion of subvolcanic basic and acid magma. Gabbroic and granitic massifs (including granophyre and rapakivi granite) were formed at this time. Large intrusions of this type are, however, more characteristic of the final stage (Vyborgian tectono-plutonic cycle) in the formation of the sedimentary—volcanic sequence.

The sedimentary formations are principally represented by coarse clastic rocks such as boulder, pebble and granite conglomerates, arkoses and lithic and quartz sandstone, but siltstones, argillites, slates or shales are also widely distributed. There is common intercalation of clastic rocks of differing grain size. Well-rounded grains of far-travelled quartz occur together with angular clasts of local rocks. Monotonous marine quartzite sandstones are typical of some regions. The psammitic rocks are commonly cross-bedded. Ripple marks, desiccation cracks and rain-drop impressions are preserved in the fine-grained rocks. In the lower parts of some groups interbeds of high-alumina slate are present among the quartzite sandstones. Depending on the metamorphic grade this material takes the form of chloritoid, kyanite or sillimanite. The quartzites themselves also contain high-alumina minerals. Metamorphic concretions composed of diaspore are present in such rocks. Probably high-alumina sedimentary rocks formed by reworking of material derived from ancient soils. Such paleosols are reported locally at the base of the volcano-sedimentary groups.

The clastic strata mainly formed by disintegration of rocks of neighbouring uplifts and of local volcanic rocks, but at times clastic material came from more distant sources in the peneplaned parts of platforms.

It has been suggested that these sedimentary and volcanic rocks are a volcanogenic molasse of the Karelian cycle. Many facts, however, contradict this idea. Firstly, the Lower Neoprotozoic strata commonly subhorizontally overlie an eroded and peneplaned surface of Karelian (and older) formations, including highly metamorphosed and folded Mesoprotozoic strata and abyssal (Karelian) granites that intrude them. Also, from a tectonic viewpoint the areas of accumulation were isolated grabens or fault-bounded troughs, the orientation of which was independent of structures of the Karelian cycle. In certain regions (such as the Baikal fold belt), the sedimentary and volcanic strata accumulated in a marginal trough—graben that cut Mesoprotozoic fold structures at an acute angle (Salop, 1964—1967). Finally, the sequence of rock types differs from that typical of molasse. The latter is characterized by increasing amounts of coarse-grained rocks up-section, but the Lower Neoprotozoic sections grade upwards into well-sorted marine platform deposits.

These sedimentary and volcanic strata were formed in a special type of subplatform taphrogenic environment formed by tectonic subsidence of certain zones of the stable platforms or in near-platform zones of geosynclines stabilized after the Karelian folding. The environment was characterized by alternation of periods of tectonic quiescence and periods of strong oscillatory movements with development of radial tectonics and intensive volcanism. The acid volcanics are thought to have formed by melting of the granitic layer of the earth's crust, and the basic volcanics from the simatic layer (Salop, 1964—1967).

In some areas tectonic movements related to fractures caused folding and metamorphism, and schistosity developed in zones of differential movements.

Shallow-marine (platform) formations of the Lower Neoprotozoic are known only in the Siberian platform. These rocks lie almost horizontally on older basement rocks and are largely represented by mature red terrigenous rocks. The lower part consists largely of quartz sandstones, locally glauconitic and including structures typical of shallow-water deposition. The upper part of the section is made up of monotonous, commonly stromatolitic, dolomites. These strata are closely related to the Middle and Upper Neoprotozoic. Stratigraphic breaks at the boundary of the lower and middle sub-erathems are only of local importance.

Lower Neoprotozoic geosynclinal strata are more widespread than the present evidence indicates. These strata may also be present in the lower parts of certain thick continuous mio- and eugeosynclinal complexes, but this is not certain because of a lack of isotopic dates. Miogeosynclinal terrigenous strata of the Sukhoy Pit Group of the Yenisei Ridge are of this type. The age of the Sukhoy Pit Group was established only by the fact that it contains glauconitic sandstone that was dated by the K—Ar method.

The end of the Lower Neoprotozoic was also characterized by intrusive magmatism, usually associated with taphrogenic sedimentation and vulcanism. The intrusions of this cycle are part of a peculiar magmatic association of subplatform (taphrogenic) or undeveloped orogenic type. In this case the rapakivi granites are associated with basic rocks: anorthosite, gabbro—norite, gabbro, alkali gabbros and alkaline syenite. These rock types make up parts of certain massifs or separate bodies which formed during a single plutonic cycle (but at different phases of this cycle). The association of rapakivi granite, anorthosite and gabbroic rocks is very typical and is reported from almost every region (e.g. in the Korosten pluton of the Ukraine, Ahvenisto Massif in Finland, Riga pluton in the Soviet Baltic region and in a number of massifs of the Labrador Peninsula — Nain, Sept Iles — etc.). Rapakivi granite and alkaline syenite or monzonite are associated in the following plutons: Korsun'—Shevchenko in the Ukraine, Berdyaush in the Southern Urals, Sept Iles in Canada, and in many more.

The sequence of rock units in many major massifs is the same or very similar. The basic magma was first intruded and, by differentiation, stratiform bodies of gabbro—norite and anorthosite were formed. These were followed by intrusions of acid magma which gave way to the rapakivi massifs, granophyre, micropegmatite and other granites. Next came derivatives of alkaline magma which formed alkaline and nepheline syenite, monzonite. The final plutonic event was intrusion of diabase, dolerite, gabbro—norite, and at times, ultrabasic dikes. In some composite massifs (Ahvenisto, Berdyaush, etc.) some older diabase dikes are present, separating two phases of (rapakivi) granite magma intrusion. Dawes (1970) established that dolerite and gabbro—norite dikes in the Tasiussaq rapakivi massif of Southern Greenland were intruded before complete crystallization of the granite, for the basic rocks contain porphyroblasts of potassic feldspar. In addition to the peculiar orbicular structure (due to oligoclase fringing potassic feldspar ovoids) all of the rapakivi granites have many common features. These include: similarities in chemical composition, iron-enrichment of dark-coloured minerals, common occurrence of fluorite, in some cases the presence of "exotic" minerals such as "armoured" olivine, abundance of xenoliths and inclusions of basic rocks of the early phases, common presence of granophyre or micropegmatite differentiates, high initial $^{87}Sr/^{86}Sr$ ratios, and the fact that potassic feldspar is usually represented by orthoclase.

The rapakivi and anorthosite plutons are commonly very large. Geophysical data (principally gravimetric) indicate that they are asymmetrical, laccolith-like bodies or flat, almost horizontal and lens-like. The "roots" appear to go down to about 18—20 km. They are usually situated close to the Conrad discontinuity (Lauren, 1970).

Rapakivi massifs and associated rocks are known from many areas of both hemispheres. The distribution of these rocks in the continents of the Northern Hemisphere is shown in Fig. 25. Most of these plutonic bodies are situated in the eastern part of the Canadian—Greenland shield, in the Baltic and Ukrainian shields, and also in the western part of the Russian plate (under a cover of younger strata). Detailed maps of these regions indicate that many intrusive bodies are grouped in linear zones, possibly related to deep basement fractures. Almost all of these intrusions are situated in platform regions. Some single massifs occur in Phanerozoic mobile belts but are confined to zones including basement stabilized at the end of the Karelian folding. The radiometric age of the rapakivi granite and associated rocks falls into the 1,600 (1,570)—1,800 m.y. range, but the most common and accurate dates are in the 1,600—1,700 m.y. range. These values are taken as the time of the tectono-plutonic cycle that concluded the Early Neoprotozoic Sub-era. The name "Vyborgian" is suggested, after the huge Vyborgian rapakivi pluton in the Karelian Isthmus. Danish geologists attribute intrusions of this cycle, in the south of Greenland, to the Sanerutian cycle (Berthelsen and Noe-Nygaard, 1965) but some older plutonic intrusions appear also to have been assigned to this cycle. Older intrusions (1,700—1,800 m.y.) were probably comagmatic with abundant acid lavas and are of the same age as the sedimen-

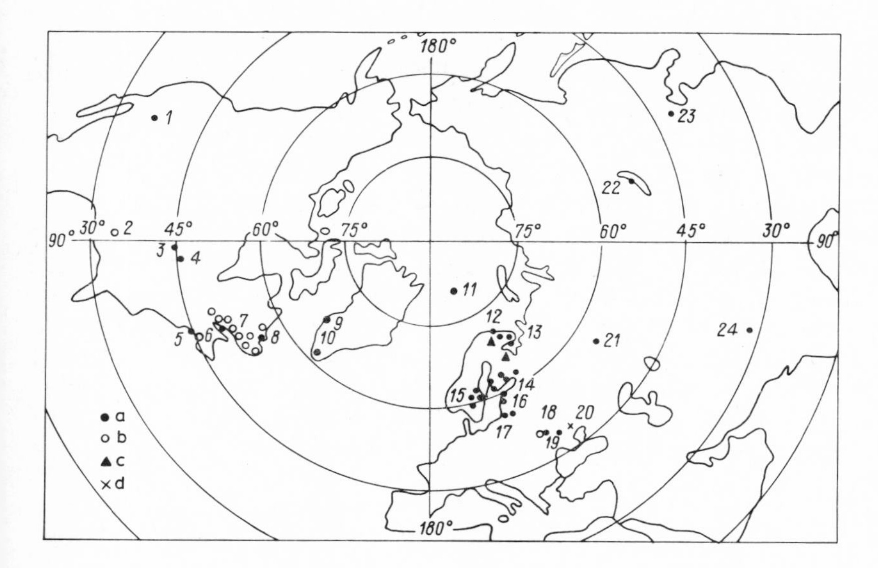

Fig. 25. Location of the major intrusive massifs of the Vyborgian tectono-plutonic cycle in the continents of the Northern Hemisphere.
a = rapakivi and related rocks; *b* = anorthosites; *c* = gabbro norites and alkaline gabbro; *d* = alkaline and nepheline syenites. Intrusive massifs: in *North America*: *1* = Gold Butte—Bonelli Peak (rapakivi); *2* = Wichita (anorthosites); *3* and *4* = a group of rapakivi massifs in Wisconsin; *5* = Head Harbour (rapakivi); *6* = Sept Isles (rapakivi and anorthosites); *7* = a group of large anorthosite massifs in the Grenville province, Labrador Peninsula; *8* = a group of anorthosite and rapakivi massifs in the Nain area, Labrador Peninsula; in *Greenland*: *9* = Egedesmine (rapakivi); *10* = Tasiussaq and others (rapakivi); in *Europe*: *11* = Lapponia, Spitsbergen (rapakivi); *12* = Western Litsa (rapakivi); Gremyakha—Vyrmes (alkaline gabbro), and Ura—Guba (rapakivi); *13* = massifs of the Lebyzhya River (rapakivi), Turiy Mys (rapakivi), and Yelet'-lake (alkaline gabbro); *14* = Salma, Vyborgian, Ahvenisto, Laitila, Satakunta rapakivi massifs; *15* = Evie, Nordingro, Ragunda, Reden, Filipstad rapakivi massifs (associated with basic rocks); *16* = Tallin and Riga (rapakivi); *17* = Polish (rapakivi); *18* = Korosten (rapakivi and anorthosites); *19* = Korsun'—Novomirgorod (rapakivi, anorthosites, alkaline syenites); *20* = Priazov massifs (alkaline syenites); *21* = Berdyaush (rapakivi and alkaline syenite); in *Asia*: *22* = Pribaikal (rapakivi); *23* = Miyun near Peking (rapakivi); *24* = Dhauladhar (rapakivi).

tary and volcanic strata of the lower Neoprotozoic.

In geosynclinal areas thermal events associated with the Vyborgian diastrophism, were rather weak and only rarely accompanied by formation of large intrusive massifs of gabbro or granite. However, at a number of localities these events caused partial mobilization of certain elements in the older rocks. It was this phenomenon that caused "rejuvenation" of ages obtained from these rocks. The common "rejuvenation" of K—Ar ages of the Karelian granite and of metamorphic rocks in the 1,800—1,700 m.y. range may be due to remobilization during the Vyborgian cycle.

The reasons for development of the unique gabbro—anorthosite—rapakivi—alkaline syenite complexes at this particular point in geological time are not certain. For further discussion of this problem see Chapter 11.

Economic deposits of lead, zinc and silver occur in a number of places in Canada in Lower Neoprotozoic sedimentary—volcanic sequences. High-alumina (chloritoid and kyanite) schists (35—38% of Al_2O_3) in the lower parts of the Teptorgo Group in the Patom Highlands, and in the Anay Group in the Western Baikal area are potential economic deposits. In the Teptorgo Group diaspore-bearing rocks of concretionary type (metabauxite) are also reported. The alumina content in these rocks reaches 60%. In the lowermost part of the Tolkachev Formation in the Ovruch ridge region, pyrophyllite-bearing slate occurs. It is used as raw material for various purposes. In many places uranium and fluorite mineralization are associated with acid volcanic porphyries.

The Middel and Upper Neoprotozoic: Regional Review and Principal Rock Sequences

Europe

As in other continents, the Middle and Upper Neoprotozoic sequences of Europe are generally more extensive than those of the Lower Neoprotozoic. Both miogeosynclinal and platform-type sequences are distinguished. In Eastern Europe (Urals fold belt) the miogeosynclinal sequences dominate and are the most complete. Platform sequences are best preserved in the Russian plate.

The best sections of miogeosynclinal aspect are in the *Southern Urals*, in the western limb of the Bashkir anticlinorium. These continuous sections have been studied in detail and may serve not only as European miogeosynclinal stratotypes of the Middle and Lower Neoprotozoic, but even as world stratotypes.

The Yurma—Tau Group, consisting of the Zigalga, Zigazino—Komarovsk and Avzyan Formations, is the Middle Neoprotozoic stratotype. The Zigalga Formation is made up of grey quartzites and phyllitic slates. Phyllites and slates are the most common rock type in the Zigazino—Komarovsk Formation, together with subordinate carbonates and sandstone—mudstone sequences. The Avzyan Formation consists of dolomite with subordinate sandstones and mudstones.

The carbonate rocks of this formation contain stromatolites: *Collenia frequens* Walc., *C. columnaris* Fent., and *Conophyton* Masl., but the most common form is *Baicalia baicalica* Kryl.; these are typical of the second (Middle Riphean) phytolite complex. Microphytolites include *Osagia tenuilamellata* Reitl. and *Vesicularites flexuosus* Reitl. (Krylov, 1963;

Zhuravleva, 1968). Thickness in the Yurma—Tau Group in continuous sections exceeds 4,000 m. An age of 1,260 m.y. was obtained from the Avzyan Formation by the K—Ar method on glauconite. This value is probably much too young, considering the geosynclinal character of the strata. A more reliable age was obtained from a phosphatic sandstone of the Zigalga Formation (1,430 m.y. by Pb-isotope analysis; Ershov et al., 1969). In addition, the Yurma—Tau Group is cut by diabase dikes with a K—Ar age of 1,200 m.y., but even this value cannot be taken as the upper age boundary of the group for the dikes may be appreciably younger.

The Upper Neoprotozoic stratotype is the Karatau Group which includes five formations which are, from base to top: (1) the Zilmerdak (700—3,300 m thick), red arkosic sandstone, siltstone and argillite; (2) the Katav (200—600 m), variegated limestone and marl; (3) the Inzer (200—700 m), glauconitic sandstone and siltstone; (4) the Min'yar (400—550 m), limestone and dolomite; and (5) the Uka (up to 200 m), biogenic and clastic limestone, quartz sandstone and siltstone.

In the Katav and Min'yar Formations the following stromatolites were identified: *Gymnosolen ramsayi* Steinm., *Katavia karatavica* Kryl., *Minjaria uralica* Kryl., *Jurusania cylindrica* Kryl.; and also the microphytolites: *Asterosphaeroides serratus* Z.Zhur., *A. humilis* Z.Zhur., *Radiosus elongatus* Z.Zhur., *R. praerimosus* Z.Zhur., *Osagia crispa* Z.Zhur., *Nubecularites uniformis* Z.Zhur., *Glebosites gentilis* Z.Zhur., *Vermiculites anfractus* Z.Zhur., *Vesicularites ovatus* Z.Zhur., which together form the third or Upper Riphean phytolite complex.

In the Uka Formation the following microphytolites were determined: *Vesicularites bothridioformis* Krasnop., *V. concretus* Z.Zhur., *Vermiculites irregularis* Reitl., *Osagia monolamellosa* Z.Zhur., *Ambigolamellatus horridus* Z.Zhur., *Volvatella zonalis* Nar., *V. vadosa* Nar., *Vesicularites compositus* Z. Zhur., *V. flexuosus* Reitl., *V. tunicatus* Nar. These are typical of the fourth complex. Because of the differences in stromatolites and microphytolites some workers separate the Uka Formation from the Karatau Group (the Upper Riphean) and assign it to the terminal Riphean (Vendian). This separation, however, seems to be artificial because the formation is an integral part of the Karatau Group. The phytolites attributed to the fourth complex appear to be fairly common in the upper part of the Upper Neoprotozoic in many localities.

K—Ar analysis on glauconite yielded values in the 610—965 m.y. range for rocks of several formations of the Karatau Group. In some cases there is a wide scatter of ages (up to 100 m.y.) from rocks of the same stratigraphic level. These results suggest considerable argon loss and consequent "rejuvenation" of the ages.

The Yurma—Tau Group rests with angular unconformity on underlying strata of the Mesoprotozoic Burzyan Group and the Lower Neoprotozoic Mashak Group. This discordance corresponds to the Bakal phase of folding

(Garan', 1946). In places the unconformity between the Yurma—Tau and Mashak Groups is not apparent (Rotar', 1974). The unconformity between the Yurma—Tau and Karatau Groups is not very significant compared to that separating the Karatau and the overlying Epiprotozoic Krivaya Luka Formation.

In other areas of the Urals the Middle and Upper Neoprotozoic sections are not so complete. The Shagar Group of the Uraltau zone may be assigned to the Middle Neoprotozoic, and the following groups are considered to be Upper Neoprotozoic: Kedrov and Basega of the Chusovaya anticlinorium, Burkochim of the Polyudov ridge, Patok of the sub-Polar Urals and the Yenganepe Formation of the Polar Urals. The Shagar Group is typified by grey, grey-green, and pink quartzites and micaceous schists which are lithologically similar to rocks of the lower and middle parts of the Yurma—Tau Group and may be correlated with them. The other units listed above, in the Polar, sub-Polar and Middle Urals, and in the Polyudov ridge, are made up of terrigenous and carbonate rocks (Borovko, 1967; Mladshikh and Ablizin, 1967; Belyakova, 1972), which are probably correlative with formations of the Karatau Group in the Bashkir anticlinorium both on the basis of lithology and phytolites (Table I). Microphytolites typical of the third complex occur in the lower, thickest part of these groups and microphytolites of the fourth complex are present in the upper part in strata comparable to the Uka Formation of the Karatau Group. It is only in the Klyktan Formation of the Kedrov Group that stromatolites of the different complexes are associated. For example, *Gymnosolen* Steinm., typical of the third complex occurs with *Baicalia baicalica* Kryl. which is normally present in the second complex, and *Linella ukka* Kryl. of the fourth complex (Mladshikh and Ablizin, 1967).

The miogeosynclinal strata of the Middle and Upper Neoprotozoic are also exposed in the zone of the Hyperborean folding in Kanin Peninsula and in Timan and also in outcrops on the western slope of the Urals.

In *Kanin Peninsula* only the Upper Neoprotozoic strata of the Ludovaty Cape are well dated. These are carbonate rocks with the following stromatolites: *Gymnosolen giganteus* Raab., *G. ramsayi* Steinm., *Inseria djejimi* Raab., *Minjaria uralica* Kryl., *Parmites concresceus* Raab. These are typical of the third phytolite complex.

The stratigraphic position of the Kanin metamorphic complex of Kanin Peninsula is not certain. The lower formation of this complex (3,500 m thick) consists of micaceous, garnet-bearing quartzite and schist. The upper formation consists of about 3,800 m of micaceous quartzite interbedded with phyllites and limestone, together with sandy dolomite and carbonaceous slate (Nalivkin, 1962). High-grade metamorphism suggests that these rocks are much older than carbonate rocks of the Ludovaty Cape region. Possibly these rocks are Middle Neoprotozoic or older. The K—Ar dates on slates from the upper formation of the complex (480—620 m.y.) give the age of a later metamorphism.

In *Timan*, correlatives of the Yurma—Tau and Karatau Groups of the Southern Urals are rather firmly established (Volochaev et al., 1967). The Svetlin Formation (800 m) is compositionally very similar to the Zigalga Formation (light-grey quartzite and micaceous schist). The Chetlas Kamen Formation (2,600 m), which is made up of phyllite and quartzitic sandstone, is correlated with the Zigazino—Komarovsk Formation. The Dzhezhim Formation (1,000 m) which unconformably overlies these units and is made up mainly of arkose sandstone, is comparable to the Zilmerdak Formation of the Karatau Group. The Bystrinsk Formation (2,000 m), which consists mainly of carbonates with stromatolites (*Gymnosolen ramsayi* Steinm., etc.) and microphytolites of the third complex, corresponds to the overlying part of the Karatau Group (probably to the Min'yar Formation). The Oselkovaya Formation (1,800 m) which is composed of phyllites with minor dolomite interbeds similar to those of the Bystrinsk Formation, is assigned to the Upper Neoprotozoic. The Bystrinsk Formation is cut by a diabase sill with an age of 1,200—1,220 m.y. (K—Ar analysis of several samples; Mal'kov, 1969). This date is very important for it defines the upper boundary of the Upper Neoprotozoic (the Upper Riphean).

The miogeosynclinal (in part subplatform facies) strata of the Middle and Upper Neoprotozoic are present on the *northwestern and northern margins of the Baltic shield* near the Caledonian fold belt.

The Lower Sparagmite Group of Southern Norway is an example. Norwegian geologists (Holtedahl, 1953) divide it into the following units (in upward sequence): Brettum sparagmite (sandstone), Brettum slate and limestone, Biri (Biskopåsen) conglomerate and Biri limestone. The latter contains stromatolites typical of the third complex (Upper Neoprotozoic). There is a stratigraphic break between the Brettum limestone and Biri conglomerate. Possibly the units below this break are Middle Neoprotozoic. Stromatolites of the third complex associated with microphytolites typical of both the third and fourth complexes (data of M.E. Raaben and V.E. Zabrodin) occur in the Porsanger dolomite (Lower Sparagmite sequence) of Northern Norway, which is considered to be correlative with the Biri limestone. The Lower Sparagmite of Northern Norway and the Sparagmite Group of Sweden are thought to correspond to the upper part of the section in Southern Norway which is placed in the upper sub-erathem.

The Visingso Group of Southern Sweden may also be Upper Neoprotozoic. It consists of basal arkose and conglomerate passing upwards into sandstone and then into slate with carbonate interbeds. Certain problematic fossils are present in rocks of this group. Remains of *Chuaria* (phytoplankton?) were described for the first time from the Chuar Group in North America. According to available data this sequence also belongs to the upper part of the Neoprotozoic (see below). The K—Ar age of slate from the upper part of the Visingso Group is 985 m.y. (Magnusson, 1965) but this figure probably represents the age of initial metamorphism.

On the Barents Sea coast (Sredny Peninsula, Kil'din Island) the Kil'din Group of clastic and carbonate rocks is Upper Neoprotozoic. These rocks are transitional in character between miogeosynclinal and platform facies (V.Z. Negrutza, 1971). The group is characterized by abundant stromatolites of the third complex (e.g. *Gymnosolen ramsayi* Steinm.). Contained glauconite was dated by the K—Ar method, but the values are probably "rejuvenated" for similar figures were obtained from both the upper and lower parts of the group. There is also a wide scatter at a single stratigraphic level (710—1,015 m.y.).

Middle and Upper Neoprotozoic platform strata are present in the *Russian plate*. They generally occur in troughs of limited extent, as shown in Fig. 30. These strata rest unconformably on the rapakivi granite and are cut by gabbro diabase dated at 1,100—1,200 m.y. This complex is correlated with the Yurma—Tau and Karatau Groups of the Urals. Definition of the boundary between the Middle and Upper Neoprotozoic is problematic in the stable platform region. This boundary is taken at a distinct break separating principally terrigenous detrital rocks of the lower part of the complex from terrigenous—carbonate strata of the upper part. This break corresponds to the stratigraphic unconformity between the Yurma—Tau and Karatau Groups. Generation of widespread gabbro—diabase dykes and sills (1,300—1,420 m.y. old) in the eastern part of the platform in the area around Kama, may be related to epeirogenic movements corresponding to the break (Prikamian or Avzyan cycle of M.I. Garan').

In the Middle Neoprotozoic, red terrigenous rocks are abundant; carbonates and volcanics are subordinate. The thickest and compositionally most varied Middle Neoprotozoic rocks occur in the southern Bashkir region (Serafimovka-377 drill hole). The following formations (Timergazin, 1959; Solontsov et al., 1966) are present (in ascending sequence): (1) Troitsk (130 m), red quartzo-feldspathic sandstone overlying basement rocks; (2) Mizgirevo (80 m), variegated siltstones and clayey rocks; and (3) Malokamysh (80 m), pink dolomite. These three formations make up the Kidash Group, which is overlain by the Serafimov Formation (250 m), consisting of red and grey quartzo-feldspathic sandstone with interbeds of pink and red silty and clayey rocks (Fig. 26).

In other bore holes in Southern Bashkir the above rocks are overlain by red quartz sandstones of the Leonidovo Formation (700 m). Recently the Serafimov and Leonidovo Formations were combined into the Solov' yevskinsk Group (Solontsov et al., 1966).

In a drill hole (Shkapovo-740) southeast of the village of Serafimov, the Middle Neoprotozoic section includes grey silicified sandstones similar to those of the Zigalga and Zigazino—Komarovsk Formations of the Urals. They occur together with red sandstones. Both the geographic location and rock types of the Shkapovo-740 bore-hole section are transitional between Serafimov and the Urals. Eastward replacement of red beds by grey-colored rocks probably reflects a change in the environment from continental to marine, and a change in tectonic setting from that of a platform to a miogeosynclinal one.

In Northern Bashkir (area around Kama) the section has the same sequence as in the Serafimov area (Fig.26). In this region the lower part of the Middle Neoprotozoic is red sandstone (>60 m), the middle part is red dolomite (60—160 m), and the upper part consists of red sandstone (up to 660 m) associated with red-brown argillite interbedded with dolomite. The lower and middle parts, which together constitute the Nadezhdinsk Formation, correspond to the Kidash Group. The upper part (Gozha and Shtandinsk Formations) is considered to be correlative with the Solov'yevkinsk Group of Southern Bashkir. The Middle Neoprotozoic rocks are cut by two gabbro—diabase dike complexes which have yielded ages of 1,300—1,420 m.y. and about 1,200 m.y.

Microphytolites of the first complex have been identified in the Kidash Group and of the fourth complex — in the coeval Nadezhdinsk Formation (Morozov and Revenko, 1969). If these identifications are correct, then the vertical stratigraphic range of the various complexes of microphytolites is considerably greater than is currently accepted.

In the western part of Northern Bashkir, sandstones, mudstones and carbonates of the Middle Neoprotozoic are replaced by coarser-grained, probably continental deposits of the Tyuryushevo Formation. This formation was formerly considered to be the base of the sedimentary cover, below the Mesoprotozoic Arlan Formation. However, the contacts between the Tyuryushevo Formation and other units are not exposed so that its stratigraphic position is uncertain. The Tyuryushevo Formation is typified by red, quartzo-feldspathic sandstone that differs from the Gozha sandstone, only in grain size. Beds of brown-red argillite, together with siltstones (typical of the Gozha Formation) are also reported from these rocks. The Tyuryushevo sandstones, as well as the Gozha rocks are characterized by complete absence of any metamorphism. Some samples are extremely indurated due to cementation by secondary quartz overgrowths. Sutured grain contacts are quite common in the Arlan sandstone. If in fact the Tyuryushevo Formation is older than the Arlan Formation, then it should show the same (or an even higher) degree of metamorphism.

At Povolzhye in the area of Kuybyshev adjoining the Southern Bashkir area, the Middle Neoprotozoic is represented by continental red sandstone and conglomerate of the Borovsk Formation. The stratigraphic sequence of the formation is highly variable. In the Baytugan area the whole section is typified by quartz sandstone. In the vicinity of the village of Borovka it is made up of quartzo-feldspathic sandstone; in the Sernovoda area the lower part is quartz sandstone but the upper part consists of quartzo-feldspathic sandstone capped by granule conglomerate. All three varieties are intercalated near the village of Yakushkino. Around the village of Chesnokovka,

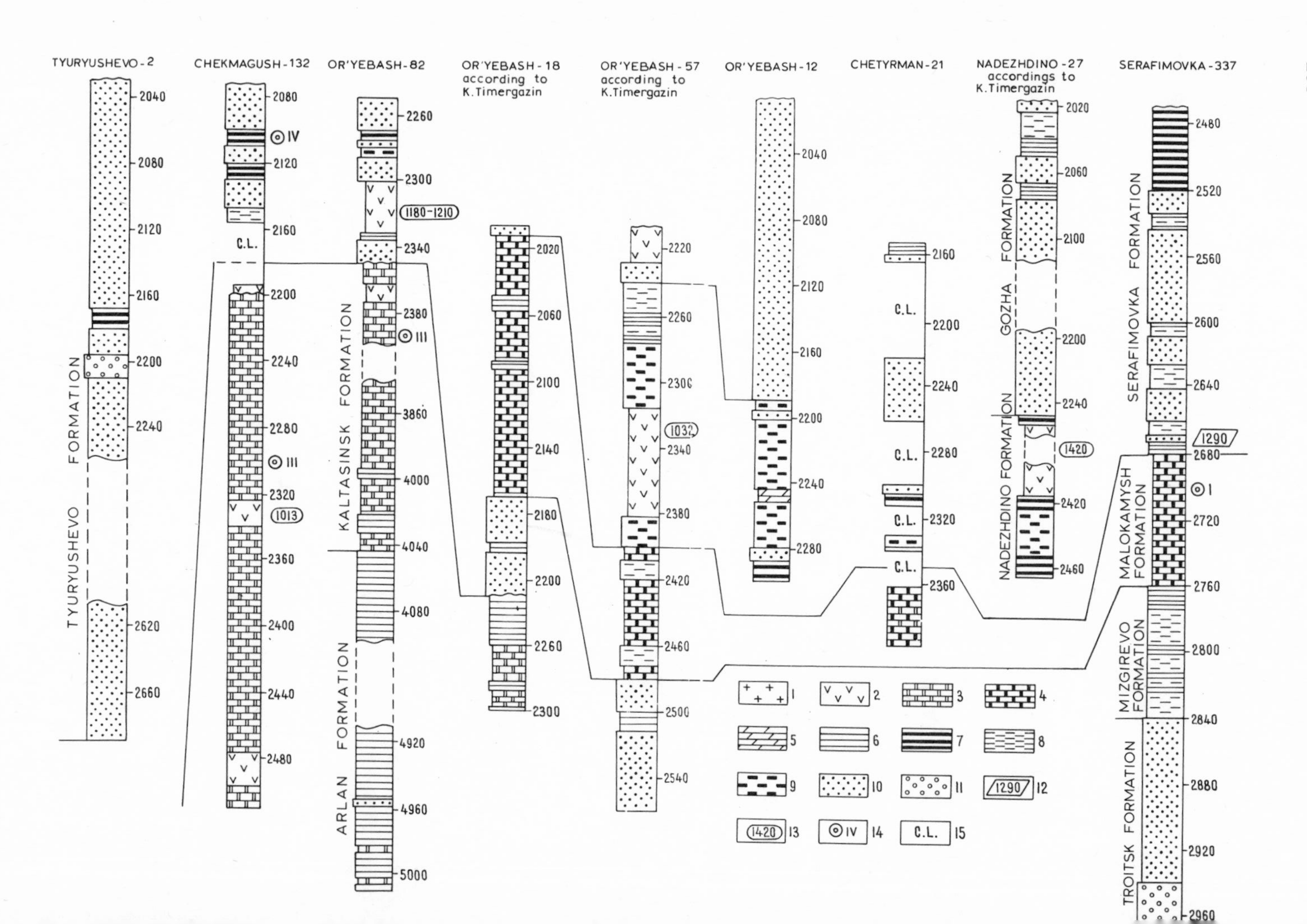
TYURYUSHEVO-2
CHEKMAGUSH-132
OR'YEBASH-82
OR'YEBASH-18
according to
K.Timergazin
OR'YEBASH-57
according to
K.Timergazin
OR'YEBASH-12
CHETYRMAN-21
NADEZHDINO-27
accordings to
K.Timergazin
SERAFIMOVKA-337
TYURYUSHEVO FORMATION
KALTASINSK FORMATION
ARLAN FORMATION
GOZHA FORMATION
NADEZHDINO FORMATION
SERAFIMOVKA FORMATION
MALOKAMYSH FORMATION
MIZGIREVO FORMATION
TROITSK FORMATION
C.L.
1180-1210
1013
1032
1420
1290
1 2 3 4
5 6 7 8
9 10 11 12
13 14 15

quartzo-feldspathic sandstone occurs in the lower part of the section and quartz sandstone in the upper part (Egorova, 1964).

The Middle Neoprotozoic is principally represented by continental rocks in the Pachelma, Kazhima, Orshansk and Belomorian troughs (Zoricheva, 1956; Bessonova, 1968; Yatskevich, 1970). Red sandstone is dominant, with subordinate granule conglomerate and brown-red silty mudstones. The ratio of quartzarenite to arkose is different in different sequences. Bessonova (1968) defined the boundary between Middle and Upper Riphean at the contact of arkosic sandstone of the Rogachev Formation with the Orshansk Formation (principally quartzarenites) sandstone in the Orshansk trough of Byelorussia. This subdivision does not seem reasonable when compared with data from other regions of the Russian plate. In particular this interpretation seems unlikely in view of the data of Povolzhye from the Kuybyshev area. Arkose and quartz-rich varieties of sandstone are not considered as separate stratigraphic units but rather as facies of a single terrigenous unit.

In Polesye and Volyn' the Polessian Group is typified by red sandstone with subordinate brown-red mudstones. This group has been studied in detail (Bruns, 1957). It is more than 800 m thick and includes rock types that are very characteristic of the Russian plate. Thus, the Polessian Group may serve as an East European platform stratotype for the Middle Neoprotozoic. It is cut by gabbro—diabase (1,180 m.y. old) which is probably the youngest Neoprotozoic dike complex related to volcanics of the Volyn' Group.

Middle Neoprotozoic terrigenous rocks associated with volcanics are present in the Kresttsy graben and in the area around Ladoga. In the Kresttsy graben variegated, poorly sorted sandstones of the Kresttsy Formation, together with units of diabasic porphyrite and tuff in its middle part, may be attributed to this sub-erathem. The porphyrite is 1,250 m.y. old (Geisler, 1966). In the southern part of the Ladoga area sandstones similar in composition to those of the Kresttsy Formation were found in the 908—273 m interval of the Kondrat'yeve bore hole. In the upper part of the succession two flows of basalt porphyrite are present among the sandstones (Tikhomirov and Yanovsky, 1970). In the eastern part of the area around Ladoga the Middle Neoprotozoic strata are only about 147 m thick. These strata rest unconformably on rapakivi granite. The lower part of the sequence is typified by a variety of terrigenous rocks (from conglomerate to mudstone). Basalt

---

Fig. 26. Correlation of pre-Eocambrian sections in the Bashkir—Urals area (compiled by K.E. Jacobson).
1 = basement rocks; *2* = gabbro diabase; *3* = grey dolomite; *4* = red dolomite; *5* = marl; *6* = argillite; *7* = brown-red argillite; *8* = siltstone; *9* = brown-red siltstone; *10* = fine- and medium-grained sandstone; *11* = coarse-grained sandstone and gritstone; *12* = glauconite age (m.y.); *13* = gabbro-diabase age (m.y.); *14* = microphytolite complexes (by number); *15* = core not available.

porphyrite and tuff (the Salma Formation) make up the upper part. The age of the porphyrite is 1,350—1,500 m.y. (Kayrak and Khazov, 1967).

Platform-type strata of the Upper Neoprotozoic are developed in the eastern extremity of the Russian plate, in the Pachelma trough, in Byelorussia and in Volyn'.

Rocks in the eastern part of the Russian plate may be directly correlated with the Zil'merdak and Katav Formations of the western slopes of the Urals. Here the marker horizon is a variegated red and grey limestone and marl which is perfectly analogous to rocks of the Katav Formation (Bekker, 1966). These strata are about 650 m thick and are underlain by red sandstones that correspond to the Zil'merdak Formation.

In the Pachelma trough the Peresypkino Formation is Upper Neoprotozoic. It rests unconformably on Middle Neoprotozoic red sandstone (Solontsov and Aksenov, 1969). The Peresypkino Formation is proposed as a stratotype of the Upper Neoprotozoic platform facies. Its lower part consists of glauconitic sandstone about 740—996 m.y. old. Dolomitic and carbonate-rich mudstones predominate in the upper part. Variegated and banded rocks of Katav type occur among these dolomites. The formation is 255 m thick. An Upper Neoprotozoic section similar to the Pachelma section is present in the Pugachev graben (Kondrat'eva, 1962).

In Byelorussia the Lapichi (Osipovichi) Formation (30 m of variegated carbonate—terrigenous rocks) is assigned to the Upper Neoprotozoic. Phytolites of both the third and fourth complexes are present in the Peresypkino and Lapichi Formations (Narozhnykh and Postnikova, 1971).

At Volyn' the Berestovets Formation (the Volyn' Group), which is made up of basalt and tuff, is also thought to be the Upper Neoprotozoic. It has a known thickness of about 480 m. There is an irregular boundary between the Polessian Group and the volcanic strata. In some places a basal unit of coarse-grained sandstone is present at the base of the volcanic sequence. Locally the volcanics seem to pass gradationally into the Polessian Group.

The volcanic rocks at Volyn' were formerly included in the Vendian, but recent data favour a Late Neoprotozoic age for these rocks. The Volyn' volcanics are separated from the overlying Vendian s.str. (Eocambrian) rocks by a sharp unconformity. These may be Epiprotozoic or Upper Neoprotozoic. The latter seems probable because pebbles of Volyn' effusive rocks occur in the Epiprotozoic tillite, and numerous dates on volcanics (K—Ar method) yielded values of about 1,000 m.y. (Jacobson, 1971b). Recently Veretennikov reported the presence of tillite under the Volyn' basalts in the Kremenets and Brody bore holes near Lvov (communication at a meeting in Kishinev, March 1973). If this is correct then the Epiprotozoic age of the Volyn' Group may be considered certain. Probably this tillite corresponds to the Lower tillite of Norway. However, it is also possible that there are two volcanic sequences.

In Europe, Middle and Upper Neoprotozoic rocks extend to the Spitsbergen Archipelago, the British Isles, France and Czechoslovakia, and are probably present in the Caucasus, in addition to occurrences in the East European platform and the Urals.

Such rocks are widely distributed in *Spitsbergen* in particular. The rocks there are of miogeosynclinal type and differ from place to place. On North-East Land the Upper Neoprotozoic Murchison Bay Group consists of the following formations (in ascending sequence): (1) Flora (up to 2,000 m), which consists of light and pink quartzite sandstone, siltstone, argillite and dolomite; (2) Norvic (280 m) of quartz sandstone (locally cross-bedded) and siltstone; (3) Raudstup (250 m) of variegated siltstones and argillite; (4) Sälodd (250 m), siltstone, dolomite and dolomitic limestone; (5) Hunnberg (250 m), dark dolomite and limestone, commonly biogenic; and (6) Ryssö (up to 800 m), light dolomite and limestone, biogenic in places. The three lower formations are essentially terrigenous and the three upper ones are mainly carbonates. In the lowest formation (Flora) oncolites of *Osagia* type of "Middle Riphean aspect" are locally present. In the Hunnberg Formation stromatolites and microphytolites of the third, second and partly of the first complexes have been found (*Gymnosolen* Steinm., *Inzeria* Kryl., *Kussiella* Kryl., *Conophyton* Masl., *Tungussia* Semikh., *Osagia columnata* Reitl., *Vesicularites flexuosus* Reitl.). In the carbonate rocks of the Ryssö Formation the stromatolites (*Gymnosolen* Steinm. and others), and microphytolites of the third complex have been identified (Krasil'shchikov, 1973).

The Backaberg Formation rests unconformably on the Ryssö Formation. It is composed of siltstone, argillite, and dolomite with microphytolites of the fourth complex. Krasil'shchikov (1973) and Norwegian geologists assign this formation to the younger Gotia Group, but it is here included in the Murchison Bay Group because it has a conformable contact with the underlying units and a great break separates it from the overlying tillite of the Gotia Group. If this interpretation is correct then microphytolites of the fourth ("Vendian") complex occur in the Upper Neoprotozoic section in Spitsbergen.

The Early and Middle Neoprotozoic age of the Murchison Bay Group, including the Backaberg Formation, is well determined not only on the basis of phytolite occurrences but also because of its stratigraphic position in relation to other Precambrian strata. It unconformably overlies the Lower Neoprotozoic Kapp Hansten (volcano-sedimentary) Group and is unconformably overlain by the Sveanor tillite of the Gotia Group (Epiprotozoic).

In the northeastern part of West Spitsbergen Island (New Frisland Peninsula) the Lumfjorden Group corresponds to the Murchison Bay Group (Harland, 1961; Krasil'shchikov, 1973). Its four lower formations (Kortbreen, Kingbreen, Glasgowbreen and Oxfordbreen) are combined into the Veteranen Subgroup (4,300 m thick). It consists principally of terrigenous rocks; greywacke, quartzite and shale, with rare interbeds of dolomite and limestone. These correlate, on the basis of lithology, with three lower formations of the Murchison Bay Group. The Kingbreen Formation contains stromatolites (*Inzeria* Kryl.) and microphytolites of the third complex. The four upper formations of the group (Grusdievbreen, Svanbergfjellet, Draken and Baklundtoppen) constitute the Akadamikerbreen Subgroup (1,400–2,500 m), composed of carbonate rocks. Those may be compared with the three upper formations of the Murchison Bay Group. Two lower formations of this subgroup contain stromatolites and microphytolites of the third complex, and the two upper formations are characterized by microphytolites of the fourth complex. The Elbobreen Formation conformably overlying the upper formation (Baklundtoppen) of the Lumfjorden Group is similar to the Backaberg Formation of the North-East Land. Harland and Krasil'shchikov attribute the Elbobreen Formation to the overlying Polarisbreen Group. The Elbobreen Formation (like the Backaberg Formation) contains microphytolites of the fourth complex and is separated from the Epiprotozoic tillite by a large break. It seems more reasonable to include it in the Lumfjorden Group.

In the western part of West Spitsbergen Island the Upper Neoprotozoic is made up of rocks of different composition. These rocks belong to the Sofiebogen Group, the basal part of which (Slyngfjellet conglomerate, up to 500 m thick) rests unconformably on the Mesoprotozoic Deilegga Group, and is overlain by the Höferpynten dolomite and limestone (300 m), followed by Gåshamna phyllite (1,500 m). The carbonate rocks of the Höferpynten Formation contain microphytolites of the second complex, on the basis of which they are assigned to the Middle Neoprotozoic together with the underlying conglomerate. The overlying phyllite is probably Upper Neoprotozoic. The phyllite is unconformably overlain by the Epiprotozoic tillite of the Kapp Linney Formation.

In *Britain*, strata of presumed Middle and Upper Neoprotozoic age are extensively developed in Scotland. These are dominantly rocks of subplatform and miogeosynclinal aspect. The first type includes the Torridonian Group, distributed on the older platform in the Northwestern Highlands. The second type includes the Moine Group which is present in the Northern and Grampian Highlands. The synchroneity of these two groups is not certain, for they differ in composition, in grade of metamorphism and are separated by a large zone of tectonic dislocation — the Moine thrust zone. Correlation of the groups is based on certain similarities in the stratigraphic sequence and also on the fact that the thickness of the Torridonian strata (together with the grade of metamorphism) increases toward the Moine thrust and the area of development of the Moine Group. The normal Torridonian section is as follows (in upward succession): (1) the Diabaig Formation (from 150 to 2,200 m), red, locally grey, sandstone, grey and green argillite, shale, conglomerate–breccia and conglomerate in the lower part; (2) the Applecross Formation (from 300 to 2,400 m), red and brown arkose with interbeds of conglomerate, local brown shale, conglomerate–breccia in the lower part; (3) the Aultbea Formation (from 75 to 1,300 m), intercalated sandstones (red near the base), dark shales and calcareous rocks with rare layers of conglomerate (Pavlovsky, 1958; J.G.C. Anderson, 1965). In the zone of the Moine thrust, red beds pass into grey and greenish rocks. Shallow-water structures are common in the rocks of the group; cross-bedding, ripple marks, sun cracks and rain-drop casts. Much of the sequence is interpreted as alluvial valley sediments.

The Torridonian sandstone rests subhorizontally on a rough surface of Archean (Lewisian) gneiss and is transgressively overlain by the Epiprotozoic Dalradian Group which includes tillite. The Torridonian Group has a wide range of possible ages. Clasts in the Torridonian conglomerate include boulders of sedimentary and volcanic rocks that are undoubtedly younger than the Lewisian complex. Oolitic (oncolitic?) siliceous and carbonate-rich rocks with microscopic algal remains analogous to those of the Mesoprotozoic Ketilidian complex of Southern Greenland (with *Vallenia erlingi* Ped. — globular structures also encountered in the Ketilidian Graensesø Formation), and to Mesoprotozoic rocks of the Labrador trough and the Penokean fold belt of Canada. Muir and Sutton (1970) assumed that these rocks were transported from Greenland which, according to them, was previously situated close to the British Isles. K—Ar age determinations on micas from boulders of metamorphic and igneous rocks in Torridonian conglomerates is 1,650—1,800 m.y. All of these data suggest that the Torridonian Group may be assigned to the Neoprotozoic and, in particular, to the middle and upper suberathems.

The Moine Group is largely composed of quartzo-feldspathic metasandstone and micaceous schist with subordinate calcareous metasandstone, dolomite and limestone. A thin conglomerate lies unconformably on the Archean gneiss and on rocks of the Mesoprotozoic (?) sub-Moine Group. Shallow-water structures are locally present in the metasandstones. The Moine Group rocks are preserved in complex, commonly overturned folds that involve basement rocks. These strata are regionally metamorphosed in the epidote—amphibolite and amphibolite facies, and, in some places, are migmatized and granitized. Both the Moine Group and the Torridonian are transgressively overlain by the Epiprotozoic Dalradian Group (J.G.C. Anderson, 1965). K—Ar age determinations on micas from metamorphic rocks of the group and from the granite cutting it usually yield "rejuvenated" values of about 420—740 m.y., related to the younger thermo-tectonic events (J.A. Miller and Brown, 1965).

In *France*, strata in the Armorican Massif (Lower and Middle Brioverian), are assigned to the Middle and Upper Neoprotozoic. These strata are probably of eugeosynclinal type. The Lower Brioverian consists largely of basic volcanics (the Erquy Formation) and the Middle Brioverian of phyllite and siliceous slate (the Land-de-Waard Formation) overlain by dark, "coaly" slate, quartzite and sandstone (the Villiers Fossar Formation (Cogné, 1962). These strata are unconformably overlain by the Upper Brioverian sequence (including tillite) that is here considered to be Epiprotozoic. In siliceous slate of the Vardes Formation there are peculiar fossil remains which were formerly considered to be radiolaria and are now a separate group of *Cayeuxidae* of unknown affinity (Graindor, 1957). Oncolites (*Osagia tenuilamellata* Reitl.) mostly typical of the Middle Neoprotozoic are present in limestones that are probably Middle Brioverian (Keller and Semikhatov, 1968). Granite cutting the Upper Brioverian and underlying the Lower Cambrian yielded an age of 545 m.y. by the Rb—Sr method on biotite (M.J. Graindor and Wasserburg, 1962).

In *Czechoslovakia* in the area around the Bohemian Massif the Neoprotozoic is also represented by a thick (up to 8,000 m) typically eugeosynclinal volcano-sedimentary assemblage which is usually divided into two groups (or formations): pre-Spilitic, largely composed of greywacke, rhythmically bedded siltstone, slate, and intraformational conglomerate, locally with abundant siliceous slates; and Spilitic, formed of volcanics of spilite—keratophyre association — pillowed spilites, variolite, diabase, keratophyre, albitophyre, and siliceous rocks (lydites), and carbonaceous slate. The siliceous rocks, as was the case in the Brioverian, contain remains of *Cayeuxidae* (Holubek, 1966). The geosynclinal complex is unconformably overlain by sandstone with possible tillite horizons (Fiala, 1964) of Epiprotozoic age. The oldest K—Ar datings of spilite do not exceed 647 m.y. (Holubek, 1966). It is likely that they were strongly "rejuvenated" by thermal processes during the Hercynian orogeny.

Asia

In Asia, Middle and Upper Neoprotozoic strata are widely developed in fold belts around the Siberian platform and partly on the platform itself. In these areas detailed studies of these strata have been carried out. Both geosynclinal and platform successions are present.

The best section of miogeosynclinal strata for a stratotype in Siberia occurs in the northern part of the Patom Highlands and in adjacent areas of the North Baikal Highlands in the Lena region in the *Baikal fold belt* (Salop, 1964—1967). There, the Middle and Upper Neoprotozoic Patom Group is about 10,000 m thick. It is subdivided into the Ballaganakh Subgroup and Kadalikan Subgroup which includes the Dzhemkukan, Barakun, Valyukhta, Zhuya and Chencha Formations. Detailed discussion of these units and of correlatives in different zones and sub-zones of the miogeosynclinal belt has been made in a number of papers by the author (Salop, 1964—1967) and others (Chumakov, 1956; Golovenok, 1957; Zhadnova, 1961; A.K. Bobrov, 1964). Each subgroup represents an incomplete transgressive—regressive sedimentary cycle. The Ballaganakh Subgroup is principally composed of terrigenous rocks which show decreasing grain size and better sorting up the section. These rocks are overlain by the Mariinskaya Formation (carbonates). The Kadalikan Subgroup is typified by intercalation of carbonate and terrigenous rocks. Thick terrigenous detrital beds (Dzhemkukan Formation) are present in the basal part of the subgroup. They are locally separated by an unconformity from the underlying rocks and are overlain by shallow-water oncolitic limestones of the Chencha Formation.

The Ballaganakh Subgroup and the major part of the Kadalikan Subgroup (with the exception of the uppermost Zhuya and Chencha Formations) belong to the Middle Neoprotozoic on the basis of numerous stromatolites and microphytolites of the second complex. The Zhuya and Chencha Formations (Zhuravleva, 1964; Dol'nik, 1969) contain stromatolites and microphytolites of the Upper Neoprotozoic (third complex). The boundary of the suberathems is gradational and may be established only by a change in the phytolite complexes at the boundary of the Valyukhta and Zhuya Formations.

The Patom Group unconformably overlies the older Precambrian strata, including the Akitkan Group. Its relationship with the Lower Neoprotozoic Teptorgo Group is more complex. In the external parts of the miogeosynclinal belt of the Lena region where there is widespread evidence of Vyborgian orogeny, it lies with disconformable to unconformable relations on the Teptorgo. In the internal zones of the belt these groups are conformable (Salop, 1974a).

Quartzitic sandstones and conglomerates of the Eocambrian Zherba Formation disconformably overlie the Chencha limestone and define the upper boundary of the Patom Group in the Lena region. The thick Epiprotozoic Bodaibo Subgroup greatly increases the thickness of the Patom section (see below) in the internal part of the miogeosynclinal belt. In the Baikal fold belt, at least in the miogeosynclinal belt, the Neoprotozoic—Epiprotozoic boundary is marked only by a change in sedimentary environment and by initiation of a new sedimentary cycle. The carbonate rocks of the Zhuya and Chencha Formation are overlain by thick terrigenous strata of the Bodaibo Subgroup which were in turn overlain by carbonates. Strong orogeny took place there only at the end of the Epiprotozoic so that the "marker" intrusions, which could help to establish the upper age boundary, are absent.

Miogeosynclinal and closely related subplatform strata of the Middle and Upper Neoprotozoic are extensive in the Baikal region, Mama—Vitim, Delyun—Uran and Lena zones of the Baikal fold belt, and also in the Sayan area and Taymyr fold belts.

In the western part of the Baikal area rocks of this age are represented by the Baikal Group which includes the Goloustnaya (dolomite and quartzite), Uluntuy (slate, siltstone, and dolomite that is locally stromatolitic and oncolitic) and Kachergat (siltstone, sandstone and slate) Formations, which correlate reliably with the Barakun, Valyukhta and Zhuya Formations of the stratotype, on the basis of physical continuity (Salop, 1964—1967). This conclusion is also supported by the presence of various stromatolites and microphytolites of the second complex and in rare cases, of *Minjaria* Kryl., a stromatolite typical of the third complex in the Uluntuy Formation

(Korolyuk, 1966; Khomentovsky et al., 1968; Dol'nik, 1969). The Baikal Group rests unconformably on rocks of the Lower Neoprotozoic Akitkan and Anay Groups and also on older strata. It is unconformably overlain by the Epiprotozoic Ushakovka Formation.

In the Mama—Vitim and Delyun—Uran zones of the Baikal fold belt the Patom Group sections differ from the stratotype in having a more complex sequence and being highly variable from place to place. Certain formations of the stratotype are easily recognized but others are so different that they are given different names. Correlation of these sections with the stratotype can, however, be done with some certainty (Fig.27). A detailed description of these sections and proposed correlation with the stratotype were given by Salop (1964—1967; 1974a).

In the *area around Eastern Sayan* the Zuntei Group of clastic rocks, including the Yermosokha and Ingashin Formations (up to 4,500 m thick), belongs to the lowest Middle Neoprotozoic. It may be correlated with the lower terrigenous formations of the Ballaganakh Subgroup of the stratotype (Mats and Taskin, 1971; Taskin, 1971). The Karagas Group (up to 2,500 m) represents higher parts of the Middle Neoprotozoic and the Upper Neoprotozoic. It includes the Shengulezha, Bogatyr, Ipsit and Techa Formations. This group is lithologically very similar to the Baikal Group of the Western Baikal area, though its lower (Shengulezha) formation of coarse-grained clastics is probably older, and corresponds to the Dzhemkukan Formation of the stratotype. The Bogatyr and Ipsit Formations correlate with the Goloustnaya, Uluntuy and partly with the Kachergat Formations of the Baikal Group, both on the basis of lithology and on the presence of microphytolites of the second complex in the first formation, of the third-complex stromatolites (*Inzeria* Kryl., *Minjaria* Kryl.) in the second one, and by the occurrence of beds of phosphorite. The Techa Formation probably corresponds to the upper part of the Kachergat Group. The red colour of many rocks of the Karagas Group differentiates them from the Baikal Group. The strata of the Karagas Group are transitional in character between miogeosynclinal and platform facies.

The Zuntei and Karagas Groups unconformably overlie different older Precambrian rocks including the Lower Neoprotozoic Kalbasyk (volcanogenic) Group. The Karagas Group is cut by dikes and sills of the Nersa complex which do not penetrate the overlying Epiprotozoic Oselochnaya Group. The K—Ar age of these diabases is 1,170—1,200 m.y. Since the analysed rocks are not altered, the ages may date the basic magmatism that preceded the orogenic cycle at the end of the Neoprotozoic. Some small (undated) intrusions of granite in the Karagas Group were probably intruded during this orogenic cycle.

In the eastern part of the Sayan area, adjoining the Baikal fold belt the Olkha Formation (up to 650 m thick) of clastic and carbonate rocks represents the Middle Neoprotozoic. It rests unconformably on Archean strata.

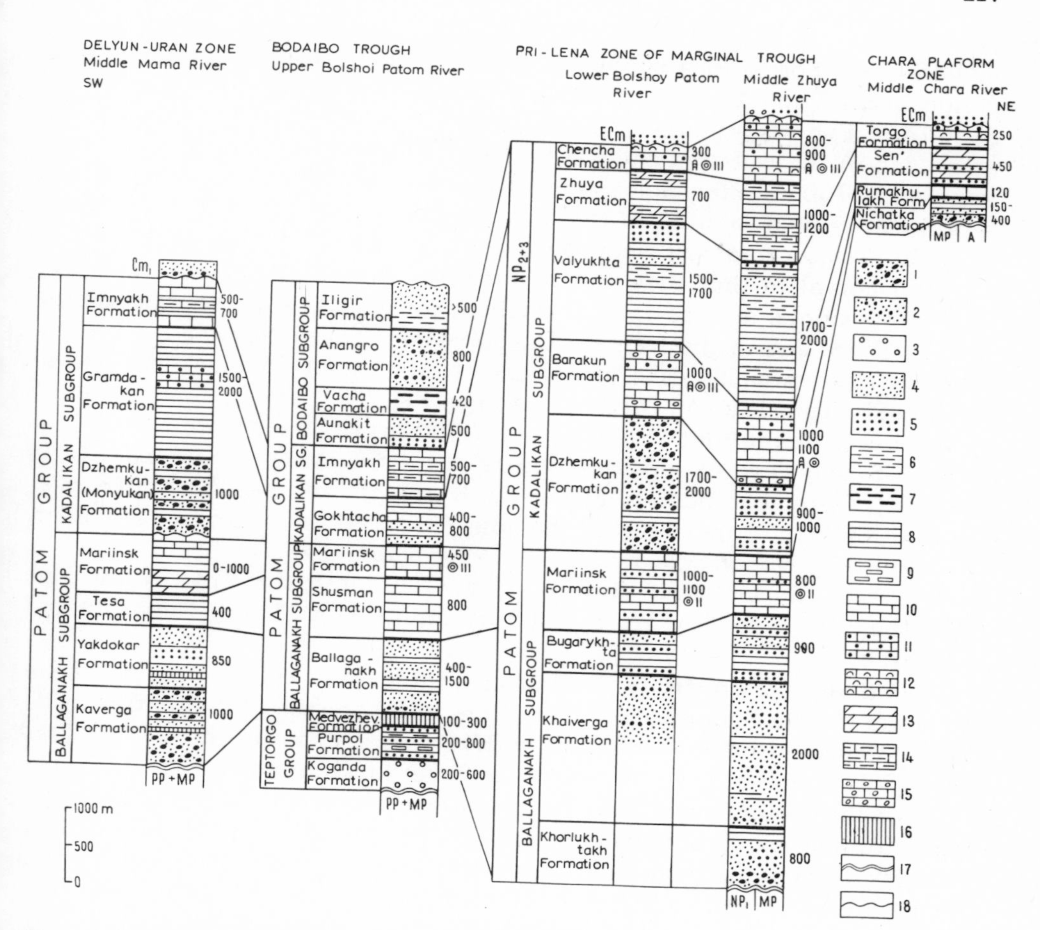

Fig. 27. Correlation of Neoprotozoic sections in different zones of the miogeosynclinal belt and near-platform margin of the Baikal fold zone.
1 = conglomerate; *2* = granule conglomerate (among sandstones); *3* = arkose; *4* = quartzo-feldspathic and lithic sandstone; *5* = quartzite sandstone (quartzite); *6* = siltstone; *7* = bituminous siltstone and quartzite; *8* = slate (phyllite), partly siltstone; *9* = chloritoid schist and quartzite; *10* = limestone; *11* = oncolitic limestone; *12* = stromatolitic limestone; *13* = dolomite; *14* = marl, calcareous slate; *15* = carbonate breccia and conglomerate; *16* = metadiabase (amphibolite); *17* = angular unconformity; *18* = erosional unconformity; *PP* = Paleoprotozoic; *MP* = Mesoprotozoic; *NP* = Neoprotozoic; *EP* = Epiprotozoic; *ECm* = Eocambrian.

This formation is transitional in type between miogeosynclinal and platform facies. V.D. Mats and V.S. Isakova (personal communication, 1970) compared it to the closely similar Uluntuy Formation of the Baikal Group. Oncolitic limestones composed of *Osagia* (second complex) are a special characteristic of both formations. According to V.D. Mats both the Olkha and Uluntuy Formations contain highly phosphatic rocks. The underlying Goloustnaya Formation of the Baikal Group may wedge out in the Sayan region for it shows a marked decrease in thickness from the Baikal to Sayan areas, and in the valley of the Angara River does not exceed 120—150 m (as opposed to 700 m in the Goloustnaya River basin). This decrease in thickness is probably related to uplift of the Archean basement.

In the *Yenisei fold belt* Neoprotozoic miogeosynclinal strata are so extensive and complete that they can be proposed as a Siberian parastratotype. The best sections are the miogeosynclinal rocks in the eastern part of the Yenisei Ridge and in the Irkiney—Chadobets aulacogen. Three groups, separated by stratigraphic breaks, are assigned to the Neoprotozoic: these are the Sukhoy Pit, Tungusik and Oslyansk Groups (Fig.28).

The lower Sukhoy Pit Group (up to 7,000 m) lies unconformably on the Mesoprotozoic Teya Group and consists of the following formations: Korda (Maroka), Gorbylok, Uderey, Pogoryuy, Kartochki and Alad'insk, of which the four lower ones are largely terrigenous rocks and the upper two are mainly carbonates. No organic remains have been found so far, but their age is known from the fact that they lie on the Mesoprotozoic Teya Group and from the fact that they are overlain by strata with Middle Neoprotozoic stromatolites.

The Uderey and Pogoryuy Formations contain phosphatic rocks like those of certain Middle Neoprotozoic formations of Siberia (e.g. the Uluntuy, Ipsit, Olkha, Mana and other formations with organic remains of the second complex). This is one of the reasons for assigning these strata and the overlying strata of the group to the Middle sub-erathem. A reliable correlation of two upper formations (the Kartochki and Alad'insk) with the eugeosynclinal Sosnovsk Formation (which has stromatolites of the second complex) also favours this conclusion (Semikhatov, 1962). Also, glauconite from the Pogoryuy Formation is dated in the 750—1,630 m.y. range by the K—Ar method (ten determinations from different localities). This wide scatter indicates partial loss of argon from glauconite due to later thermal and other processes common in old geosynclinal strata which are somewhat deformed and locally highly metamorphosed. Probably the oldest value (1,630 m.y.) is the only one that approaches the true age of the strata. This date was reported by Kozakov for rocks of the Pogoryuy Formation sampled along the Chapa River, upstream from its tributary, the Almaninan brook. Two lower formations of the Sukhoy Pit Group are accordingly considered to be Lower Neoprotozoic on the basis of these facts and also the break between the Sukhoy Pit Group and the underlying Mesoprotozoic Teya Group to be relatively insignificant.

The Tungusik Group (up to 5,000 m) is composed of carbonates and terrigenous rocks of the Krasnogorsk and Dzhur Formations (sometimes combined into one formation called the Potoskuy) together with the Shuntar Formation which is locally unconformable on the Dzhur Formation and the

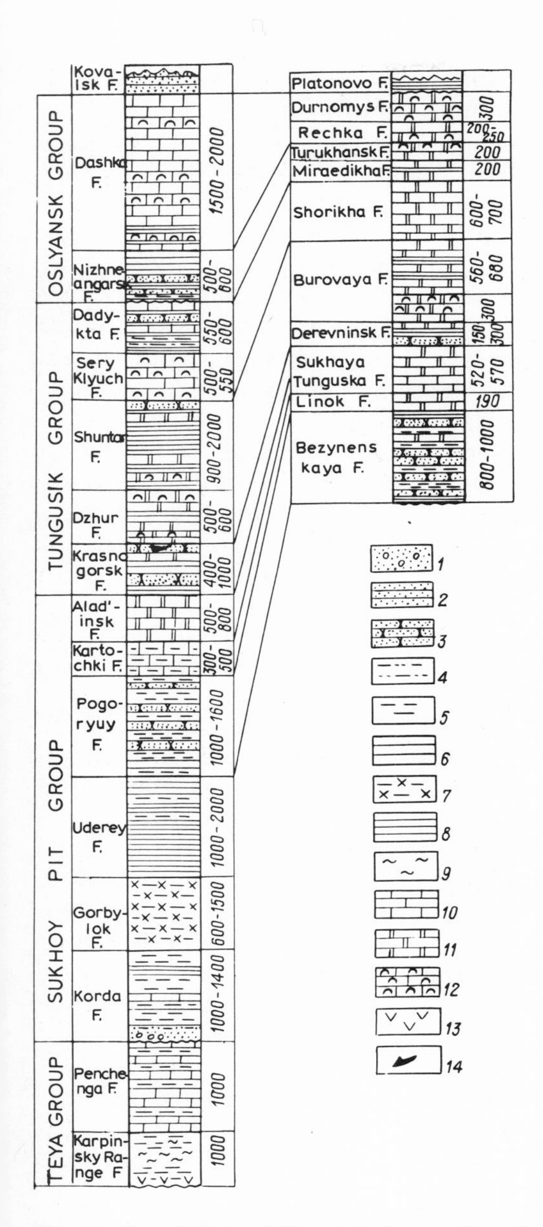

Fig. 28. Correlation of Precambrian sections in the Yenisei Ridge and Turukhansk region (after Semikhatov, see Keller et al., 1967).
1 = conglomerate and granule conglomerate; *2* = sandstone; *3* = quartzite; *4* and *5* = siltstones; *6*, *7* and *8* = different slates and phyllites (*7* = phyllite with magnetite); *9* = schist; *10* = limestone; *11* = dolomite; *12* = stromatolitic limestone; *13* = volcanics; *14* = iron formation. There are no current names for the formations of the Turukhansk region. In Table II the names proposed by Dragunov (1967) are given. The names shown in the figure correspond to Dragunov's names in the following way: Besymenskaya Formation: to Strel'nogorsk Formation; Linok Formation: to Il'yushkino Formation; Derevninsk Formation: to Vtorokamensk Formation; Turukhansk Formation: to Pervoporog Formation.

Sery Klyuch and Dadykta Formations (commonly combined under the name of Kirgitey Formation). The Dzhur and Shuntar Formations contain stromatolites of the second complex (*Conophyton lituus* Masl., *C. cylindricus* Masl., *C. frequens* Masl., *Baicalia ampa* Semikh., *Tungussia nodosa* Semikh., etc.), and the Sery Klyuch Formation contains stromatolites of the second and third complexes (*Conophyton baculus* Kiric., *Baicalia unca* Semikh., *Tungussia nodosa* Semikh., *Minjaria uralica* Kryl., *M. nimbifera* Kryl., *Gymnosolen confragosus* Semikh.). Thus the boundary between the Middle and Upper Neoprotozoic must lie within the Tungusik Group (Semikhatov, 1962).

The uppermost group of this complex — the Oslyansk Group (up to 4,000 m) includes the Nizhneangarsk Formation (shale, sandstone and limestone) and the Dashka Formation of clayey limestone and dolomite. This group belongs entirely to the Upper Neoprotozoic, for both its formations contain stromatolites of the third complex (*Minjaria nimbifera* Semikh., *Gymnosolen confragosus* Semikh.), and even fourth-complex microphytolites appear in the upper part of the Dashka Formation. Thus, the Neoprotozoic complex in the Yenisei Ridge area is also capped by strata bearing phytolites of "Vendian type".

Epiprotozoic post-geosynclinal groups containing microphytolites of the fourth complex and rare medusoid casts rest above the Oslyansk Group, either with a great break or with angular unconformity.

In the Yenisei fold belt, early-kinematic gabbroic intrusions and synkinematic and late-kinematic granitoid intrusions were formed in relation to orogeny that terminated the Neoprotozoic. In different zones granitoid intrusions are combined into different complexes. In the Yenisei Ridge area two Neoprotozoic granitoid complexes are subdivided as follows: the Teya, consisting of leucocratic tourmaline granite, gneissic granite and pegmatite (emplaced in the Sukhoy Pit rocks); and the Tatarsk—Ayakhta, composed of granite, adamellite and diorite, and cutting strata of the Tungusik Group. K—Ar analysis on biotite from granite of the Teya complex yielded values in the 910—1,100 m.y. range (Pb-isotope analysis on zircon and allanite yielded 950 m.y.) and 650—880 m.y. for granitoid of the Tatarsk—Ayakhta complex (Volobuev et al., 1964).

Some workers (Volobuev et al., 1964) think of the Teya and Tatarsk—Ayakhta granitoids as belonging to different tectono-plutonic cycles. This concept is probably erroneous for the breaks between the Neoprotozoic groups are not angular unconformities and the groups appear to belong to a single stratigraphic complex. An angular unconformity is present only between the Neoprotozoic and Epiprotozoic. Possibly, emplacement of the granitoid complexes was related to the orogeny that caused this unconformity. The fact that the intrusions vary in composition is probably due to the fact that they are different phases (the synkinematic — Teya and late-kinetic — Tatarsk—Ayakhta) and may also be related to different depths of emplacement. Discrepancies in the K—Ar ages of the intrusions (a wide scattering of values in the case of granitoids of the Tatarsk—Ayakhta complex) indicate "rejuvenation" due to argon loss. Probably the oldest age values for the Teya granite indicate the time of the Grenville orogeny which terminated the Neoprotozoic Era. Recently Moscow geochronologists concluded that the Grenville orogeny was important in the Yenisei Ridge area (Volobuev et al., 1973). This conclusion was based on new Pb-isotope and Pb-isochron dating of granitoids of the Teya complex. The values reported were of the order of 1,000 m.y.

The various groups of the Yenisei Ridge area may be correlated with the miogeosynclinal rocks of the *Igarka and Turukhansk regions* of the Yenisei fold belt (in the Khantai—Rybninsk uplift of the Siberian platform). These correlations are made on the basis of lithology and organic remains. The correlations were made by a number of workers (Dragunov, 1967; Keller et al., 1967; Kirichenko, 1968). The correlations used are similar to those used here (see Fig.28). Particularly strong similarities exist between the Burovaya and Shuntar Formations, the Vtorokamensk and Dzhur Formations and the Il'yushkino (Linok) red-bed and Kartochki Formations. The Middle—Upper Neoprotozoic boundary lies within the complex, between the Burovaya and Shorikha Formations, as exposed in the Yenisei Ridge. The Shorikha Formation contains stromatolites of the second complex associated with microphytolites of the third complex, and higher in the section there are strata with stromatolites of the third complex. V.I. Dragunov's find of sabelliditids (pogonophores?) represented by *Paleolina* Sokol. (determination by B.S. Sokolov) in the Vtorokamensk (Derevninsk) Formation (which contains stromatolites and microphytolites of the second complex) is of great importance, for it indicates the relatively ancient occurrence of organisms with an outer tubular skeleton.

In the Igarka and Turukhansk areas the lower units of the Neoprotozoic (lowest middle sub-erathem and the whole lower sub-erathem) are not exposed, and Eocambrian strata rest unconformably on the Upper Neoprotozoic, so that the Epiprotozoic is missing. Stratigraphic breaks are absent or rare in Neoprotozoic sequences of these regions (some slight erosional features are present at the base of the Shorikha Formation). This differentiates these strata from the synchronous rocks of the Yenisei Ridge area.

Sedimentary deposits of limited extent in the Siberian platform (the Chadobets uplift) are very similar to the strata of the Yenisei miogeosynclinal parastratotype. These strata are of subplatform type for they formed in the Irkineye—Chadobets aulacogen which branches from the miogeosynclinal belt of the Yenisei system. However, these rocks are discussed here because of their strong similarities to the Yenisei parastratotype. According to Sklyarov (1968) and Zabirov (1966) these strata may be divided into several formations as shown in Table II, column 10. The Terina Formation and part of the Chuktukon Formation are lithologically similar to the Shuntar Formation of the parastratotype. The overlying (locally with a stratigraphic break) Brus, Medvedkovo and Bezymyannaya Formations correspond to the Upper Neoprotozoic Sery Klyuch and Dadykta Formations. This correlation is confirmed by the presence of stromatolites (*Compactocollenia cf. sarmensis* Korol.) in the Terina Formation. These are typical of the Goloustnaya Formation (Middle Neoprotozoic) of the Baikal region, and by the presence of oncolites (*Radiosus limpidus* Z.Zhur.) in the Bezymyannaya Formation. These belong to the third complex — Upper Neoprotozoic.

Correlation of the lower parts of the Neoprotozoic sections in both areas is not certain. Correlations proposed by Blagoveshchenskaya (1959), Kirichenko (1967) and Zabirov (1966) are all different. However, they all agree that most of the strata of the Chadobets Uplift correspond to the Teya and Sukhoy Pit Groups of the Yenisei Ridge. Thus, Zabirov correlates the lowermost Chadobets Formation of the Uplift with the Pechenga Formation of the Teya Group, and the overlying Semenovskaya Formation with the four lower formations of the Sukhoy Pit Group. The overlying Dolchikov Formation is thought to correspond to the Kartochki and Alad'insk Formations, the Chuktukon and Terina to the Potoskuy Formation, and the remaining formations (Brus, Medvedkovo and Bezymyannaya) to the upper part of the Tungusik Group. It is considered more reasonable to correlate the Chuktukon and Terina Formations with the Shuntar Formation (containing stromatolites of the second complex) and the overlying three formations with the Kirgitey Formation of the ridge. Correlation of the Semenovskaya Formation with the middle part of the Sukhoy Pit Group section was based mainly on K—Ar dates on glauconite from these rocks (835—1,290 m.y., in the Semenovskaya Formation). More recent dates provided older ages (1,630 m.y.). These are from glauconite of the Pogoryuy Formation.

Considering all the features of the rocks it seems more realistic to correlate the Chadobets Formation (dolomite) with the similar Alad'insk Formation of the Sukhoy Pit Group, the terrigenous rocks of the Semenovskaya Formation with the Krasnogorsk Formation and the carbonate-rich Dolchikov Formation with the Dzhur Formation of the Tungusik Group. According to this proposal, the older Precambrian formations of the Chadobets uplift would be Middle Neoprotozoic. The Chadobets Formation is unconformably overlain by the Semenovskaya Formation, which has conglomerates at its base (according to R.Ya. Sklyarov). Thus, in the Chadobets uplift area (as in the Yenisei Ridge) there is a stratigraphic unconformity at this level.

In the *Taymyr fold belt* the Neoprotozoic strata are also of transitional character between typical miogeosynclinal and platform assemblages (Sobolevskaya and Mil'shtein, 1961; Golovanov and Zlobin, 1966). These rocks are represented by the terrigenous red-beds of the Stanovsk Formation and the Kolosovka Formation (carbonates). In the eastern part of the Taymyr Peninsula the Neoprotozoic includes the overlying Krasivaya and Fomina Formations which are of mixed terrigenous—carbonate composition. In the

North Land Islands the Fjord Formation closely resembles the Stanovsk Formation (Egiazarov, 1959). These strata rest unconformably on Mesoprotozoic rocks and are in turn overlain unconformably by sedimentary rocks that are considered to be Epiprotozoic (Sovinsk and Canyon Formations). Many stromatolites and microphytolites of the second complex are present in the Kolosovka Formation, so that this formation and probably also the underlying Stanovsk Formation, are Neoprotozoic. Golovanov and Zlobin (1966) recognized various stromatolites and microphytolites of the third (Upper Neoprotozoic) complex in the Krasivaya and Fomina Formations.

Eugeosynclinal-type strata of the Middle and Upper Neoprotozoic are extensively developed in the Baikal, East Sayan and Yenisei fold belts. The best sections are situated in the Katera zone of the eugeosynclinal part of the *Baikal fold belt*. The Katera Group consists of an extremely thick sequence (up to 11,500 m) of volcano-sedimentary rocks and may be chosen as a eugeosynclinal stratotype for Middle Siberia (Salop, 1964—1967).

This group includes the following formations (in upward sequence): (1) Ukolkit (3,000—5,000 m) — metamorphosed sandstones and granule conglomerates which are locally tuffaceous, with phyllite interbeds and acid and intermediate lavas, rare basic volcanics and with a basal conglomerate at the base of the succession; (2) Nyandoni (1,500—3,000 m) — various slates and metasiltstones with interbeds of metasandstone, carbonaceous phyllite and limestone and also with some acid and intermediate volcanics; (3) Barguzin (500—4,000 m) — interbedded limestone, slate and commonly carbonaceous phyllite; and (4) Yanchuy (>1,500 m) — phyllite and porphyroblastic slate. Stromatolites and oncolites of the second complex are reported from limestones of the Barguzin Formation.

Other stratigraphic units of the same eugeosynclinal belt (the Uakit, Selenga and Ikat Groups) correlate with the Katera Group (Salop, 1964—1967). All these groups overlie the Mesoprotozoic rocks with an angular unconformity. The Katera Group rests on the Lower Neoprotozoic Padra Group which has a close similarity to the Akitkan Group. Strata of Eocambrian to Early Cambrian age unconformably overlie the Katera Group (the Kholodnaya and overlying formations, the Turik Formation etc.). The Ikat Group is unconformably overlain by the Epiprotozoic (?) Tochera Formation. Oncolites (mainly *Osagia tenuilamellata* Reitl. and *O. columnata* Reitl.) typical of the second phytolite complex, are reported from carbonate rocks of the middle (and in rare cases of the lower parts) of these groups (Barguzin, Mukhtun, Nerunda and other formations). Stromatolites are also locally reported from these rocks. They are usually poorly preserved (strongly recrystallized) but closely resemble forms of the *Baicalia* group. The upper Ikat (Parenga) Formation of the group of the same name contains stromatolites of the second complex and oncolites of the second and third complexes (Korolyuk et al., 1961). Rocks similar to the Ikat Formation are present in other eugeosynclinal groups of the Baikal fold belt (Yanchuy and Dabata

Formations), so that it may be concluded that only their uppermost units are Upper Neoprotozoic, the major part being older (Middle Neoprotozoic). Correlation of these groups with the miogeosynclinal Patom Group supports this conclusion and is based on lithological comparisons and on data concerning the geological development of the geosynclinal system (Fig. 29).

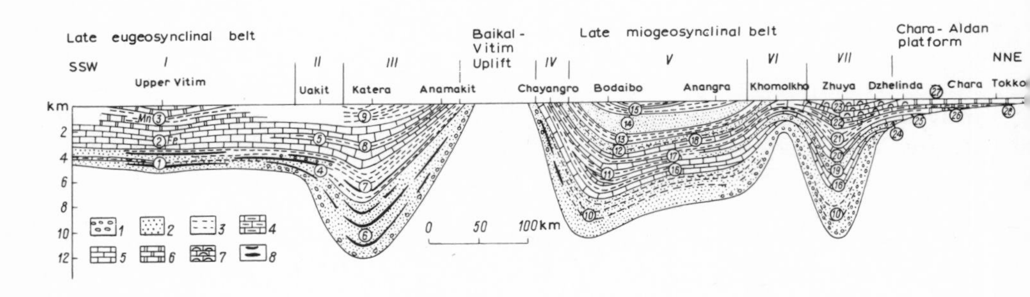

Fig. 29. Schematic section of the Middle and Upper Neoprotozoic and Epiprotozoic strata across the strike of the Baikal geosynclinal system (after Salop, 1964—1967).
1 = conglomerate; *2* = sandstone; *3* = siltstone and slate; *4* = calcareous slate and marl; *5* = limestone; *6* = dolomite; *7* = biogenic limestone; *8* = volcanics. Structural-formational zones: *I* = Ikat; *II* = Uakit; *III* = Katera; *IV* = Delyun—Uran; *V* = Mama—Vitim; *VI* = Zyuya; *VII* = Lena zone of the Baikal—Patom marginal trough. *Formations* (figures in the section): *1* = Suvanikha; *2* = Tilim; *3* = Ikat; *4* = Mukhtun; *5* = Nerunda; *6* = Ukolkit; *7* = Nyandoni; *8* = Barguzin; *9* = Yanchuy; *10* = Ballaganakh (subgroup); *11* = Kadalikan Subgroup, undifferentiated (flysch); *12* = Aunakit; *13* = Vacha; *14* = Anangro; *15* = Iligir; *16* = Mariinsk; *17* = Gokhtacha; *18* = Imnyakh; *19* = Dzhemkukan; *20* = Barakun; *21* = Valyukhta; *22* = Zhuya; *23* = Chencha; *24* = Nichatka; *25* = Kumakhulakh; *26* = Sen'; *27* = Torgo; *28* = Lower Tolba.

These groups correspond to the lower and middle (Ballaganakh and Kadalikan) subgroups of the Patom Group (Salop, 1964—1967). Neoprotozoic eugeosynclinal strata of the Baikal fold belt are typified by a dearth of volcanics, especially in comparison with Meso- and Paleoprotozoic rocks of similar facies in the internal zones of geosynclinal systems. Volcanic rocks are particularly abundant in the lower part of the section, but even there they do not constitute more than 30% of the section. A decrease in volcanic activity in the middle stage of development is a general characteristic of the geosynclinal zones of many mobile belts of different age.

The Bitu—Dzhida and Zun—Murin (Iroy, Temnik) Formations of the *Khamar-Daban Mountains* may correlate with the eugeosynclinal strata of the Baikal area. The Bitu—Dzhida Formation is very similar to the Ukolkit and the Nyandoni Formations, both lithologically and in stratigraphic sequence. The latter formations make up the lower part of the Katera Group. This correlation is also supported by the stratigraphic position of these units in the Precambrian section of Khamar-Daban. The Bitu—Dzhida Formation rests unconformably on the Irkut Formation or on the Paleoprotozoic Khangarul strata and is itself overlain by terrigenous and carbonate rocks of the Zun—Murin Formation. The Zun—Murin Formation is closely similar to the Barguzin Formation with the exception of the uppermost part, which probably corresponds to the Yanchuy Formation of the stratotype. Many workers (Dodin et al., 1968) correlate the Iroy and Temnik Formations in the eastern part of Khamar-Daban with the Zun—Murin Formation. These strata are almost identical to the Burlya and Dabata Formations of the Selenga Group, developed in adjacent areas of the southeastern part of the Baikal region. These strata are essentially the same but were given different names because they were studied by different authors. The Burlya and Dabata Formations are equivalent to the Barguzin and Yanchuy Formations of the stratotype section (Fig.29).

In *Eastern Sayan*, the Kuvay Group of the Uda—Derba zone and the Orlik Group of the Irkut—Oka zone belong to the erathem under discussion. Both groups generally lie unconformably on the Derba and Kitoy Groups (Paleo- and Mesoprotozoic) and are unconformably overlain by Epiprotozoic rocks (the Izyk Group and correlatives). These groups have the same succession and lithology as Neoprotozoic eugeosynclinal groups of the Baikal fold belt. The lower part of these groups (Urman and Oka Formations) consists of clastic rocks with metavolcanic horizons. The middle parts (Mana and Mongosha Formations) are largely composed of carbonates that contain stromatolites and oncolites of the second complex, and the upper parts are composed of metavolcanics, with some interbeds of terrigenous rocks (the Bakhta Formation) or principally of terrigenous rocks with subordinate metavolcanics (the Dibin Formation; Dodin et al., 1968). Comparison is strengthened by the fact that manganese occurs only in the Mana Formation of the Kuvay Group and in the Ikat Formation of the Ikat ridge. In the Ikat Formation there are manganese-bearing siliceous and carbonate-rich slates, whereas in the Mana Formation manganese ores occur in a band characterized by interbedded limestones and quartzites. In the latter case mineralization is probably related to redeposition of manganese from disseminated primary sedimentary stratiform ores.

Correlation of Neoprotozoic rocks in Eastern Sayan and in the internal parts of the Baikal fold belt, on the basis of lithological comparison, is confirmed by occurrences of similar or identical organic remains in the corresponding carbonate rocks. Thus, both the Mongosha Formation of Eastern Sayan and the Tilim Formation of the Ikat ridge contain *Newlandia* remains. If this correlation is accepted, the upper (regressive) parts of the Kuvay and Orlik Groups (the Bakhta and Dibin Formations) which contain no organic remains, are probably Upper Neoprotozoic.

In the *Siberian platform*, Middle and Upper Neoprotozoic platform deposits are widespread. Strata in the northern part of the platform are probably of this age. These include the Yusmastakh Formation of the Billyakh Group in the Anabar uplift, the major (upper) part of the Solooli Group of the Olenek uplift, the Ukhta, Eselekh, Neleger and Sietachan Formations of Northern Kharaulakh. The Yusmastakh Formation of the Anabar uplift may be regarded as the stratotype for the Siberian region. Its lower member (up to 210 m) consists of brown dolomite with stromatolites of the *Anabaria*, *Baicalia*, *Tungussia*, *Conophyton*, *Stratifera* and *Gongilina* groups and also oncolites typical of the second, and partly of the third phytolite complexes. Its upper member (up to 400 m) is composed of grey biogenic dolomite with phytolites of the third and fourth complexes (stromatolites of the *Nucleela* and *Irregularia* groups and oncolites including *Osagia grandis* Z.Zhur., *Radiosus limpidus* Z.Zhur. and others). The change from stromatolites of the third to those of the fourth complex takes place approximately in the middle of the upper member. On the basis of organic remains, the boundary between the Middle and Upper Neoprotozoic is either near the contact between the two members or in the upper part of the lower member.

The Billyakh Subgroup is cut by diabase that does not penetrate into the Epiprotozoic Starorechenskaya Formation. Its K—Ar age (whole rock) is 912 m.y. (L.P. Belyakov, personal communication, 1972). This value is probably somewhat young, but is nevertheless close to the time of basic magmatism at the end of the Neoprotozoic and provides an upper age limit for the Billyakh Subgroup.

Reliable correlation of the Billyakh Subgroup (by formations) with Neoprotozoic units of other regions of Northern Siberia may be carried out on the bases of lithology, rock sequence and contained stromatolites and microphytolites (Table II). The unconformity between the Lower and Middle Neoprotozoic reported in the western part of the Anabar uplift is also strongly expressed in the Udzhin and Olenek uplifts. According to the stromatolites and microphytolites the boundary between the Middle and Upper Neoprotozoic occurs in homogeneous beds that form the upper part of the section (as also is the case in the stratotype). The upper boundary of the Neoprotozoic is marked by a significant unconformity.

In the Olenek uplift several formations of the Solooli Group have been dated by the K—Ar method on glauconite. The age values decrease up-section from 1,380 to 920 m.y., but, as is commonly the case, all of the values seem to be "rejuvenated". The "rejuvenation" is probably greater in older units because higher temperatures are related to depth of burial. The age of 920 m.y. obtained from glauconite in the upper part of the group is probably also "rejuvenated", but is quite close to the time of sedimentation of these rocks. An upper age limit is also established on the basis of isotopic dating of diabase dikes and sills in the Solooli Group and in strata of the same age in the Udzhin uplift. Whole rock K—Ar dating of these igneous rocks yielded values of about 1,100—1,130 m.y. These correspond to the period of basic magmatism that preceded the Grenville orogenic cycle in many parts of the world.

In the *Aldan shield* the platform facies of both Middle and Upper Neoprotozoic are represented by the Maya Group (eastern part of the shield) and the unnamed group, including the Nichatka, Kumakhulakh, Sen' and Torgo Formations (northwestern region). Both of these groups have been studied in detail and contain abundant organic remains.

The Maya Group consists of five formations which, in upward sequence are: (1) Enna — cross-bedded quartz sandstone with interbedded stromatolitic dolomite in the central part. Forms present include *Stratifera* Korol., *Baicalia* Kryl., and *Omachtenia* Nuzh.; (2) Omnya — argillite and siltstone; (3) Malgin — variegated limestone; (4) Tsipanda — grey dolomite with stromatolites, including *Stratifera* Korol., *Baicalia* Kryl., *Tungussia* Semikh. and *Conophyton* Masl.; and (5) Lakhanda — argillite intercalated with carbonates bearing stromatolites of the *Conophyton* Masl., *Baicalia* Kryl. and other groups. In the Uchur—Maya area of the Aldan shield, typical platform deposits of the Maya Group are about 1,200—1,500 m thick and pass laterally into subplatform facies of the same name. These are transitional into a miogeosynclinal sequence which is present in the Yudoma—Maya aulacogen. These rocks are thicker (up to 3,500 m), more highly altered, and more intensely deformed.

The Maya Group overlies Archean gneisses with a sharp angular unconformity or disconformably overlies the Lower Neoprotozoic Uchur Group. Stromatolites of the lower unit (Enna Formation) are typical of the first and second phytolite complexes. Stromatolites and oncolites of the second complex occur in the Tsipanda Formation. In the lower part of the Lakhanda Formation there are stromatolites and microphytolites of the second complex and in the upper part, stromatolites of the third complex occur (Nuzhnov, 1967). Thus, according to the phytolites, the major part of the Maya Group is Middle Neoprotozoic, and only the topmost part is Upper Neoprotozoic. The boundary between the sub-erathems is within the Lakhanda Formation. Glauconite ages of 1,190 and 1,270 m.y. were obtained from the lower part of the group. These ages must have been "rejuvenated". Age determinations from the upper part of the group (Lakhanda Formation) gave values of 890—1,000 m.y. The age of 1,000 m.y. probably approximates the true age of sedimentation (at the end of the Neoprotozoic).

Neoprotozoic strata on the northwestern margin of the Aldan shield are similar to the miogeosynclinal stratotype — the Patom Group of the Lena region in the Baikal fold belt. In the northeastern margin of the Patom Highlands rocks of this group show a strong facies change. There is a sudden decrease in thickness with loss of the Ballaganakh Subgroup and certain formations of the Kadalikan Subgroup, with transition into platform deposits of the Aldan shield (Figs. 27 and 29). Stratigraphic continuity indicates that the Nichatka Formation of the Aldan shield region corresponds to the Dzhemkukan Formation, the Kumakhulakh Formation to the Barakun Formation, the Sen' Formation to both the Valyukhta and Zhuya Formations and the Torgo Formation may be correlated with the Chencha Formation (Zhuravleva et al., 1959; Reitlinger, 1959). These correlations are also supported by organic remains (Zhuravleva, 1964). Thus, platform analogues of the Patom Group include the upper part of the Middle Neoprotozoic and most of the Upper Neoprotozoic.

Outside the Siberian platform and its surrounding fold belts, Lower and Middle Neoprotozoic rocks are present in many Phanerozoic fold belts of Central and Eastern Asia. They are known from Kazakhstan, Middle Asia, Altai, Western Sayan, Tuva, the Far East, the northeastern part of the U.S.S.R. and also from China and Korea. In some of these areas these rocks are extensive and well studied. In many areas, however, their stratigraphic subdivision and position in the succession are poorly known.

In *Kazakhstan*, Lower and Middle Neoprotozoic sequences include the Bozdak and overlying Kokchetav and Koksuy Groups of Ulutau (Zaytsev and Filatova, 1971).

The Bozdak Group (up to 3,000 m) is composed of coarse clastics, including boulder conglomerates, together with slates interbedded, in the lower and middle part, with volcanics and acid and basic tuffs. The upper part of the group is mostly dolomite, quartzite and phyllite with several intraformational breaks. This group lies unconformably on the Paleoprotozoic Karsakpay Group and the Lower Neoprotozoic Maytyube Group and is separated by a stratigraphic break from the overlying Kokchetav Group. The Kokchetav Group consists of monotonous light-coloured quartzite with a maximum thickness of 1,200 m. This homogeneous composition is typical of the group, not only in Ulutau but also in many areas of Kazakhstan, so that it may be used for correlation of widely separated sections in different structural zones.

The Koksuy Group (up to 4,000 m) is largely composed of acid porphyry, ignimbrite and tuff. Basic volcanics (diabase and porphyrite) are present in the middle part of the group. Its relationships with the Kokchetav Group of Ulutau are not certain, but it is assumed to occupy a higher stratigraphic position by analogy with similar rocks in the Precambrian section of the Atasu—Mointa watershed. The K—Ar age of porphyry of the Koksuy Group is 840 m.y. Granite cutting this group gave 650 m.y. by Pb-isotope analysis on zircon.

In the Ulutau region conglomerate, sandstone and tuff of the overlying Akbulak Group are either Neoprotozoic or Epiprotozoic. They underlie an Epiprotozoic succession that includes tillites.

Thus, in Western Kazakhstan (Ulutau) the Middle and Upper Neoprotozoic sequences are mainly composed of subaerial terrigenous and volcanic rocks. However, well-sorted mature marine strata are present in the middle part of the succession (upper part of the Bozdak Group and the Kokchetav Group).

In *Tien-Shan* the Middle and Upper Neoprotozoic strata are highly variable from place to place. In the Kirgiz—Terskey zone the Karadzhilgin, Ortotau and Kenkol Groups are Middle Neoprotozoic and the Terskey and Uchkoshoy

Groups Upper Neoprotozoic. All of these groups are separated by stratigraphic breaks (Kiselev and Korolev, 1964; Korolev, 1971; Krylov, 1971).

The Karadzhilgin Group rests unconformably on Archean rocks. It is made up of shale and siltstone with some limestone interbeds in the middle part. Stromatolites from the limestones have not yet been identified. The Ortotau Group (3,000 m) consists of three sedimentary cycles, each beginning with quartzite, followed by carbonate rocks overlain by shale. The carbonate rocks contain stromatolites of the second complex (*Baicalia* Masl., and *Stratifera* Korol.). The Kenkol Group (1,500—3,000 m) is divided into two formations. The lower one (Kurgantash Formation) is made up of clastic rocks with local andesite flows. The overlying Ov Formation is composed of finely bedded shaly carbonates intercalated with limestones that contain abundant stromatolites (mainly *Baicalia baicalica* Masl. and *B. kirgisica* Kryl.). These forms are also typical of the second phytolite complex.

The Terskey Group begins with a thin (50—100 m) unit of cross-bedded quartzitic sandstone (the Dzheldysuy Formation) containing various shallow-water structures. This is succeeded by a thick (up to 2,400 m) sequence of greenstone (altered basic lava) with shale, limestone and jasper-like interbeds (the Terek Formation). The Uchkoshoy Group (1,700 m) consists of three sedimentary cycles separated by erosional surfaces. The lower part of each cycle is made up of quartzitic sandstone and conglomerate. These are overlain by sandstone or shale and siltstone, followed by limestone. The Uchkoshoy Group is assigned to the Upper Neoprotozoic on the basis of correlation with the Karagain Group which contains stromatolites of the third complex.

In the Karatau—Talass region, the Middle Neoprotozoic is represented by the Ichkeletau Group and the unconformably overlying Uzumakhmat Formation. The unconformably overlying Karagain Group contains stromatolites. It is assigned to the Upper Neoprotozoic (Maksumova, 1967).

The Ichkeletau Group consists of two formations. In ascending sequence these are: the Bakair (700—1,000 m) which consists of marble and slate, with interbeds of cobble conglomerate in the lower part, and the Karaburin (up to 500 m) which is made up of slate and phyllite interbedded with quartz sandstones and carbonate rocks. These groups are in fault contact with Paleoprotozoic (?) rocks. Organic remains are not known so that its age is not well established. It may be Lower Neoprotozoic, or even Mesoprotozoic. The Uzunakhmat Formation (200 m) is composed of lithic sandstone, phyllitic slate and marble. It overlies the Karaburin Formation and has conglomerate at its base. Strong resemblance to the Kenkol Group of the Kirgiz—Terskey region suggests that it may be Middle Neoprotozoic (the Kenkol Group contains stromatolites of the second complex). The lowest unit of the Karagain Group is the Sarydzhon Formation (700 m) which is composed of slate and clayey-carbonate rocks. It is conformably overlain by the Chatkaragay Formation (900 m) which is typified by the presence of flysch-like interbedded sandstones, silty and clayey limestones, and marls. The carbonate rocks contain stromatolites of the third complex (including *Minjaria*, *Inzeria* and *Gymnosolen*). This formation is unconformably overlain by the Malokaroy Group, which probably belongs to the lower part of the Epiprotozoic.

In the Chatkal—Naryn zone of Tien-Shan an unnamed succession of dolomite, slate, conglomerate and basic volcanics (up to 1,700 m), rests with sharp unconformity on metamorphic rocks known as the Kassan Group (Paleoprotozoic?), and may be correlated with the Ichkeletau Group. The Bolshoy Naryn Group of that region is probably Upper Riphean. It is up to 3,000 m thick and largely composed of schistose acid volcanics overlain by phyllites and calcareous and dolomitic marbles. The Epiprotozoic Dzhetym Formation (including tillite) overlies it with angular unconformity.

In contrast to the Middle and Upper Neoprotozoic strata of Western Kazakhstan which are of subplatform (taphrogenic) and platform type, rocks of the same age, throughout the Tien-Shan region, are of miogeosynclinal facies.

In the *Altai—Sayan fold belt* the Neoprotozoic strata are, in some cases, rather arbitrarily subdivided. In the Altai region the Terektin Range Group is probably Neoprotozoic. It is up to 5,500 m thick and is composed of slate and metasandstone with bands of marble and quartzite. The base of the group is not seen but it is unconformably overlain by Cambrian sedimentary and volcanic rocks. K—Ar dating of granite cutting rocks of this group gave "rejuvenated" values (560—620 m.y.). In Western Sayan the Terektin Range Group may be correlated with the thick Dzhebash Group (up to 10,000 m) of para- and orthoschist, quartzite and marble. Clasts of these rock types are present in Lower Cambrian conglomerates. K—Ar dating of micas from metamorphic schists of the group usually give "rejuvenated" values related to Paleozoic thermal events, but some ages up to 950—1,000 m.y. have been obtained. These probably reflect the time of the Grenville orogeny terminating the Neoprotozoic Era.

In Western Tuva the Dzhebash Group (or at least its upper part) may be correlated with the Akkol Formation which includes oncolitic marble. In Eastern Tuva the Aylyg, Kharal and Okhem Formations, with a total thickness of up to 11,000 m, are probably of the same age. The Aylyg Formation lies, without obvious break, on metamorphic rocks that are comparable to the Derba Group (Paleo- to Mesoprotozoic) of Eastern Sayan. It consists of limestone with interbedded slate. The limestone contains *Osagia*. The overlying Kharal Formation is made up of various ortho- and paraschists including calcareous bands and metasandstones which grade upwards into calcareous slate and sandy limestone of the Okhem Formation. This unit is unconformably overlain by Lower Cambrian strata. Part of this thick volcano-sedimentary complex in the Tuva region may be Epiprotozoic. If this is so, then the Precambrian sections of Eastern Tuva are typified by the absence of large breaks between erathems. Continuous Precambrian sections are typical of many regions of adjacent Eastern Sayan. The Eastern Tuva rocks are of eugeosynclinal type but have subordinate volcanics. In this respect these rocks are similar to eugeosynclinal complexes of Eastern Sayan and to the internal zone of the Baikal fold area.

In the *Soviet Far East* the Middle and Upper Neoprotozoic strata can be arbitrarily subdivided in the Khankai Massif and also in the Bureya and Khingan—Bureya regions (Shatalov, 1968; Bersenev, 1969). In the Khankai Massif of the Lesozavodskaya region these strata are represented (in ascending order) by the Kabarga Formation which is composed of chlorite—sericite slate and phyllite interbedded with metasandstone (1,100 m), followed by limestones and dolomites (80 m), sericitic carbonaceous slate and chert (up to 600 m), and capped by limestones with slaty interbeds (> 360 m). In the Voznesensk zone these strata were formerly correlated with the Dal'zavodsk, Pervomaysk, Berezyansk, Novoyaroslavsk, Vokushino and Kovalenkovo Formations which contain microphytolites of the second complex. However, these strata are not lithologically very similar and archaeocyathids and trilobites have been found in the Pervomaysk Formation so that the Voznesensk zone strata are now assigned to the Lower Cambrian. The possibility remains, however, that Cambrian strata are tectonically emplaced among Neoprotozoic units, for exposure is poor, and the tectonic setting is complex. On the other hand, Lower Cambrian rocks of the Okhotsk coast appear to contain microphytolites of the second complex (Stepanov and Shkol'nik, 1972).

In the Bureya anticlinorium three formations belong to the Neoprotozoic. In ascending sequence these are: (1) Tokur — sandstone with interbeds of slate and limestone; (2) Ekimchan — rhythmically bedded phyllites and slate, siltstone, sandstone; and (3) Amnus — rhythmically bedded slate and sandstone with lenses of granule conglomerate and sedimentary breccias. The total thickness is 4,600 m. In the Khingan—Bureya region their stratigraphic analogues are the Igincha Formation, which is made up of sandstone with slate and siltstone interbeds, and the overlying Murandava Formation which is composed of dolomite, limestone and slate.

In the Far East all these complexes lie unconformably on Mesoprotozoic metamorphic rocks (see Chapter 6) and are unconformably overlain by Lower Cambrian strata. In the Khingan—Bureya region they are also overlain by the ferruginous rocks of the Eocambrian. These strata are of miogeosynclinal aspect.

In the *Northeastern U.S.S.R.* Middle and Upper Neoprotozoic strata are reliably identified in an anticlinorium in the Kolyma region and in the Okhotsk and Omolon Massifs. In other areas they are arbitrarily subdivided (e.g. Koryaksk Ridge, Tas—Khayakhtakh, Omulev Mountains). In the Kolyma anticlinorium (near the middle tributary of the Kolyma River) the following succession has been determined (V.P. Rabotnov et al., 1970): the Tyyuryuyakh Formation (1,000 m) made up of quartzitic sandstone with conglomerate at the base, overlying the Paleoprotozoic (?) gneissose and schistose rocks, the Oroyek Formation (600 m) composed of slate and phyllite with quartzite interbeds, and the Chebukulakh Formation (750 m) which is limestone, dolomite and calcareous slate. The latter is unconformably overlain by Epiprotozoic rocks containing tillite. Stromatolites of the second complex are present at various levels in the thick lower part of the Chebukulakh Formation (*Conophyton cylindricus* Masl., *C. metula* Kiric., *Tungussia* sp., *Baicalia* sp.) and in its uppermost bed, phytolites of the third complex are present (stromatolites: *Inzeria tjomusi* Kryl., microphytolites: *Osagia grandis* Z.Zhur., *Vesicularites raabenae* Zabr., etc.). These rocks, both on the basis of lithology and paleontology, are similar to the Maya Group of the Yudoma—Maya aulacogen (V.P. Rabotnov et al., 1970).

Rocks of the Okhotsk and Omolon Massifs are also similar. In the Okhotsk Massif, at the base of the section, conglomerates and quartzites lie unconformably on Archean gneisses. They pass successively upwards into shale, quartzitic sandstone with stromatolitic limestone interbeds containing the following forms: *Conophyton cylindricus* Masl., and *Collenia* sp., siltstone interbedded with slate, and finally quartzite. The total thickness is 2,600 m. In the Omolon Massif red granule conglomerates in the lower part of the section rest on Archean gneiss, and grade upwards into red and grey sandstone (in some places glauconite-bearing) with various shallow-water structures. These are overlain by quartz sandstones intercalated with carbonaceous shale, followed by siltstone with desiccation cracks. A thick sequence (up to 900 m) overlies these terrigenous strata (up to 350 m) and is characterized by intercalated limestones and sandstones. Stromatolites are present in the limestone (*Conophyton baculus* Kiric. in the lower part and *Gymnosolen ramsayi* Steinm. in the upper half). These forms suggest a Middle and Late Neoprotozoic age for these terrigenous and carbonate strata.

In *China* the Middle and Upper Neoprotozoic deposits of the Sinian complex are distributed in the China—Korean platform. The best sections are situated in the Peking area. Kao Chen-hsi and others (1962) divided the Sinian complex into Lower, Middle and Upper parts.

The Lower Sinian includes the following formations (in upward sequence): (1) Chancheng (650 m) — quartzite, unconformably overlying older gneiss; (2) Chuanlingou (480 m) — dark bituminous shale with sandstone interbeds; (3) Tahunui (60—400 m) — light-coloured cross-bedded quartzite and feldspathic quartzite with sheets of amygdaloidal porphyry (andesite) lava flows in the upper part of the formation; and (4) Kaoyuchuang (up to 1,300 m) — limestone interbedded with quartzite and slate. The limestone contains the stromatolite *Collenia*. The following formations are Middle Sinian: (5) Yangchuang (400 m) — red shale and argillite overlying the Kaoyuchuang limestone with an erosional contact; (6) Umishan (up to 1,500 m) — siliceous limestone; (7) Hunshuichuang (200 m) — black shale with quartzite interbeds in the upper part; and (8) Tiehling (350 m) — limestone, commonly stromatolitic. The Upper Sinian which has an erosional contact with the underlying Tiehling limestone is composed of the following formations: (9) Hsiamaling (360 m) — carbonaceous shale which has oolitic hematite interbeds in the lower part and passes upwards into cross-bedded coarse-grained sandstone and quartzite; and (10) Chinehryü (150 m) — finely bedded siliceous limestone with a basal unit of variegated shale. This formation is unconformably overlain by the red slate of the Manto Formation which contains the trilobite *Redlichia chinensis* Grab. which is characteristic of the upper part of the Lower Cambrian.

The Sinian formations were formerly regarded as the youngest Precambrian strata. However, K—Ar dates on glauconite from the upper formation (Chinehryü) of the Sinian and from the overlying Lower Cambrian Manto Formation provided a wide range of ages: 870—890 m.y. for the Chinehryü Formation and about 500 m.y. for the Manto Formation. Thus, there appeared to be a large gap before accumulation of the Cambrian strata. It has been suggested that the glauconite age from the Chinehryü Formation is too great due to formation of glauconite from clastic micas. However, this is not likely because formation of glauconite causes complete destruction of the crystal lattice of the micas with loss of the contained argon. Some Chinese authors reported *Redlichia* remains in the Chinehryü Formation, but in one case the fossiliferous beds appear to be part of the unconformably overlying Manto Formation, and in another case the fossils attributed to the Chinehryü Formation appear to be dubious. Stromatolites from the Sinian complex have not as yet been studied by modern methods. On the basis of descriptions and illustrations of stromatolites from Chinese publications M.A. Semikhatov considers columnar stromatolites from the Tiehling Formation to resemble some forms typical of the third phytolite complex. The upper formation of the Middle Sinian may therefore be Upper Neoprotozoic. In Southern China stratigraphic equivalents of the Sinian probably occur in a metamorphic complex that is cut by granite dated at 900—1,000 m.y. (by the K—Ar method). Younger platform strata of that region include Epiprotozoic sedimentary rocks including tillite.

Correlatives of the Sinian complex are widely distributed in the eastern part of the Sino—Korean platform in the Phennam trough of *North Korea*. These rocks are subdivided into the Sanvon Group with two formations, which are, in upward sequence: Chikhen (500—1,700 m) — terrigenous rocks including quartzite and micaceous and mica—quartz slate with interbeds of marble in its upper part; and Sadan'u (up to 2,000 m) consisting of dolomite and limestone with layers of slate, phyllite, and local quartzite. The Sadan'u Formation is transgressively overlain by the Epiprotozoic Kuhen Group (including tillite) which is in turn unconformably overlain by the Lower Cambrian Yandok Group with trilobites (Masaitis, 1964). I.N. Krylov and M.A. Semikhatov identified stromatolites of the *Tungussia* Semikh. and *Baicalia* Kryl. groups in the lower part of the Sadan'u Formation. In the upper part of the formation stromatolites of the *Inzeria* Kryl. and *Jurusania* Kryl. groups were identified. The former are typical of the second phytolite complex and the latter, of the third one. The upper carbonate part of the Sanvon Group thus appears to be Middle and Upper Neoprotozoic. Probably the lower terrigenous part of the group is not older than Middle Neoprotozoic, but the possibility that it is Lower Neoprotozoic cannot be ruled out. In one locality in the southern part of the Phennam trough there is another thick terrigenous—carbonate sequence (more than 2,000 m) which was given the name Kore Group by V.V. Russ. It is situated above the Sanvon Group and probably beneath the Kuhen Group. The Mukchen Formation (sandstone, argillite and clayey limestone) and the overlying Meraksan Formation (limestone) as established by Korean geologists, may be correlatives. The group as a whole is assigned to the Upper Neoprotozoic.

Terrigenous rocks of the Chikhen Formation may be correlated with Lower Sinian terrigenous rocks, and carbonates of the Sadan'u Formation and the Kore Group may correspond to the Middle and Upper Sinian of China. In this case the Sinian complex also corresponds to the Middle and Upper Neoprotozoic. In Northern China Cambrian rocks lie directly on Upper Neoprotozoic rocks. The Upper Sinian rocks are not correlative with the Epiprotozoic tillite-bearing sequences of Korea because there are great differences in lithology and succession, and because a glauconite age from the Chinehryü Formation contradicts its assignation to the Eocambrian.

In Southern Asia, Middle and Upper Neoprotozoic rocks are present in several areas of the eastern and northern parts of the *Hindustan Peninsula*. The Cuddapah Supergroup in Madras state outcrops in an arcuate band (convex to the west). It is probably Middle Neoprotozoic. It is thick (up to 5,500 m) and consists of four cycles (groups) separated by small stratigraphic breaks. Interbedded quartzite and conglomerate are present at the base, overlain by slate (or phyllite), and locally dolomite. The quartzite—sandstone facies is dominant in the uppermost group. Some diabase sills are present among the sedimentary rocks together with basic lavas. The rocks are slightly deformed and have undergone low-grade metamorphism. Folding and metamorphism are strongly developed only on the concave side of the arc. Many of the rocks are red-coloured and cross-bedded and have suncracks and other features typical of a shallow water environment. Stromatolites are common in dolomite.

The supergroup overlies Archean gneisses and charnockites which were deformed and metamorphosed in Paleoprotozoic times and partly in the Mesoprotozoic (K—Ar dating reveals that the latest thermal events happened about 1,600—2,700 m.y. ago). The sedimentation age of the supergroup is 1,450 m.y. on the basis of a glauconite date from the sandstone (Vinogradov and Tugarinov, 1964). Rb—Sr isochron dating of weakly altered lava from the lowest part of the supergroup yielded 1,580 m.y. (Crawford and Compston, 1973). The upper age boundary of the Cuddapah strata is based on

K—Ar dating of intrusive diabase with an age of 1,160 m.y. (Aswathanarayana, 1964). The Cuddapah Supergroup is unconformably overlain by the Upper Neoprotozoic (?) Carnool Group.

The Pakhal Group which is present in the Pranhita—Godovary (Andhra Pradesh) tectonic depression is very similar to the Cuddapah Group. Abundant stromatolites are present in carbonate rocks of this group (A. Chaudhuri, 1970). The Sullaway Group rests unconformably on the Pakhal Group. It is equivalent to the Carnool Group.

The Kaimur Group, the second lowest group of the Vindhyan Supergroup in Northern India, may correspond to the Cuddapah Supergroup or possibly only the upper (quartzitic) Kistna Group. The Kaimur Group, which has a thickness of about 300 m, is made up of quartzitic sandstones interbedded with slate and porcellanite. It unconformably overlies the Semri Group (Lower Vindhyan) or various older formations. The K—Ar age of glauconite from rocks of this group is 910—940 m.y. (Tugarinov et al., 1965a)., but this value may be strongly "rejuvenated". The Kaimur sandstone is cut by a pipe-like body of kimberlite. Phlogopite from the kimberlite yielded an age of 1,140 m.y. by Rb—Sr analysis (Crawford and Compston, 1970). However, interpretation of this value is complicated by the fact that the kimberlite is composed of rock and mineral fragments of different age and genesis. Thus, the true age of the group is not certain and it may possibly be Upper Neoprotozoic.

The Upper Neoprotozoic in India is more widely represented. The above-mentioned Carnool and Sullaway Groups and also the Chattisgarh Supergroup and the two upper groups of the Vindhyan Supergroup are probably Upper Neoprotozoic.

The Carnool Group is rather thin (about 500 m) and includes two sedimentary cycles. The lower one unconformably overlies the Cuddapah Supergroup. It starts with a horizon of diamond-bearing conglomerate (the Banganapalli), which passes upwards into a unit of stromatolitic limestone with beds of shale and calcareous slate. The upper cycle consists of a basal quartzite, overlain by bituminous limestone, followed by red sandstone interbedded with limestone. Salt casts, desiccation cracks and ripple marks are present in the red beds.

The Chattisgarh Supergroup may be correlated in detail with the Carnool Group. According to Schnitzer (1969a, b) it is composed of two groups: the Chandarpur (up to 500 m) and the overlying Raipur (up to 1,200 m). The first is largely comprised of sandstone and granule and coarser conglomerates. The second group is composed of limestone and dolomite (commonly stromatolitic) intercalated with red calcic sandstones. The red quartzites, which display cross-beds and ripple marks, occur in the upper part of the group. The stromatolites locally form reefs up to 1 km^2.

The Rewa and Bhander Groups of the Vindhyan Supergroup (Upper Vindhyan) may be correlative with the above-mentioned units. The Rewa Group (up to 3,000 m) is principally composed of coarse, cross-bedded sandstone with diamond-bearing conglomerate at the base. Slate is of subordinate abundance. The most likely source of the diamonds is in the kimberlite cutting the underlying Kaimur Group. The Bhander Group (up to 1,000 m) begins with diamond-bearing conglomerate (in some places they are also present in the uppermost part of the Rewa Group), followed by the Ganurg slate, the Bhander limestone (or the Nagol limestone), sandstones ("the Bhander lower sandstone"), the Sirbu slate, and the topmost unit is sandstone ("the Bhander upper sandstone"). The limestone contains stromatolites which Valdiya (1969) identified as *Collenia baicalica* Masl. (= *Baicalia baicalica* Kryl.), a form typical of the second (Middle Riphean) complex. However, the validity of these determinations is questioned on the basis of examination of the published material. Some of the forms illustrated resemble stromatolites of the third (Upper Riphean) complex. The Bhander Group includes worm trails (Verma and Prasadk, 1968) and disc-like problematica such as *Fermoria* (B.F. Howell, 1956) which closely resemble *Chuaria*, reported from Upper Neoprotozoic rocks of Sweden (the Visingso Group) and North America (the Chuar Group).

Correlation of the upper groups of the Vindhyan Supergroup with the Carnool Group is based on the presence of diamond-bearing conglomerate, on the similar position in the upper part of the Precambrian section and on the presence of similar rock types (particularly the stromatolitic limestone or dolomite groups).

North America

Middle and Upper Neoprotozoic strata are fully developed and widely distributed in North America. Rocks of both platform (including subplatform) and miogeosynclinal facies are present. No truly eugeosynclinal sequences of this age are known from North America. In the *Canadian—Greenland* shield the Hornby Bay Group may be regarded as a stratotype for the Middle Neoprotozoic, and the Coppermine River Group for the Upper Neoprotozoic. Both of these groups occur in the Bear Lake province. The lower part of the Hornby Bay Group (2,650 m) is made up of sandstone and conglomerate, lying with angular unconformity on Mesoprotozoic rocks (the Echo Bay and Cameron Bay Groups). These are followed by dolomite (commonly stromatolitic), then siltstone and sandstone. The group is capped by red sandstones and siltstones which show evidence of a small break between their deposition and sedimentation of the underlying rocks (Donaldson, 1968; Fraser et al., 1970). These rocks are unconformably overlain by the Dismal Lakes Group (1,200 m) which is made up of a lower unit (430 m) of sandstone, siltstone and shale, overlain by a thick dolomite sequence, containing stromatolites. The Dismal Lakes Group is separated by an unconformity from the underlying Hornby Bay Group. Intrusion of the Muskox complex took place during the period of development of the unconformity. The overlying Coppermine River Group (4,700 m) is largely composed of thick basaltic lava flows with red sandstone in the upper part and with development of dolomite and sandstone in the lower part. The age of basalt is 1,200—1,250 m.y. according to K—Ar and Rb—Sr isochron methods (W.R.A. Baragar, 1969), and the age of the Muskox is 1,200 m.y. by K—Ar analysis (W.R.A. Baragar and Donaldson, 1973). In adjacent areas contemporaneous diabase dikes were intruded. The rocks of these groups have a dip of about 5—10° to the north. The Coppermine River Group is cut by diabase dikes with an age of 1,200 (1,150) m.y. (K—Ar analysis) and is unconformably overlain by sandstone, shale and dolomite of the Rae Group.

In the Bathurst Inlet area, rocks corresponding to the Hornby Bay Group include the following formations: Tinney Cove (sandstone and conglomerate), Parry Bay (dolomite with stromatolites) and Kanuyak (red sandstone and dolomite). These formations correlate well with the lower and middle parts of the stratotype (Table III). The overlying basaltic volcanics correlate with the Coppermine River Group. The stratigraphic succession of all these formations is very similar to that of the stratotype: clastics of the Tinney Cove Formation lie unconformably on the Mesoprotozoic Goulburn Group and the basalt is unconformably overlain by the Epiprotozoic sandstone and slate (Fraser and Tremblay, 1969).

In the Churchill province there are two Middle Neoprotozoic groups: the Athabasca developed around Lake Athabasca, and the Upper Dubawnt Group which is more widely distributed in more northerly parts of the province (Eckelmann and Kulp, 1956; Donaldson, 1965, 1967 (1968), Fraser et al., 1970). Both groups are similar in composition: the basal part is quartzitic sandstone, conglomerate and siltstone (red in the Lower Dubawnt) which unconformably overlie the Lower Neoprotozoic Martin Lake and Lower Dubawnt Groups. Stromatolitic dolomite is present higher up the section. The Upper Dubawnt Group is cut by diabase of 1,415 (1,360) m.y. old. It is unconformably overlain by basaltic lavas, comparable to those of the Upper Neoprotozoic Coppermine River Group. The basalt is cut by diabase dated at 1,150 (1,100) m.y. Diabase dikes of similar age (1,200—1,250 m.y.) cut the Athabasca Group. In the Lake Athabasca region the same age (1,100 m.y.) is characteristic of the second generation of hydrothermal uranium mineralization. These events may be related to thermal processes at the end of the Neoprotozoic (Koeppel, 1968).

In the Superior province the Middle and Upper Neoprotozoic are represented by the Lower Keweenawan Group (Lower and Middle Keweenawan). This group is thick and extensive in the area of Lake Superior where its rocks unconformably overlie older Precambrian rocks, including the Mesoprotozoic Animikie Supergroup (Grout et al., 1951; Hamblin, 1961; etc.). The lower part of the group is principally composed of conglomerate, sandstone and slate with some stromatolitic dolomite (the Pakwage—Sibley—Nipigon Formation). These rocks are probably Middle Neoprotozoic. The major upper part is made up mainly of thick basic volcanics (with some acidic rocks), together with subordinate sedimentary units (Osler—North Shore—Portage Lake). This sequence is comparable with the Upper Neoprotozoic Coppermine River Group. The stromatolites of the Sibley Formation were described in detail by Hofmann (1969), but the methods he used differ from those of Soviet workers, so that comparison with Neoprotozoic stromatolites of Eurasia is difficult. Acid volcanics of the Portage Lake Formation yielded an age of 1,100 m.y. by the Rb—Sr isochron method (S. Chaudhuri and Faure, 1967). This value may be too young, for the group is cut by diabase dikes and the large intrusion of the Duluth gabbro with an age of 1,180—1,250 m.y. by both K—Ar and Rb—Sr methods (Faure, 1964). A small body called the Melen granite cuts the volcanics. It is dated by the K—Ar method at 1,150 (1,100) m.y. (Goldich et al., 1961).

Middle and Upper Neoprotozoic platform deposits are also reported in the Canadian Arctic. The Darnley Group is probably Middle Neoprotozoic. It occurs around Darnley Bay, near the northwestern margin of the Canadian—Greenland shield (Yorath et al., 1969). Sedimentary rocks in Boothia Peninsula and Somerset Island may also belong to this sub-erathem (Blackadar, 1967b). These rocks are of similar composition to rocks of the Hornby Bay Group, the stratotype of the sub-erathem (rows 39,40, Table III). Stromatolites are present in carbonates in the upper part of these sequences. Thick deposits of the Ikvalulik—Uluksan Group are probably also Middle and Upper Neoprotozoic. These rocks are in the northwestern part of Baffin Island (Blackadar, 1970). The base of the group is made up of andesite and basalt, but the rest of the section is typified by interbedded terrigenous, partly red-coloured, quartzites, sandstones, argillites etc., and dolomite or limestone (row 42, Table III).

All these rocks in the Arctic region of Canada lie subhorizontally on old Archean (?) crystalline basement rocks. The Ikvalulik—Uluksan Group is cut by diabase dikes dated at 1,190 (1,140) m.y. Thus, if these dates are reliable the group is older than the Epiprotozoic. Basalt from the base of the group yielded an age of 940 (900) m.y. by the K—Ar method, but this age is probably too young for it is less than the age of the diabase dikes that cut the whole group (Blackadar, 1970).

In Greenland Middle and Upper Neoprotozoic deposits include the Gardar Group (Eriksfjord) which is developed in the Nanortalik area in the southern part of the island (V. Poulsen, 1964; Berthelsen and Noe-Nygaard, 1965; Bridgwater, 1965). These rocks lie subhorizontally on the Mesoprotozoic Ketilidian complex and on the intrusive rocks of the Lower Neoprotozoic Sanerutian complex, including the rapakivi granite with an age of 1,750 m.y. They are cut by an alkaline syenite which is 1,245 m.y. old according to Rb—Sr isochron analysis (Van Breemen and Upton, 1972) and 1,150 (1,100) m.y. old by the K—Ar method. These results give an accurate estimate of the age of the group. The group is typified by intercalation of sandstone and quartzite (commonly red) with basaltic lavas (row 51, Table III). In the lower part of the group there is a local stratigraphic break similar to that in the Ikvalulik—Uluksan Group. This probably indicates some tectonic movements at the boundary of the Middle and Upper Neoprotozoic.

Neoprotozoic platform deposits (probably Upper Neoprotozoic) are also well represented in Northern Greenland. In the Thule area (Cape York) the Lower Thule Group is of this age. It is composed of the Wulstenholm quartzite—sandstone (500 m) and an overlying shale with interbedded dolomites and sandstones of the Dundas Formation (650 m). The arkosic conglomerate at the base of the group lies unconformably on metamorphic basement rocks which are cut by diabase dikes dated (K—Ar method) at 1,623 (1,563) m.y. old. Sedimentary rocks of the group are also cut by sills of younger diabase dated at 1,220 (1,172) m.y. by the same method (Dawes et al., 1973). These dates define the age limits of the group. The red beds of the Epiprotozoic Upper Thule Group unconformably overlie the Dundas Formation.

The Lower Thule Group may be correlated with platform deposits of Inglefield Land. At the base of this succession red sandstone and conglomerate of the Renseler Bay lie unconformably on the crystalline basement and are overlain by stromatolitic Cape Leiper dolomite and the Cape Ingersoll dolomite. Diabase sills in the Renseler Bay sandstone are 1,240 (1,190) m.y. old according to K—Ar analysis (Dawes et al., 1973). The Cape Ingersoll dolomite is unconformably overlain by the Wulff River Formation. The formation contains fossils (brachiopods, gastropods and trilobites) typical of the upper part of the Lower Cambrian (J.W. Cowie, 1961).

In Northeastern Greenland, in the south part of Peary Land the Midsummer sandstone corresponds to the Wulstenholm and Renseler Bay Formations. It is cut by diabase dikes with an age of 1,045 (1,000) m.y. (Jepsen, 1971). In the Danmarks Fjord area the Nordsemandale red sandstone is probably correlative with these formations. In both regions the sandstone is unconformably overlain by Epiprotozoic strata with local tillite.

Within the limits of the *North American plate* the Neoprotozoic platform deposits are reported in the area of the Grand Canyon of the Colorado River (North Arizona). These strata include the Unkar Group which lies almost horizontally on the Mesoprotozoic (?) Vishnu metamorphic complex, the overlying Nankoweap and Chuar Groups which are separated by a stratigraphic break and are also unconformable on the underlying rocks. All of these rocks form part of the Grand Canyon Supergroup.

The Unkar Group consists of a thin (60 m) stromatolitic dolomite overlain by red sandstone and argillite (1,800 m). Basic lavas are interbedded with the sedimentary rocks; the thickest is the Cardenas basalt at the top of the group. It lies on an eroded sandstone surface. K—Ar dating of basalt from the upper flow gave 845 m.y. (Ford and Breed, 1972) but this age is probably too young. Schopf et al. (1973) assume that the true age of the lava is of the order of 1,200 m.y.

The Nankoweap Group (100—150 m) is mostly red cross-bedded sandstone with a few interbeds of argillite and stromatolitic limestone. On the bedding surface of a sandstone unit a problematic trace fossil was reported. Some people regard it as the imprint of a medusoid (Van Gundy, 1951). Others consider it to be inorganic (Cloud, 1968). According to data by Ford and Breed (Ford and Breed, 1972) the Chuar Group is divisible into three formations. The lowest one — Galeros (1,320 m) consists of massive grey dolomite with subordinate limestone, slate and sandstone in the upper part; the middle unit — Kwagunt (675 m) is red sandstone and argillite with dolomite interbeds; the upper formation — Sixty Miles is a thin unit (36 m) composed of brecciated rocks and granule conglomerate with clasts derived from the underlying rocks. The Galeros Formation contains *Inzeria*, *Stratifera*, and *Baicalia aff. rara* Semikh., stromatolites typical of the third phytolite complex. The Kwagunt Formation has *Boxonia* stromatolites, typical of the fourth complex. The Kwagunt Formation contains *Chuaria* (problematica) like those reported for the Upper Neoprotozoic (?) Visingso Group in Sweden. On the basis of contained organic remains the Chuar Group is attributed to the Upper Neoprotozoic. The occurrence of *Boxonia* in the upper part of the group does not contradict this idea because the fourth complex phytolites are quite common in the upper part of the Neoprotozoic in certain regions of Eurasia. Recently well-preserved microscopic remains of blue-green algae (filaments and spherical bodies) were reported in siliceous oolites from the Kwagunt Formation (Schopf et al., 1973). The Sixty Miles Formation may be part of the Epiprotozoic glacial complex. The two lowest groups (Unkar and Nankoweap) of the Grand Canyon Supergroup are Middle Neoprotozoic.

These strata are widespread in the adjacent region of Central Arizona (in the Mazatzal Mountains) close to the boundary between the Colorado Plateau and the Cordilleran fold belt. They rest subhorizontally and unconformably on the Mesoprotozoic Pinal Group or on red porphyry of the Lower Neoprotozoic Red Rock Group which has an age of 1,640—1,690 m.y. At the base of the section there is a thick (up to 1,200 m) succession of the Mazatzal Quartzite (which is only locally developed) and above this the varied Apache Group (400—530 m) lies unconformably, and in some areas, directly, on Mesoprotozoic rocks. It consists of the following formations (in upward succession): (1) Pioneer (50—170 m) — greyish-red shale, siltstone and argillite, underlain by quartzite or quartz conglomerate (the Scanlan conglomerate); (2) Dripping Spring (180—250 m) — cross-bedded quartzite passing into siltstone; in the lower part a local quartz-pebble conglomerate (Barnes) lies disconformably on the underlying shale; and (3) Mescal (80—140 m) — dolomite with local stromatolitic bioherms, passing upwards into orange-red argillite; a basalt flow is present in the upper half of the formation. The group is crowned by a thick (up to 120 m) sheet of amygdaloidal basalt (Shride, 1967). The stromatolites *Conophyton cylindricus* Masl. and

Tungussia Semikh. are present in the Mescal Formation (Cloud and Semikhatov, 1969). These forms are characteristic of the second complex of the Middle Neoprotozoic.

The Apache Group is overlain unconformably by the "Troy quartzite" which is about 400 m thick. As well as quartzite it includes arkose and conglomerate. The conglomerates have ventifacts with a desert varnish. The quartzite is cut by a diabase with an age of 1,250 (1,200) m.y. (Wasserburg and Lanphere, 1965). This group is probably Upper Neoprotozoic. The Bolsa quartzite—sandstone of the Lower (?) Cambrian unconformably overlies it.

All the Neoprotozoic strata of Central Arizona (including the Lower Neoprotozoic Red Rock Group) are reliably dated and subdivided. These strata are therefore regarded as a parastratotype of the Neoprotozoic Erathem in the southern part of North America. Unfortunately, the stratigraphic position of the Mazatzal Quartzite and of the Apache Group in the Neoprotozoic section is not certain. The Mazatzal Quartzite is cut by monzonite with an age of 1,420—1,480 m.y. by the Rb—Sr and K—Ar methods (Livingston and Damon, 1968). It has been suggested that the monzonite intruded during a separate orogeny called the Mazatzal orogeny. However, these intrusive rocks appear to be of anorogenic character, the result of platform magmatism. The Apache Group unconformably overlies the Mazatzal Quartzite and the monzonite. Thus, it is younger than 1,420 m.y. and for this reason is attributed to the Upper Neoprotozoic. Stromatolites in the Mescal Formation suggest that it belongs to the Middle Neoprotozoic. However, the ages obtained from the monzonite may be slightly "rejuvenated", and the few determinations of stromatolites are not sufficient to be reliable. The vertical range of the stromatolite forms is not yet known.

Miogeosynclinal strata of Middle and Upper Neoprotozoic age are mostly present in Phanerozoic fold belts surrounding the North American craton. Within the craton itself they are only reported from the Grenville province of the Canadian—Greenland shield. The Belt Supergroup of the Cordilleran fold belt in Montana and Idaho is well studied (Ross, 1963, 1970; Mac Mannis, 1964; Smith and Barnes, 1966) and is suggested as a stratotype for these rocks. The stratotype is in the Big and Little Belt Mountains of Central Montana, but in this area the upper part is not fully exposed. In Western Montana and Idaho the section is more complete but the base is not certain. Thus, the sections in the two regions differ somewhat and for this reason several correlations have been proposed. In this publication (Table III, rows 61 and 62) the proposal of Smith and Barnes (1966) is accepted. This correlation is based on the marker beds of stromatolitic carbonates (Helena Formation and the Piegan Group).

In Western Montana and Idaho the supergroup includes four conformable groups as follows: (1) pre-Ravalli (Prichard) composed of siltstone, argillite and sandstone with some limestone interbeds; (2) Ravalli — shale, phyllitic slate, argillite, siltstone, sandstone that is red in the upper part; (3) Piegan —

marl and dolomite, and (4) Missoula — thick and very complex succession (includes eight formations), typified by intercalation of siltstone, quartzite and shale (commonly red-coloured and containing salt casts); the Purcell lavas are confined to this group. The La Hood (coarse-grained facies) is distinguished in the lower part of the supergroup in Central Montana. It originated in margin zones of uplifts (MacMannis, 1964) and roughly corresponds to the pre-Ravalli Group of more westerly areas. The thickness of the supergroup and of its constituent units varies greatly, with a general increase from east to west (maximum 13,000 m). Stromatolites are present in carbonate rocks of the pre-Ravalli Group (Newland Formation), the Piegan and Missoula Groups (Snowslip Formation). Semikhatov (see Cloud and Semikhatov, 1969) identified *Collenia frequens* Walc., *Baicalia* sp., and *Conophyton cylindricus* Masl. in the Missoula Group. These are typical of the second phytolite complex or of the Middle Neoprotozoic.

Rocks of the Belt Supergroup are folded and cut by granite with associated contact metamorphism. In the contact aureole saline deposits of the Missoula Group are transformed into scapolite-bearing schist (Hietanen, 1967).

The Belt Supergroup overlies a Paleoprotozoic metamorphic complex, mobilized and intruded by granites at the end of the Mesoprotozoic, some 1,900 m.y. ago according to dating by K—Ar, Pb-isotope and Rb—Sr isochron methods (Obradovich and Peterman, 1962; Smith and Barnes, 1966). The upper age limit of the supergroup is determined as 1,100—1,200 m.y. by Pb-isotope dating of hydrothermal uranium mineralization in rocks of the Ravalli Group (Eckelmann and Kulp, 1957). K—Ar dating of glauconite from shale of the Ravalli Group (Empire Formation) and of the Missoula Group (McNammara Formation) gave rejuvenated ages of about 1,150 (1,100) m.y. (Obradovich and Peterman, 1962). The Rb—Sr isochron age of granophyre (granodiorite) from the Hellroaring Creek stock that cuts lower units of the Belt Supergroup, and from a diabase sill that intrudes these rocks in British Columbia (near the U.S.A. border) is 1,260 m.y. (Ryan and Blenkinsop, 1971). The available data, particularly the high initial ratio of $^{87}Sr/^{86}Sr$ (0.81), suggests that the granodiorite formed by melting of older rocks (by heat generated by basic magma?). The magma may be represented by the diabase which is widely distributed in British Columbia. The upper age boundary of the Belt Supergroup in the U.S.A. is fixed by the unconformably overlying Epiprotozoic Pokatello tillite and the Middle Cambrian Flathead quartzite.

These data do not preclude a Lower Neoprotozoic age for the lower part of the Belt Supergroup. In many parts of the Cordilleran fold belt the Lower Neoprotozoic is, however, represented by rocks of quite different character (acid volcanics of taphrogenic type), so that a Lower Neoprotozoic age is improbable.

The Belt Supergroup extends into British Columbia (Canada) where it is called the Purcell Supergroup. Because of facies changes, detailed correlation poses problems. The most important marker horizon is the Purcell basalt. The underlying part of the Purcell Supergroup is usually correlated with all other units of the Belt Supergroup that underlie lava (Reesor, 1957; Price, 1964). This correlation is not well founded. Carbonates (with stromatolites) in the lower part of the Purcell Supergroup may be correlated with the Piegan Group (principally carbonates) which is in the middle part of the Belt Supergroup. If this correlation is correct then correlatives of the Ravalli and pre-Ravalli Groups are not known from the Purcell Supergroup (Table III).

The Purcell Supergroup is younger than basement rocks which were metamorphosed about 1,900 m.y. ago. These older rocks are exposed in several areas of Northern British Columbia (e.g. Tuchodi Lake area). However, the contacts between the Purcell rocks and the older rocks are nowhere exposed. An upper age limit is fixed by intrusion of numerous diabase dikes with an age of 1,250 (1,200) m.y. The age of this diabase is the same as that of Hellroaring Creek granodiorite. This tends to favour the proposed origin for the granodiorite by crustal melting related to the presence of basic magma. The Epiprotozoic Windermere Group (with tillite) overlies these successions unconformably.

In more southerly areas of the North American Cordillera the Harrison Group (Nevada, Utah and Idaho) and the Big Cottonwood Group (Utah) are probably correlatives of the Belt Supergroup or perhaps only of its lower part (Blackwelder, 1932; Crittenden et al., 1952; K.C. Condie, 1966, 1967; Woodward, 1967). Both of these groups consist of quartzite, sandstone, siltstone and dolomite. A few basic lava flows are also present. The Harrison Group is generally more highly metamorphosed. Sandstone and siltstone are commonly transformed into micaceous slate, dolomite into marble, and volcanics are amphibolitized. Both groups overlie an Archean gneiss complex mobilized in the Paleoprotozoic and Mesoprotozoic. The Big Cottonwood Group is overlain unconformably by the Epiprotozoic Mineral Fork tillite (which is sometimes erroneously assigned to the upper part of the group).

In Eastern California (Death Valley) the two lower formations of the Pahrump Group are probably also of the same age. These include the Crystal Spring (up to 1,200 m) Formation composed of sandstone and granule conglomerate (lower part), stromatolitic dolomite (middle part) and argillite and siltstone (upper part), and Back Spring (400 m) composed of dolomite with shaly interbeds (Hunt and Mabey, 1966). The upper formation (Kingston Peak) is considered to be Epiprotozoic. It unconformably overlies the Back Spring Formation and contains tillite. Dolomite of the group contains *Baicalia* sp. and *Jakutophyton* sp. (D.G. Howell, 1971) stromatolites typical of the second phytolite complex. The Crystal Spring Formation lies with sharp unconformity on the Mojave metamorphic complex which was altered during the Karelian orogeny (Silver et al., 1962; Stern et al., 1966). All of these data define the Middle Neoprotozoic age of the lower formations of the Pahrump Group.

In the *East Greenland fold belt* the Eleonore Bay Group is considered to be Middle and Upper Neoprotozoic. It outcrops in the Kong Oskars Fjord and Scoresby Bay areas and also the lower part of the Hagen Fjord Group in Kronprins Christians Land. The first is made up of weakly altered sandstone, quartzite and argillite. Carbonate rocks with stromatolites and microphytolites are present in the upper part of this group. Among the stromatolites tungussids are dominant, represented by new forms called *Poludia* and *Eleonora*, but *Inzeria* and *Jurusania*, typical of the third phytolite complex, are also present. Microphytolites include *Vesicularites aff. lobatus* Reitl. *V. aff. compositus* Z. Zhur., *Osagia* sp. (Bertrand-Sarfati and Caby, 1974). These phytolites favour assignation of the upper part of the group to the Upper Neoprotozoic. Certain rocks and the Brogetdal Formation in particular (row 72, Table III), are characterized by red colour, cross-bedding and ripple marks. The base of the group is not well defined, but it probably lies above crystalline basement rocks that underwent metamorphism during both the Paleoprotozoic (2,600—2,800 m.y.) and the Mesoprotozoic (1,900—2,000 m.y.). The group is unconformably overlain by the tillite-bearing strata of the Epiprotozoic Merkebjerg Group (Katz, 1954, 1961).

The lower part of the Hagen Fjord Group consists of terrigenous and carbonate rocks metamorphosed to greenschist facies. It lies unconformably above metamorphic rocks of the so-called pre-Carolinian Mesoprotozoic (?) complex. A tillite that is assigned to the upper part of this group should probably be considered as a separate group, for a stratigraphic break separates it from the underlying rocks. The contact of the Hagen Fjord Group with the pre-Carolinian complex is everywhere tectonic, but the younger age of the former is clearly indicated by an abrupt change in metamorphic grade (Berthelsen and Noe-Nygaard, 1965).

In the Queen Louise Land region the Zabra terrigenous Group correlates with the lower part of both groups. It lies with angular unconformity on the Lower Neoprotozoic (?) Trekant Group which is cut by diabase (Berthelsen and Noe-Nygaard, 1965).

In the *Innuitian fold belt* the Kennedy Channel Formation (Group) is conventionally assigned to the Middle Neoprotozoic. It consists of phyllite, sandstone and limestone and is unconformably overlain by the Epiprotozoic (?) subplatform Ella Bay dolomite (Kerr, 1967). In the Churchill tectonic province of the *Canadian—Greenland shield* the Seal Lake Group is also assigned to the Middle Neoprotozoic. It is present in the Labrador Peninsula in the area adjacent to Nain province.

According to Brummer and Mann (1961) this group includes the following formations (in upward sequence): (1) Bessie Lake or Majoqua (1,400—4,000 m) — light-grey and pink quartzite that is locally feldspathic, intercalated with amygdaloidal basic lavas and underlain by granule conglomerates and conglomerates; (2) Wuchusk Lake (6,500 m) — interbedded cherty slate, phyllite, argillite, fine-grained quartzite and finely interbedded carbonates with stromatolites and oncolites; (3) Whisky Lake (1,000 m) — red shale and grey phyllite with subordinate red argillite and quartzite; (4) Salmon Lake (1,000 m) — interbedded red shale and green-grey slate with thick flows of amygdaloidal basalt; (5) Adeline Lake (450 m) — grey and red shale and quartzite; and (6) Upper red quartzite (>650 m) — red quartzite with quartz pebbles. All of these formations are intruded by diabase and gabbro—diabase sills. The K—Ar age of basalt from the Salmon Lake Formation is about 1,250 (1,200) m.y. For this reason the upper part of the group is regarded as Upper Neoprotozoic. Important hydrothermal copper deposits (related to basalts) are present in the Salmon Lake Formation.

The Seal Lake Group resembles platform complexes in some ways, but it differs in its great thickness (>11,000 m), rather strong deformation and by the fact that it is locally metamorphosed. Thus, the group is transitional in character, between platform and miogeosynclinal deposits.

The Seal Lake Group unconformably overlies porphyry of the Lower Neoprotozoic Letitia Lake Group. Its upper boundary is fixed by the fact that it was affected by Grenville movements (1,000—1,100 m.y.).

The Seal Lake Group may possibly be correlated with the Ramah Group (up to 1,500 m) which forms a narrow syncline (striking NNW) in the Torngat Mountains on the Labrador coast (Knight, 1973). The lower part of the group is white and pink quartzites including an amygdaloidal andesitic lava. A regolith is present at the base. The oldest formation (800 m) lies unconformably on gneiss. The second unit is black sandstone and quartzite (60—140 m) followed by variegated argillite, siltstone, limestone and dolomite of the third formation (450 m), intraformational dolomitic breccias, dolomite, limestone (with local stromatolites), and sandstone with gypsiferous interlayers (fourth formation, 160 m), black and grey argillite with interbedded quartzites, sandstones and limestones (the fifth formation, 130—150 m) and finally greywacke sandstones, black shales, and siltstones (the sixth formation, 200 m).

Possible correlatives of the lower part of the Seal Lake Group are present in the Wakeham Group in the northern part of the Grenville province, near the north shore of the Saint Lawrence River, opposite Anticosti Island. This group consists of terrigenous rocks which are folded and have undergone low-grade metamorphism. These rocks are principally quartzite and arkose resting on Archean-type crystalline rocks, which suffered multiple mobilization and are cut by Early Neoprotozoic intrusions of anorthosite and rapakivi-type granite. The Wakeham Group is cut by small granitic intrusions that are probably related to the Grenville orogeny.

General Characteristics

There is great similarity among the Middle and Upper Neoprotozoic sequences of different continents of the Northern Hemisphere. In contrast to the rocks of the Lower Neoprotozoic, platform (and partly miogeosynclinal) strata are much more extensive. Red-beds are particularly widespread and may indicate a significant increase in free oxygen in the atmosphere. Glauconite is common in these rocks. Hematitic and sideritic iron ores are present in some areas among the sedimentary rocks of both platform and miogeosynclinal type. These formations are similar to the Urals-type iron ores that first developed and were widespread in the Mesoprotozoic (see Chapter 6). However, these ores are relatively rare in the rocks discussed above. Jaspilitic iron formation (including the Superior type) are absent in the Neoprotozoic. This type of iron formation did not form after the Mesoprotozoic.

Among the carbonate rocks, limestone becomes more abundant in comparison to dolomite. This may be related not only to a change in atmospheric composition (decrease in CO_2), but may also be related to the development of a relatively high relief which could have produced a rapid flow of fresh water into the seas causing an overall decrease in alkali content (Strakhov, 1963).

Coarse-grained rocks are important in many sedimentary complexes of this age. Rhythmically bedded strata of flyschoid type are widely distributed in miogeosynclinal complexes (for example, the Kadalikan Subgroup of the Patom Group in the Mama—Vitim zone and the Kachergat Formation of the Baikal Group). In some sequences the coarse-grained rocks resemble, but are not identical to, molasse. A great number of stratigraphic breaks are present in both miogeosynclinal and platform successions.

There is a great increase in abundance and variety of phytolites. They became very important as rock formers and built up larger bioherms. The typical Middle Neoprotozoic phytolites are of the second ("Middle Riphean") complex containing numerous stromatolites of the groups *Baicalia*, *Tungussia*, *Jakutophyton*, *Gongilina*, *Anabaria*, *Colonella* and others; oncolites such as *Osagia columnata* Reitl., *Ȯ. undosa* Reitl., *O. composita* Z. Zhur.; and katagraphs: *Vesicularites flexuosus* Reitl., *Vermiculites angularis* Reitl., *Glebosites glebosites* Reitl. In the Middle Neoprotozoic of the Turukhansk area remains of pogonophora (?) identified as *Paleolina* Sokol. were found. However, this find is only a single one and *Sabellidites*-allied forms are known from much younger strata (Eocambrian and Cambrian). More details of the identification are needed, together with better documentation of the stratigraphic position. In some areas microscopic remains of algae appear in Middle Neoprotozoic strata. The cells of these algae appear to have an inner nucleus (eukaryota).

The Upper Neoprotozoic is typified by phytolites of the third complex ("Upper Riphean") with abundant representation of the *Gymnosolen*, *Inzeria*, *Minjaria*, and *Jurusania* groups, together with various older and rarer forms, and oncolites such as *Osagia grandis* Z. Zhur., *Asterosphaeroides serratus* Z. Zhur., *Radiosus limpidus* Z. Zhur. and other forms such as the katagraphs *Nubecularites uniformis* Z. Zhur., *Vermiculites anfractus* Z. Zhur. etc.

In many areas of Northern Eurasia the paleontology of the Neoprotozoic rocks has been studied in detail and it has been established that various forms typical of the fourth phytolite complex are present in the upper part of the Upper Neoprotozoic. Some workers contend that the fourth-complex phytolites are typical of rocks attributed to the Vendian or Yudomian. This is taken as evidence for drawing the boundary line between the Upper Riphean and Vendian within these units. In this case the stratigraphic subdivisions are not reflected by a marked change or break in sedimentary conditions, but rather are based simply on the change from phytolites of the third complex to those of the fourth. Those who consider the phytolite complexes to be of prime importance in subdivision of the Precambrian put the lower age boundary of the Vendian at about 650 m.y. The available facts indiindicate that the fourth-complex phytolites occur in rocks as old as 1,100 m.y., and in some cases in strata older than 1,420 m.y. (the Shtandinsk and Gozha Formations of the Kama region).

In Western Europe remains of *Cayeuxidae* (problematic, earlier erroneously considered to be radiolarians) were reported from the cherty rocks associated with Neoprotozoic eugeosynclinal volcanics.

Metamorphism of these sub-erathems is different in different tectonic zones. On platforms the rocks are usually altered as a result of diagenesis or epigenesis. Only rarely do the rocks reach the lowest greenschist facies. This is in marked contrast to the Mesoprotozoic platform (protoplatform) strata which have commonly undergone greenschist-facies metamorphism. In geosynclinal areas metamorphism is of a zonal character and ranges from greenschist to amphibolite, but in most areas the rocks exhibit greenschist facies.

Extensive volcanism of platform-type is characteristic of the Middle and Upper Neoprotozoic. The most intensive volcanism occurred near the Middle and Late Neoprotozoic boundary, and at the beginning of the Late Neoprotozoic in particular. In contrast to the subplatform taphrogenic strata of the Lower Neoprotozoic which were dominated by acid porphyritic volcanics, Middle and Upper Neoprotozoic platform sequences are dominated by basic lavas. In most cases acid volcanics are minor. Dikes and sills of diabase and gabbro—diabase are associated with these volcanics. In many areas however, such intrusive bodies appear without any associated volcanics, and locally they are accompanied by small stock-like intrusions of diorite and granodiorite. There are minor occurrences of geosynclinal volcanics of the spilite—keratophyre clan, but in many eugeosynclinal complexes the sedimentary rocks are more abundant than volcanics.

The Grenville orogeny, which terminated the Neoprotozoic, was accompanied in geosynclinal areas by numerous rather small intrusions of ultrabasic, basic and acid magma. Granitization and migmatization were not extensive and occurred mainly in contact aureoles around granite intrusions.

The Middle and Late Neoprotozoic have strikingly well differentiated tectonic elements. In geosynclinal systems both miogeosynclinal and eugeosynclinal belts may be distinguished. These are subdivided by internal uplifts and deep faults into zones and sub-zones of different tectonic character. In some geosynclinal systems deep marginal troughs were distinctly isolated (Salop, 1964—1967). Many peculiar aspects of the composition and structure of Neoprotozoic sedimentary strata are explained by increased tectonic differentiation and speeding up and increase in scale of tectonic movements.

This stage in geological development is also characterized by considerable consolidation of the older platforms. It was during this period that the major platforms of the Northern Hemisphere (East European, Siberian, Hindustan and North American) were formed. Paleotectonic reconstructions also show that, during the Neoprotozoic, these platforms were even greater than in the Phanerozoic.

Fig.30 shows a paleotectonic reconstruction of Eastern Europe in Neoprotozoic times. The major part of the area was occupied by a large craton, which, both in shape and size is rather similar to that of the modern East European platform. On the northwest the craton was bounded by a deep trough in which thick miogeosynclinal strata of "sparagmite" type were accumulating. This trough was not broad, for the sediments were derived from both the southeast and the northwest at the same time (Holtedahl, 1953). These rocks are now largely covered by formations of the Caledonian fold belt. Because the tectonic movements that led to the formation of the trough and the folds developed in it, preceded the Caledonian movements, and because the Neoprotozoic (and Epiprotozoic) trough and the Caledonian geosyncline are practically superimposed, the name Eocaledonian fold belt is proposed for it. In the Eocaledonian belt the most intensive folding took place at the end of the Epiprotozoic, during the Katangan orogeny.

A miogeosynclinal belt extended from Scandinavia in a southwesterly direction to Timan through Finmark Rybachy and Kanin Nos Peninsulas. That geosynclinal belt, forming the northeast margin of the craton is the Hyperborean—Timan belt. Many Neoprotozoic strata developed there are transitional in facies from miogeosynclinal to platform. This geosynclinal trough was also relatively narrow, forming a sort of canal between cratonic blocks. Another rather small craton, called the Barents platform (Salop, 1958b, 1960) was probably present to the northeast. Its presence is suggested by the general structure of the surrounding fold belts, by orientation of the structures within these belts, and by the miogeosynclinal aspect of the strata.

In the area of the Polyudov Ridge the Hyperborean—Timan belt joined the great Riphean geosynclinal belt bordering the East European craton on the east. Neoprotozoic strata extend along the length of this belt from the Polar Urals to the Mugodzhary Mountains. All of these strata are of miogeosynclinal type. Certain metamorphic rocks, however, are of eugeosynclinal aspect. These are developed on the eastern slope of the Urals (Chulaksay

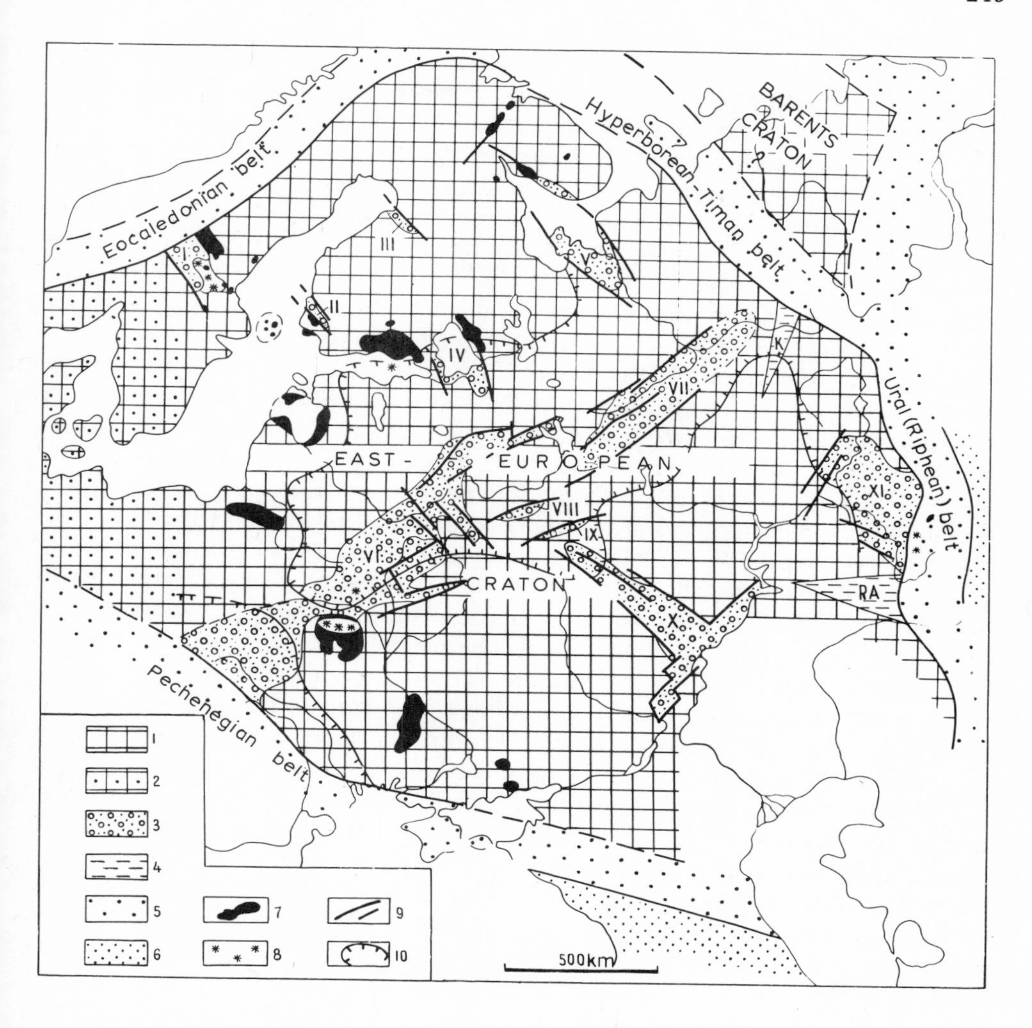

Fig. 30. Major structural elements of Eastern Europe during the Neoprotozoic and limits of the Eocambrian platform cover (Vendian s.str.). Compiled by L.J. Salop and K.E. Jacobson, using data of Klevtsova (1970), Petrov (1972) and Solontsov and Aksenov (1969).

1 = craton; *2* = parts of the craton reactivated at the end of the Neoprotozoic; *3* = platform depressions and trough-grabens (*I* = Dalarna; *II* = Satakunta; *III* = Muhos; *IV* = Ladoga; *V* = Belomorian; *VI* = Volyn'—Kreststy; *VII* = Soligalich—Yarensk; *VIII* = Gzhatsk; *IX* = Moscow; *X* = Pachelma; *XI* = Kama—Bel'sk); *4* = aulacogens (*K* = Kazhima; *RA* = Radayevo—Abdulino); *5* = miogeosynclinal belts; *6* = eugeosynclinal belts; *7* = subplatform intrusions of Vyborgian cycle (rapakivi, anorthosite, alkaline rocks); *8* = taphrogenic subaerial sedimentary—volcanogenic strata of the Lower Neoprotozoic; *9* = faults; *10* = limit of the Eocambrian platform cover — Vendian s.str. (hachures are directed towards the cover).

Group etc.). Their age is, however, not well established. In the well-studied western and central parts of the Urals, strata of the peripheral part of the geosynclinal belt are well developed. Certain Middle and Upper Neoprotozoic strata in the western limb of the Bashkir anticlinorium are of subplatform aspect so that they resemble platform strata located near the Urals region. The Neoprotozoic Riphean geosynclinal belt was probably almost the same size as the Hercynian Urals belt. Its limits differ from those of the older Burzyan fold belt that, in Mesoprotozoic times, bordered the Sarmatian craton — the embryonic East European craton.

To the southwest and south the Pechenegian geosynclinal belt bordered the East European craton (the name "Pechenegian" is from the ancient inhabitants of the Southern Steppe of Russia). Folded, meso- and epizonally metamorphosed Neoprotozoic strata of this belt are known from drill holes to the south of Moldavia, in Northern Crimea, and in the Kuban River area where these strata are unconformably overlain by Eocambrian (Vendian) rocks, which are slightly metamorphosed and faulted. Outcrops of the former are also reported in Dobrudzha. Typical rock types include phyllitic shale, micaceous slate, siltstone, schistose sandstone and in part green chloritic orthoschist. The Neoprotozoic age of these rocks is not certain, but they are certainly older than the Eocambrian and lie unconformably on metamorphic strata containing jaspilite. These metamorphic strata are present in a drill hole in the area of the village of Tuzla, south of Constantia in Roumania (A.Ya. Dubinsky, personal communication, 1973). They greatly resemble iron formation of the Mesoprotozoic Krivoy Rog Group. In the Great Caucasus Ridge, Middle and Upper Neoprotozoic strata are represented by sedimentary and volcanic rocks of eugeosynclinal type (the Khasaut and Chegem Groups).

Neither the extension of the Pechenegian belt to the southeast nor its structural relation to the Riphean belt are known, for there is a thick Phanerozoic cover in the Caspian Lowland region, where the structure of the basement is not revealed even by geophysical methods.

The East European craton was not completely stable during the Neoprotozoic. In the Early Neoprotozoic, internal grabens appeared in which subaerial red, sedimentary and volcanic taphrogenic strata accumulated. Basic, acid, and alkaline magmas were intruded along faults. These initiated formation of platelike and laccolithic bodies of gabbro—anorthosite, rapakivi granite and alkaline syenite. During the Middle and Late Neoprotozoic numerous large and small troughs were formed. In places these were bounded and transected by faults. Such troughs and trough—grabens were mainly in two directions, northwest and northeast, approximately parallel to the craton boundaries and to the existent geosynclinal troughs. In the grabens red terrigenous or carbonate—terrigenous platform strata accumulated with local extrusion of basic lavas.

Two deep aulacogens (Radayevo—Abdulino and Kazhim) projected at acute angles into the platform, in both the Urals and Timan geosynclinal belts. The Middle and Upper Neoprotozoic strata in these aulacogens are very thick (up to 6—8 km in the Radayevo aulacogen). The southwestern part of Scandinavia and adjoining areas of Jutland probably suffered partial mobilization at the end of the Neoprotozoic, during the Grenville orogeny. However, there were no major tectonic effects. This idea is based mainly on "isotopic rejuvenation" of the older crystalline basement rocks, and on the presence of small granite intrusions with an age of 1,000 m.y. (the Bohus granite, etc.) and of local folding of Neoprotozoic strata (the Almesåkra Group).

In Northern Asia during the Neoprotozoic the huge Siberian craton originated. It was much larger than the modern Siberian platform which developed from it (Fig.31). This craton formed as a result of consolidation of the Angara, Okhotsk and Kara Mesoprotozoic cratonic blocks and by

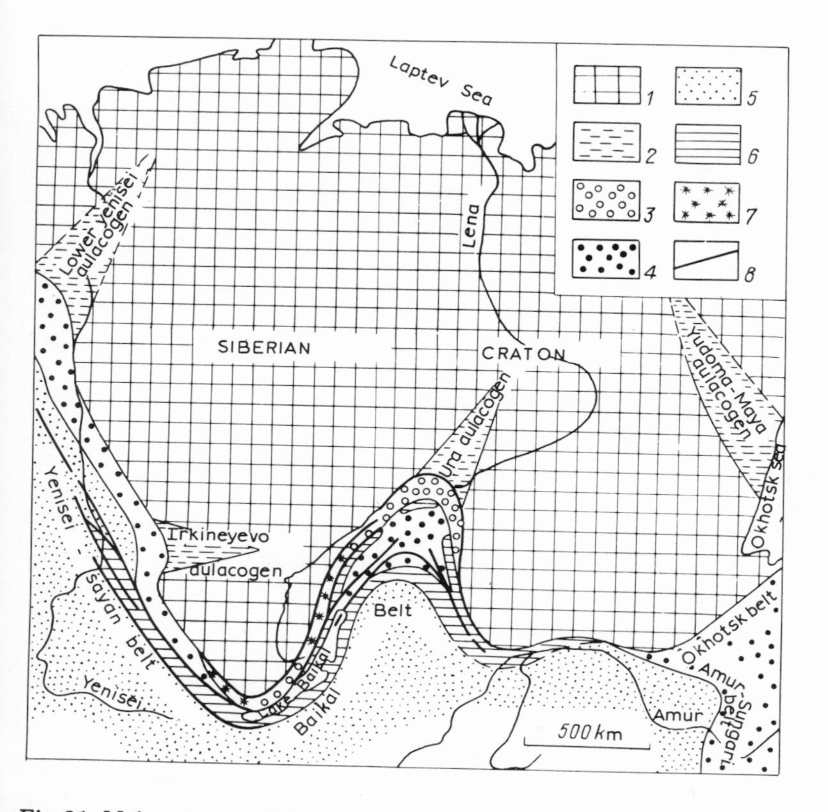

Fig. 31. Major structural elements in Siberia during the Neoprotozoic.
1 = craton (platform); *2* = aulacogens; *3* = marginal troughs; *4* = miogeosynclinal belts; *5* = eugeosynclinal belts; *6* = internal geosynclinal uplifts; *7* = taphrogenic subaerial sedimentary and volcanic rocks of the Lower Neoprotozoic; *8* = deep faults.

closure of the older Taymyr and Olenek geosynclinal systems. In Taymyr and Northern Kharaulakh (the Lower Lena River) the Middle and Upper Neoprotozoic strata are a typical platform assemblage. They unconformably overlie Mesoprotozoic rocks that are folded and metamorphosed. These facts indicate that the above-mentioned geosynclinal systems no longer existed.

To the southwest and south the Siberian craton was bordered by the Yenisei—Sayan, Baikal and Okhotsk geosynclinal systems. Aulacogens related to these systems include the Lower Yenisei (Dragunov and Smirnova, 1964), Irkineyevo (Salop, 1958b; Kirichenko, 1968), Ura (Salop, 1964—1967) and Yudoma—Maya (Nuzhnov, 1967). In the Yenisei—Sayan and Baikal systems the mio- and eugeosynclinal belts are clearly defined as also are inner troughs (intrageosynclines), and uplifts (intrageoanticlines). In the near-platform margin of the Baikal system (Fig. 29) was a deep and narrow marginal trough (Baikal—Patom). Many of these features were fault-bounded. Such faults are common at the boundary between the craton and geosynclinal systems (Salop, 1964—1967). Thus, in the Neoprotozoic the structural development of mobile belts became much more complex.

In the Middle and Late Neoprotozoic most of the basement rocks of the Siberian craton were covered by a shallow epicontinental sea in which platform sediments accumulated. Deep cratonic troughs and grabens, typical of the East European craton, were not so widely distributed in the Siberian craton. Poor development of Neoprotozoic volcanics in the Siberian craton was probably due to greater tectonic stability in that region. However, the Siberian craton is not well known because of a lack of drill-hole data. Within the limits of the craton most tectonic activity was in the aulacogen regions where Middle and Upper Neoprotozoic strata are folded, (especially near fracture zones) and faulted, with local development of imbricated thrusts.

In Neoprotozoic times the tectonic features of North America (Fig. 32) were similar to those of the present day. The huge North American craton (platform) appeared in place of the several small cratonic blocks that existed in the Mesoprotozoic. It was bounded on the west by a large geosynclinal belt that, to some extent, coincided with the Alpine belt of the North American Cordillera. This older mobile belt is called the Beltian after the geosynclinal complex developed within it.

To the northeast (east shore of Greenland) is the Carolinian geosyncline belt which developed during the Mesoprotozoic (Fig. 24). Southeast of the craton lay the Grenville belt which also existed in the Mesoprotozoic but had distinctly different outlines. Perhaps the present site of the Appalachian Mountains was formerly an uplifted block within the vast Grenville geosyncline, for Neoprotozoic strata are not present there, and Epiprotozoic rocks appear to rest directly on Archean (?) crystalline rocks. Presence of a cratonic block in that region is unlikely for isotopic dating of basement crystalline rocks in the Appalachians commonly gives "Grenville" values of about 1,000—1,100 m.y., related to thermal events during the Grenville orogeny.

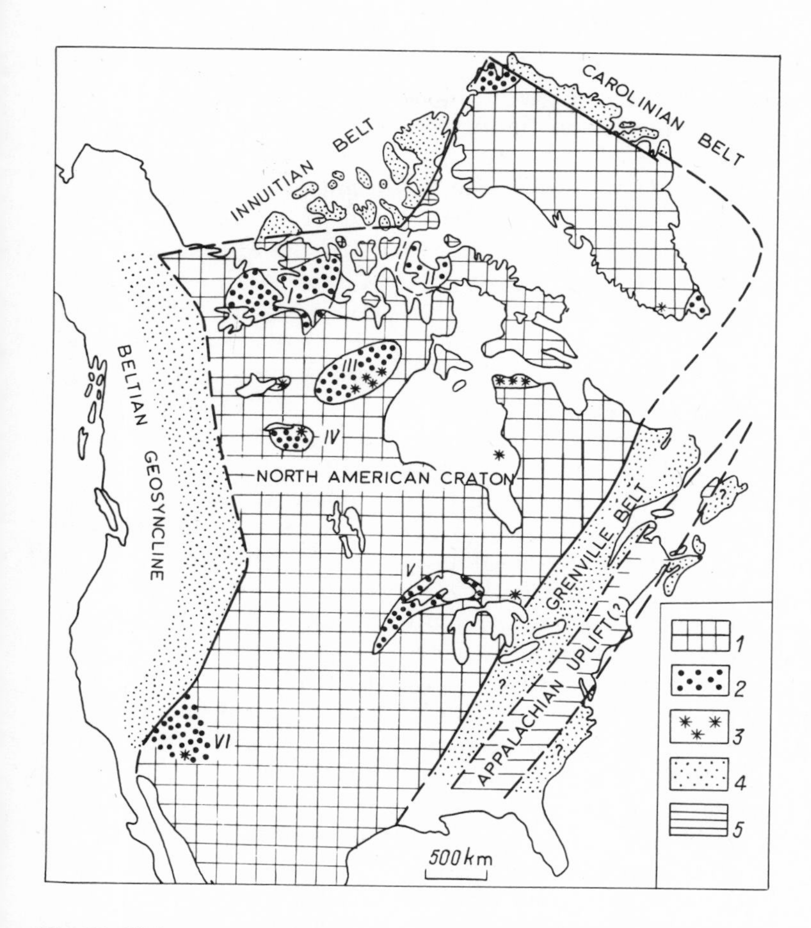

Fig. 32. Major structural elements of North America during the Neoprotozoic.
1 = craton (platform); *2* = platform troughs = areas of deep subsidence or thick platform sedimentation (*I* = Bear Lake, Darnley, Victoria; *II* = Borden; *III* = Dubawnt; *IV* = Athabasca; *V* = Keweenawan; *VI* = Colorado); *3* = areas of distribution of Lower Neoprotozoic taphrogenic subaerial volcano-sedimentary rocks; *4* = miogeosynclinal belts; *5* = internal geosynclinal uplift (?).

Though it is not ruled out that some crystalline basement rocks of the Appalachians (for example the Cranberry gneisses) may be highly metamorphosed rocks of Neoprotozoic age. In this case the site of the Appalachian Mountains was probably a deeply submerged part of the geosynclinal belt at that time.

The northern (northwestern) boundary of the North American craton cannot be defined because of a lack of data. However, in the northern part of the Canadian Arctic Archipelago, a Neoprotozoic miogeosynclinal belt may have existed in the site of the Caledonian—Early Hercynian Innuit geosyncline. Neoprotozoic sedimentary strata are of platform type in Victoria Island and Baffin Island, and also in northeast Greenland (Peary Land and Danmarksfjord). In Ellesmere Island these strata are represented by miogeosynclinal formations.

Considerable problems arise in attempting to define the southern boundary of the craton. Subplatform taphrogenic volcanic strata of Lower Neoprotozoic age are reported from the southern part of the Midcontinent. In the Colorado Plateau and adjacent areas of Arizona (later subjected to Mesozoic—Cenozoic activity) strata of all three Neoprotozoic sub-erathems are represented by platform rocks (the lower part — taphrogenic rocks; the middle and upper parts — typical platform rocks). Perhaps the North American craton extended to the south, beyond the modern boundary of the platform.

As in the East European craton, deep troughs and grabens existed in the Neoprotozoic North American craton. Thick red sedimentary and sedimentary—volcanic assemblages were characteristic. The approximate outlines of these troughs are shown in Fig. 32. The original boundaries cannot be reconstructed as yet in detail, though certain data have already been collected by Canadian workers (Fraser et al., 1970).

Most Middle and Upper Neoprotozoic platform strata occur within the limits of the modern North American platform (with the exception of the Grenville province) and miogeosynclinal strata were principally present in the surrounding Phanerozoic fold belts. The fact that miogeosynclinal strata in the overlying Paleozoic and Mesozoic rocks are, in general, similarly disposed to those of the Neoprotozoic, North American platform and also the surrounding Phanerozoic systems, suggests that the paleogeographic configuration originated in the Neoprotozoic. Highly metamorphosed eugeosynclinal sedimentary—volcanic strata of uncertain age are present in the North American Cordillera. These rocks are commonly regarded as Paleozoic or even Mesozoic. In the Innuitian, East Greenland and Appalachian fold belts Neoprotozoic eugeosynclinal strata are possibly covered by ocean waters.

Numerous mineral deposits are related to the Middle and Upper Neoprotozoic formations, but in the continents of the Northern Hemisphere these are neither large nor of unique type. The stratiform copper deposits of the Belt Supergroup might provide an exception. The biggest copper deposits are in the lower part of the supergroup, in the Ravalli Group, and in the Spokane Formation in particular. In other areas deposits of iron, manganese, magnesite and phosphorus occur in the sedimentary strata. Some large deposits of iron occur in rocks of this age. These include hematitic granule conglomerates in the Angara—Pit area of the Yenisei Ridge, which form part of the sequence of the Upper Neoprotozoic Oslyansk Group. Deposits of iron in the Southern Urals (Avzyan and others) probably formed as a result of weathering of ferruginous dolomite of the Middle Neoprotozoic Avzyan Formation.

Magnesite deposits of the Alad'insk and Dzhur Formations in the Yenisei

Ridge are of great economic importance (Talskoe, Karkadan and other deposits). Deposits in the Murondava Formation of the Khingan—Bureya area in the Soviet Far East formed as a result of redistribution of magnesium in syngenetic dolomites. The following deposits are also of economic interest: manganiferous cherty-carbonate slate of the Ikat Formation in the Vitim Plateau, manganiferous limestone of the Mana Formation of Eastern Sayan, phosphorite of the Uluntuy Formation in the Baikal region of the Pogoryuy and Uderey Formations in the Yenisei Ridge, and also bedded and concretionary hematite and siderite ores in carbonate rocks of the Tilim Formation in the Vitim Plateau and of the Omnya and Lakhanda Formations in the Uchur—Maya area of Eastern Siberia. Diamond-bearing conglomerates of the Carnool and Rewa Groups in India are probably fossil placers.

Almost all of these mineral deposits of sedimentary origin are related to the Neoprotozoic miogeosynclinal and platform strata. Only the manganese deposits occur in a eugeosynclinal sequence.

In Canada the largest hydrothermal deposits of native copper and copper sulphides are genetically and spatially related to Middle Neoprotozoic basalts. These are especially significant in the Upper Neoprotozoic. The best examples are the deposits in the Lower Keweenawan of the Lake Superior area, those in basalts of the Seal Lake Group in Labrador and those in the Coppermine River Group in Northwestern Canada. Similar mineralization is present in basalts of the Volyn Group (Berestovets Formation) in the Russian plate. These rocks are also considered to be Upper Neoprotozoic.

The boundary between the two upper sub-eras of the Neoprotozoic, as defined by orogenic cycles, appears to be of the same age in different continents of the Northern Hemisphere. The Middle and Upper Neoprotozoic boundary is defined by diabase dikes and sills of the Prikamian (Kibarian, Avzyan) second order tectono-plutonic cycle. The age of this cycle is within the 1,400—1,300 m.y. range. In the Beltian fold belt (North American Cordillera) diabase of this cycle is accompanied by small granodiorite intrusions. Folding related to this cycle was weak and local, but oscillatory epeirogenic movements were widespread. These movements led to gaps in sedimentation and to great stratigraphic unconformities at the boundary between the middle and upper sub-erathems.

The upper age boundary of the Upper Neoprotozoic and of the Neoprotozoic as a whole is everywhere drawn at a great structural and stratigraphic (on the platforms) unconformity that separates Neoprotozoic from Epiprotozoic or younger rocks. This unconformity is a manifestation of the widespread Grenville orogeny that took place about 1,100—1,000 m.y. ago. Ubiquitous intrusion of basic magma along fissures preceded the Grenville orogeny and resulted in numerous diabase dikes and sills, and large gabbro massifs with an age of 1,100—1,250 m.y.

CHAPTER 8

THE EPIPROTOZOIC

Various sequences formed after the Grenville orogeny and before the termination of the Katangan orogeny (in the 1,000—650 m.y. interval) are assigned to the Epiprotozoic. The occurrence of glacial deposits in sedimentary complexes of this erathem is its most diagnostic feature.

Regional Review and Principal Rock Sequences

Europe

In Europe the strata of this erathem are mostly of platform and miogeosynclinal facies. The first are distributed in the Russian plate and locally in the Baltic shield. Miogeosynclinal-type rocks are present in the Urals in the northern (Eocaledonian) margin of the Baltic shield, and in the Paleozoic fold belts of Western and Central Europe. Unlike the Neoprotozoic strata, the Epiprotozoic rocks are almost entirely composed of terrigenous rocks and locally of volcanics. Carbonate rocks are a minor constituent. In these areas (as in other continents) glacial formations (tillite) are present. Eugeosynclinal units occupy a small area in Central England.

In the *Russian plate* Epiprotozoic platform strata are quite distinct from the underlying and overlying rocks and lie transgressively on older rocks of Archean (Strugova Buda), Middle Neoprotozoic (Kaverino), and Upper Neoprotozoic age (Serdobsk, East Byelorussia).

A section in the northwestern part of the Pachelma trough, in a drill hole in the vicinity of the village of Kaverino, is suggested as the platform stratotype of the Epiprotozoic. The Pachelma Group in this region includes three formations: (1) Vedenyapino — greenish-grey siltstone and clay with a basal bed of quartz—glauconite sandstone about 1,000 m.y. old (50 m thick); (2) Vorona — interbedded pink and cherry-brown sandstone with silty and clayey rocks (37 m); and (3) Partsinsk — interbedded micaceous sandstone and mudstones with tillite (50 m).

The Partsinsk Formation (tillite) was considered to unconformably overlie layered rocks of the Vorona Formation. This idea was based on the assumption that the Volyn' Group (with tillite) lay unconformably on the underlying formations. Studies of the Kaverino section by K.E. Jacobson showed that the tillite there is closely related to the Pachelma Group. The tillite in the Kaverino section forms subordinate interbeds in a sequence that is predominantly siltstone and sandstone.

In the central zone of the Pachelma trough the Vedenyapino and Vorona Formations are quite distinct, as also is the Krasnoozersk Formation which may be correlated with the Partsinsk Formation (intercalation of grey argillite and siltstone with some dolomite beds). Glauconite from the Vedenyapino Formation was dated in the 765—807 m.y. range. In the southeastern part of the Pachelma trough the Epiprotozoic is represented only by the Vedenyapino and Vorona Formations. In the Pachelma trough the Epiprotozoic exceeds 700 m in thickness.

In the area of the Moscow syneclise, the Epiprotozoic sections resemble those of the Pachelma trough. They are known from bore holes. (Kirsanov, 1970; Klevtsova, 1971) in Pavlov—Posad (2,891—1,692 m interval), in Roslyatino (4,552—1,849 m interval) and in Bologoye (2,895—1,630 m interval). Recently A.A. Klevtsova and K.E. Jacobson reported the presence of Epiprotozoic strata in the Kresttsy R-1 bore hole (1,244—1,155 m interval). The thickness of the Epiprotozoic varies greatly, but it is everywhere composed of finely rhythmically bedded micaceous siltstone—mudstones and sandstones. Grey sandstones occur in the lower part (Vologda Formation) and variegated sandstones in the upper part (Bologoye Formation).

The Epiprotozoic rocks of Byelorussia have been studied and described in detail (Bessonova and Chumakov, 1969; Klevtsova, 1971). These strata are known as the "Vilchansk Formation" which is made up of tillite and sandy to clay-rich rocks. In some areas it consists wholly of tillite (Osipovichi-3 bore hole), in others the lower part is sandstone and conglomerate (Osipovichi-14 bore hole), and in others there are three tillite units and three units of mudstone (Vilchitsy bore hole). The sandstones and mudstones that occur together with tillite, are finely bedded, highly micaceous, and thus resemble the rocks of the Pachelma Group. Carbonate layers are common in the Vilchansk Formation and in the Pachelma Group. In addition to Byelorussia and the Pachelma trough, Epiprotozoic tillites are also present in the graben in the Ladoga region (Bruns, 1963).

In the northern part of the Kola Peninsula and in Scandinavia, strata transitional between platform and miogeosynclinal facies are present. In the *Kola Peninsula* the following terrigenous groups are Epiprotozoic: Volokovaya, Eina Bay and Bargout of the Rybachy and Sredny Peninsulas. These rocks were studied in detail by V.Z. Negrutza (1971). The widely separated Volokovaya and Eina Bay Groups are probably synchronous. Both unconformably overlie the Upper Neoprotozoic Kil'din Group and are overlain unconformably by the Bargout Group. The Volokovaya Group includes: (1) the Kuyakan Formation (from a few metres to 170 m) of poorly sorted, cross-bedded arkose and granule conglomerate with lenses of boulder conglomerate; and (2) the Puman Formation (>350 m) which is made up of sandstone and shale. The Eina Bay Group consists of three formations which, in upward succession are: (1) Mota (350 m) — unsorted boulder and pebble conglomerate, regarded by many as tillite; (2) Lonskaya (700 m) — polymict conglomerate, granule conglomerate, siltstone and argillite with rare interbeds of clayey limestone; and (3) Pereval (>1,000 m) — coarse-grained lithic sandstone and granule conglomerate with coarser conglomerate lenses. Rounded inclusions of sandstone with a carbonate matrix are very typical.

The Bargout Group is composed of: (1) the Maya Formation (250 m) — cross-bedded alluvial sandstone and granule conglomerate with thick (up to 30 m) beds and lenses of boulder conglomerate in the lower part; (2) the Zubovskaya Formation (500 m) — intercalated sandstone, granule conglomerate and shale; and (3) the Tsip—Navolok Formation (>200 m) — shale interbedded with carbonates. Conglomerate of the Maya Formation is probably of fluvioglacial origin. Problematic structures (formerly thought to be casts of medusoids) are present in sandstone and shale that are in tectonic contact with the Bargout Group. These structures are probably inorganic.

The K—Ar age of phyllitic slate from the Eina Bay Group falls in the 670—900 m.y. range (Polkanov and Gerling, 1961). The Volokovaya Group is cut by diabase with an age of 600 m.y. (Bekker et al., 1970).

In *Scandinavia* the most complete Epiprotozoic sections and sections most similar to those of the Kola region occur in Norway, in Eastern Finnmark (Varanger Peninsula and adjacent areas). According to studies by A. Siedlecka, S. Siedlecki and D. Roberts (Siedlecka and Siedlecki, 1967; Siedlecka and D. Roberts, 1972; Siedlecka, 1973), the Eastern Finnmark Supergroup and the unconformably overlying Västertana Group are Upper Precambrian. A. Siedlecka attributes the three groups to the Eastern Finnmark Supergroup. These are the Laksefjord, Raggo and Barents Sea Groups. Their relationships are not certain, for they generally have tectonic contacts. A. Siedlecka has suggested that all three groups may be synchronous in part. They are all composed largely of sandstone with subordinate siltstone, shale and carbonates. The Barents Sea Group is the best known. Its base is not exposed but the thickness is of the order of 9,000 m. The lowest exposed part of the group is the Kongsfjord Formation (about 3,500 m) which is composed of dark argillite, shale and feldspathic sandstone with graded bedding. This formation includes a unit (>4 m thick) of unsorted coarse-grained clastic rocks regarded as tilloid or tillite (Lökevik tillite) (Siedlecka and Roberts, 1972). The Kongsfjord Formation (about 2,500 m) is conformably overlain by the Båsnaering Formation (about 2,500 m) the lower part of which is grey quartzitic sandstone, and the upper part of which is coarse-grained and cross-bedded, and red, with interbedded red siltstone. The overlying Båtsfjord Formation (about 1,500 m) is composed of interbedded dolomite, clayey limestone, shale and sandstone. The overlying Lille Molvik Formation (200—300 m up to 1,500 m) is grey-green sandstone and sandy slate with pyrite and siderite concretions. The section is capped by the Tanafjord Formation (up to 1,300 m), characterized by interbedded quartzitic sandstone, shale and siltstone with a unit of dolomite at the top. This formation lies unconformably on the underlying one. Conglomerate is common in its base.

The Vestertana Group (Upper or Red Sparagmite) unconformably overlies the Barents Sea Group and includes three formations: (1) Smallfjord tillite

(30—60 m) which is composed of green and red-brown mudstones with boulders and pebbles of dolomite, sandstone and granodiorite; (2) Nyborg Formation (up to 400 m) made up of brown and red sandstone, interfingering with red and greenish, finely bedded siltstones and shale; and (3) Upper Tillite (60 m), similar in composition to the Lower Tillite. In some areas siltstone and sandstone interbedded with shale are present above it. The Stappugiedde shale and sandstone unconformably overlie the Vestertana Group and these are unconformably overlain by fossiliferous lowermost Lower Cambrian rocks. The Rb—Sr isochron age of the Nyborg Formation shale is 665—680 m.y. (Pringle, 1973), and of the Tanafjord Formation shale is 800 m.y. (Siedlecka, 1973).

The sedimentary rocks of Eastern Finnmark and of the Rybachy and Sredny Peninsulas are almost certainly correlative. They represent a single stratigraphic unit separated by the Varanger fjord. The Barents Sea Group (or Eastern Finnmark Supergroup) correlates with the Eina Bay and Volokovaya Groups, and the Vestertana Group with the Bargout Group. This correlation is also favoured by the presence of tillite in the lower part of both groups. Thus, in northern Fennoscandia there are two levels of glaciogenic deposits separated by a thick sedimentary succession that appears to have accumulated under warm-climate conditions. Two tillite horizons in the Vestertana Group probably belong to different stages of a single glacial epoch (the Varanger or Lapland Glaciation).

Epiprotozoic glaciogenic rocks are also extensive in Southern Norway. In the Oslo graben region the lower part of the Rene Group (the Upper Sparagmite), which lies unconformably on the Upper Neoprotozoic Biri limestone, is probably Epiprotozoic (Björlykke et al., 1967). The lower part of the group is sandstones and shales (250 m). Higher in the section the Moelv tillite is present (10—110 m). It is overlain by a thin shale (up to 40 m). Rocks of the overlying Rene Group rest unconformably on the shale and are considered to be Eocambrian for worm trails are reported and trilobite-bearing strata of the Lower Cambrian unconformably overlie them. In Sweden rocks that correspond to the Upper Sparagmite are the Varegian Group which lies unconformably on the Upper Neoprotozoic Sparagmite Group. This group is also typified by the presence of tillite (Geier, 1963).

The Serebryanka Group of the *Middle Urals* may be regarded as a stratotype of the Epiprotozoic miogeosynclinal facies (Mladshikh and Ablizin, 1967). It starts with tillites of the Taninsk Formation (up to 800 m). Up section it is followed by the following formations: Garevsk (up to 730 m), Koiva (up to 700 m), Buton (up to 350 m) and Kernos (up to 1,500 m). These units are made up of variegated, green and black shale, siltstone and lithic sandstones, that form separate stratigraphic units, or are interbedded. In this part of the section carbonate rocks, alkaline basalts and tillites (the Koiva Formation) are subordinate. Above these is the Vilva Formation which is made up of tillite, basic volcanics and shale (1,380—1,760 m). The section is capped by the Pershinskaya Formation which is shale and quartzitic sandstone (500 m).

Epiprotozoic strata similar to those of the Middle Urals are known from the Polyudov Ridge (Borovko, 1967). In that region the Upper Neoprotozoic Nizva Formation is overlain by the Churochnaya Group which has two formations, the Ust'—Churochin and Srednechurochin. The Ust'—Churochin Formation is sandstone and siltstone (1,000 m), but the rocks of the Srednechurochin Formation are more varied and include beds of tillite together with siltstone, shale and dolomite. The K—Ar age of glauconite from the upper part of the Ust'—Churochin Formation is 680 m.y., and from the Eocambrian strata which unconformably overlie the tillite, is 640 m.y. These values are probably not reliable for there is a very small difference in age between strata that must vary appreciably in age. Strong deformation of these rocks may have led to argon loss from the glauconite.

The Sablegorsk and Laptapay Formations of the sub-Polar part of the Urals are also thought to be Epiprotozoic. In the Polar Urals the Iz'yakhoy and Khaydyshor Formations are also of this age. Conglomerates of the Laptapay and Khaydyshor Formations resemble tillites (G.A. Chernov, (1948).

The Mazarino Formation of the Uraltau zone may also be Epiprotozoic. It is typified by augen schists with local disseminated pebbles and angular clasts passing into conglomerate (S.S. Gorokhov, 1964). These rocks are possibly tillites.

In the Southern Urals, subdivision of Epiprotozoic strata has been carried out only in the eastern limb of the Bashkir anticlinorium. In this area the Krivaya Luka Formation (sandstone—clay and volcanogenic rocks) is overlain by the Kurgashli Formation which includes tillite-like conglomerate. The first lies unconformably on the Upper Neoprotozoic Karatau Group and the second is overlain by the Eocambrian Asha Group.

In Western Europe Epiprotozoic strata are developed in Great Britain, France, Czechoslovakia and in the Spitsbergen Archipelago.

In the *British Islands*, viz. in Scotland, the lower part of the Dalradian Group (or Supergroup) is Epiprotozoic. It is known as the Lower Dalradian basal calcareous unit (J.G.C. Anderson, 1965; Johnson, 1969). At its base black limestones and slates, underlain by quartzite, transgressively overlie Neoprotozoic rocks (Moine and Torridonian) and locally lie directly on Archean gneisses. Higher in the section there are boulder conglomerates (Schichallion, Portaskaig, Fanad, etc.) that are regarded by many as tillites. Above the tillite, with a break, lies the upper part of the Lower Dalradian which is composed of quartzites and slates which are assigned to the Eocambrian, for they pass upwards into the Upper Dalradian with Lower Cambrian fossils.

In Central England the Epiprotozoic is represented by the Uriconian Group (in Shropshire) and the Charnian Group (in Leicestershire). Both groups are composed of eugeosynclinal sedimentary—volcanogenic rocks. The Charnian Group is of great interest, for it contains many well-preserved casts of non-skeletal animals (principally coelenterates of the Ediacara type). The group is divided into three formations: the lowest (Blackbrook) is composed of volcanic rocks but the overlying (Maplewell and Brand) formations are sedimentary and volcanic strata (Watts, 1947, Ford, 1958). The organic remains (*Charnia masoni* Ford, *Charniadiscus concentricus* Ford, *Cyclomedusa davidi* Sprigg) occur in the upper part of the Maplewell Formation (1,300 m), which is acid volcanics, agglomerate, tuff, slate and tuff conglomerate. The K—Ar age of diorite porphyry dikes cutting the Maplewell Formation is 684 m.y. Emplacement of the dikes may date the break between the Maplewell and Brand Formations. The dikes are probably closely related to the Maplewell volcanics so that the date obtained reflects the age of the formation. The entire Charnian group is cut by diorite that is older than the Cambrian strata. The age of the diorite is 547 m.y. but this age is probably slightly "rejuvenated". Thus, the coelenterates appear to be present in Epiprotozoic rocks.

In *France* the Upper Brioverian Group of the Armorican Massif is Epiprotozoic. It rests unconformably on the Neoprotozoic Lower Brioverian Group and is overlain with angular unconformity by Lower Cambrian strata with archaeocyathids. In the lower part of this group the Granville tillite occurs. The tillite passes up into finely bedded ("varved") and coarse-bedded sandstones intercalated with siltstones and slate (M. Grandor and Wasserburg, 1962).

In *Czechoslovakia* (Krušne Mountains) the so-called "Postspilitic strata" which are sometimes classified as "Eocambrian", are considered to be Epiprotozoic. They unconformably overlie Neoprotozoic eugeosynclinal sedimentary and volcanic rocks. They are thick (up to 2,600 m) and consist of finely bedded slate and sandstone with mixtite beds (Fiala, 1964). Their upper age boundary is defined by the fact that they underlie Lower Cambrian strata.

In the *Spitsbergen Archipelago* two groups (or parts of these groups) belong to the Epiprotozoic. These are the Polarisbreen which is distributed in the northwestern part of West Spitsbergen Island, and the Gothian which is in the North-East Land (Krasil'shchikov, 1973). Both these groups (excluding the lower Elbobreen and Backaberg Formations which terminate a Neoprotozoic sedimentary cycle) are considered to be Epiprotozoic (see Chapter 7).

The Polarisbreen Group is composed of two formations. The lower one, Wilsonbreen, (140—240 m) is composed of two mixtite units with thicknesses of 100 and 60 m, separated by terrigenous—carbonate rocks similar to those that overlie the upper tillite. The overlying Drakoisen Formation (280 m) comprises variegated shale, siltstone argillite and dolomite with katagraphs of the fourth complex (*Vesicularites lobatus* Reitl., *Nubecularites abustus* Z.Zhur.). The Gothian Group has a similar sequence. The lower part is the Sveanor Formation (50—130 m), composed of quartzitic sandstone, passing upwards into tillite. This formation is conformably overlain by the Klakberg Formation (250 m) which is made up of dolomite intercalated with quartz and quartzofeldspathic sandstone.

Lower Cambrian strata with a skeletal fauna (*Salterella*, *Volbortella*, etc.) unconformably overlie both groups. The upper formations of both groups may be Eocambrian, but this does not seem likely for these groups are conformable on the underlying rocks and in most parts of the world an unconformity is reported between the Epiprotozoic and the Eocambrian.

Asia

In the Siberian platform and its surrounding fold belts Epiprotozoic strata formed following important tectonic movements which terminated the Neoprotozoic Era. These were the last strong manifestations of geosynclinal orogeny in major Precambrian fold belts of Siberia. Sedimentary rocks of both platform and orogenic facies are widely distributed there. It is only in the Baikal and East Sayan fold belts that miogeosynclinal strata are present.

The Khorbusuon Group of the Olenek uplift is taken as a platform stratotype of the Epiprotozoic in Siberia. It consists of three formations which in upward sequence are the Maastakh, Khattyspyt and Turkut (Biterman and Gorshkova, 1969). The Maastakh Formation (35—80 m) unconformably overlies different units of the Middle—Upper Neoprotozoic Solooli Group. Its lower part is cross-bedded sandstone and granule conglomerate (coarser conglomerate is less common) overlain by grey biogenic dolomite. The Khattyspyt Formation (up to 190 m) consists mainly of dark-grey limestone but at the base there is a band of variegated quartz sandstone. Sokolov (1972) identified a cast from rocks of this formation as *Rangea* sp., a non-skeletal coelenterate known from the Ediacara faunal complex. Diabase sills are common in the Khattyspyt Formation. According to some workers the Turkut Formation (up to 200 m) lies unconformably on the underlying rocks. It consists of montonous grey dolomite with the stromatolites *Stratifera* Korol. and *Paniscollenia* Korol., etc. and also katagraphs of the fourth complex: *Vesicularites bothrydioformis* (Krasnop.), *V. lobatus* Reitl. and *V. irregularis* Reitl. Glauconite from the upper part of the group is dated at 670 m.y. The Turkut Formation is unconformably overlain by the Eocambrian—Lower Cambrian Kessyuse Formation.

The Khorbusuon Group may be correlated with other platform strata in the northern part of the Siberian platform such as the Starorechenskaya Formation of the Anabar uplift, the Tomtor and Turkut Formations of the Udzha uplift (Tkachenko, 1970), and the Khara—Yuettekh Formation of Northern Kharaulakh (Vinogradov and Sobolevskaya, 1958; Krylov et al., 1971). All of these units, together with the stratotype, form a distinct complex in Precambrian sections of the northern part of the platform. They have particular organic remains. The contacts with underlying and overlying strata are well known. These rocks are everywhere unconformable on Neoprotozoic sequences and are unconformably overlain by rocks of Eocambrian—Lower Cambrian age (see Chapter 9). At the base of all the units correlated with the stratotype, variegated clastic rocks of variable thickness (usually thin) are present. The thicker, upper part of these units is formed of chemical, partly biogenic, dolomite. All of these successions contain microphytolites of the fourth complex. Stromatolites are also present in the Starorechenskaya Formation of the Anabar uplift and the Khara—Yuettekh Formation of Kharaulakh. These include *Stratifera glebosata* Gol., *Paniscollenia emergens* Kom., and *Boxonia* Korol. which are also typical of the fourth complex (Vinogradov and Sobolevskaya, 1958; Tkachenko, 1970).

In addition to those of the northern part of the Siberian platform Epiprotozoic platform strata are also present in other areas of Siberia, in the eastern part of the Aldan shield, in the Irkutsk amphitheatre and in the Taymyr fold belt.

In the *Aldan shield* the Epiprotozoic is represented by the Uy Group which consists of sandstone, siltstone and argillite which is generally greenish-grey or rarely red in colour. In the shield proper the group is of platform facies and is not more than 400—500 m thick, but to the east, in the Yudoma—Maya aulacogen, it takes on a miogeosynclinal aspect, is distinctly folded, and increases in thickness up to 3,000—4,000 m (Nuzhnov, 1967). Acritarchs and medusoid casts have been found in the Kandyk Formation which is the lowest of the group. The upper formation, Ust'Kirba, contains glauconite, which was dated in the 610—720 m.y. range. The Uy Group lies unconformably on the Maya Group which contains Middle and Upper Neoprotozoic phytolites. It is cut by the ultrabasic and alkaline rocks of the Ingili Massif which have a K—Ar age of 610—680 m.y., but the most reliable values, obtained by the Pb-isotope method on different radioactive minerals, are mainly of the order of 650±20 m.y. (Tugarinov and Voytkevich, 1970). The Uy Group and intrusive rocks of the Ingili complex are unconformably overlain by the Yudoma Formation — the Eocambrian stratotype of Siberia. Thus, in the eastern part of the Aldan shield the upper boundary of the Eocambrian is well established at 650—680 m.y.

In the *Taymyr fold belt* the Sovinsk Formation of East Taymyr and the Canyon Formation of the North Land may be attributed to the Epiprotozoic (Egiazarov, 1959; Sobolevskaya and Mil'shtein, 1961). Both formations are unconformable on Neoprotozoic rocks, and the eroded surface of the Sovinsk Formation is overlain by Lower Cambrian "gastropod beds". These formations differ appreciably and their correlation and assignment to the Epiprotozoic are rather arbitrary. The Sovinsk Formation is largely made up of dark limestone. Unsorted boulder conglomerate (glaciogenic?) occurs at its base. The Canyon Formation is composed of red and variegated silty shales, interbedded with marl and with rare sheets of acid lavas.

The platform strata of the Irkutsk amphitheatre are similar to the sequences in adjacent fold belts which are discussed briefly below.

Epiprotozoic "orogenic" strata that accumulated in isolated depressions (troughs and grabens), formed after closure of major geosynclinal systems, are essentially clastic red rocks which include the following groups: Taseyeva, Vorogovka, and Chingasan (Yenisei Ridge) and the Ushakov Formation of the West Baikal region, and the Tochera Formation of the Vitim Highlands.

The Taseyeva Group of the *Yenisei Ridge* is suggested as a stratotype. This group is developed in the Angara—Kansk trough. It consists of the Aleshkin, Chistyakovka and Moshakovo Formations (with a total thickness up to 2,500—3,000 m). The upper and lower formations are largely composed of red sandstone and granule conglomerate. The middle formation is composed of grey sandstone and siltstone with dolomite bands. Medusoid casts were found by N.S. Podgornaya in the Aleshkin Formation. The Taseyeva Formation unconformably overlies Precambrian rocks of various ages (from the Sukhoy Pit Group to the Oslyandsk Group) and is unconformably overlain by the Eocambrian Redkolesnaya Formation.

Podgornaya and others correlate the Taseyeva Group with the Chingasan and Vorogovka Groups of the Yenisei and Teya troughs (rows 8, 9, Table II). Microphytolites of the fourth complex occur in the Sukhorechensk Formation of the Vorogovka Group. This unit is correlated with the Chistyakovka Formation. Correlation with the Chingasan Group is more complex. At the base of this group is the thick Suktal'ma Formation which is mainly boulder and pebble conglomerates and breccias that are probably glaciogenic. Glacial origin of the Suktal'ma conglomerates has been refuted (Grigor'ev and Semikhatov, 1958) and they have been considered to be the product of mud flows and turbidity flows, laid down on the margin of the older cordillera of the Lebyazhinsk uplift. Some of these rocks may be resedimented glaciogenic mixtites. The presence of granite boulders which are not known from the Lebyazhinsk uplift (Itskov, 1970) also tends to contradict the idea of local origin.

According to N.S. Podgornaya (personal communication, 1972), at the time of accumulation of the conglomerate, clastic material was mainly derived from the northeast and not from the Lebyazhinsk uplift. The Suktal'ma Formation has no correlatives in the stratotype section and may be slightly older. Other formations of the Chingasan Group (Suvorov, Podyemnaya, Tayezhnaya), composed of red and grey terrigenous rocks with rare carbonate units, are correlative with the Taseyeva and Vorogovka Groups. The Podyemnaya Formation contains microphytolites of the third and fourth complexes and glauconite dated at 700 m.y. The Suktal'ma Formation includes minor lava flows that yielded about 700 m.y. by K—Ar analysis. This value is probably slightly "rejuvenated".

Some workers include the underlying Kar'yernaya and Lopatino Formations (red terrigenous rocks, partly carbonate-rich) with the Chingasan

Group. These formations are unconformably overlain by the Suktal'ma tillite, and the Lopatino Formation contains microphytolites of the third complex (Gavrilenko et al., 1971) so that they are probably Upper Neoprotozoic, and comparable to the Oslyansk Group of the Yenisei Ridge and the Karagas Group of the Sayan region.

Itskov (1970) considered the emplacement of granite and syenite of the Porozhninsk complex to have occurred at the break between deposition of the Yenisei Ridge successions described above and the Eocambrian (or Lower Cambrian). The K—Ar age of the granite and syenite is about 630 m.y. but the Epiprotozoic age of the complex is not geologically demonstrable.

In the Chadobets uplift of the *Irkutsk amphitheatre* grey and red sandstones, siltstones and minor carbonate rocks of the Togon Formation are similar to the Taseyeva Group. This unit lies unconformably on the Upper Neoprotozoic Medvedkovo or Brus Formations and is unconformably overlain by the Lower Cambrian Ogon'sk Formation.

The Taseyeva Group and correlatives resemble the Oselochnaya Group of Pri-Sayan both in lithology and succession. The glauconite age of the lower part of this group is 737 m.y. The carbonate beds in the middle part of the group contain microphytolites of the fourth complex. The Oselochnaya Group rests unconformably on the Middle and Upper Neoprotozoic Karagas Group which is cut by diabase dated at 1,200 m.y. and is unconformably overlain by the Eocambrian Mota Formation.

In the western part of the *Baikal area* the Ushakovka Formation (greywacke, granule conglomerate and coarser conglomerate) is an "orogenic" facies of the Epiprotozoic. It lies unconformably on the Kachergat Formation of the Middle and Upper Neoprotozoic Baikal Group and is transgressively overlain by the Eocambrian Mota Formation. Towards the Irkutsk amphitheatre the thickness of this formation rapidly decreases (from 1,300 to 50—100 m) and the coarse clastics are replaced by fine-grained sandstones and siltstones that directly overlie the Archean basement of the Siberian platform.

The Tochera Formation is probably also an Epiprotozoic "orogenic" deposit. It is of limited extent in the upper tributary of the Vitim River, in the internal part of the Baikal fold belt. These coarse-grained clastics together with subordinate carbonates and acid volcanics, rest unconformably on the Neoprotozoic Ikat Group and are in turn unconformably overlain by conglomerates of the Burunda Formation which is arbitrarily assigned to the Lower Cambrian. The nature of the Tochera Formation suggests that large tectonic movements preceded its deposition. The Tochera Formation resembles the Ushakovka Formation of the Western Baikal area. It is also similar to the Sarkhoy Formation of Eastern Sayan, in that it includes acid volcanics.

The miogeosynclinal facies of the Epiprotozoic is represented in Siberia by the Bodaybo Subgroup of the Patom Group in the Patom Highlands, the Izyk Group and the Sarkhoy and Gorkhon Formations, which are of the same age as the Izyk Group, in Eastern Sayan. These strata are of limited extent, poorly exposed and have few organic remains, so that it is difficult to establish a stratotype.

The Bodaybo Subgroup of the Patom Group is present in the Bodaybo zone of the miogeosynclinal region of the Baikal fold belt. This sequence is typified by a great thickness (up to 4,800 m), of dominantly terrigenous rocks, presence of conglomeratic greywackes in the upper half of the section (Anangro Formation), relatively high and variable grade of metamorphism (due to large granite intrusions) and close relationships with the underlying Neoprotozoic rocks (Salop, 1964—1967). These peculiarities reflect their geosynclinal depositional environment which developed during the regressive phase of a large-scale sedimentation cycle of the Neo—Epiprotozoic, prior to the Katangan ("Early Baikalian") orogeny that terminated the geosynclinal development of the Baikal region.

The Neoprotozoic (Grenville) orogeny did not cause a significant break in sedimentation in the miogeosynclinal belt of this system. However, there was a change to deposits which were thick and mainly terrigenous (Bodaybo Subgroup) from the terrigenous and carbonate rocks typical of the Kadalikan Subgroup. The Epiprotozoic age of the Bodaybo Subgroup is defined by its occurrence between the Middle to Upper Neoprotozoic (the Kadalikan Subgroup) and the Eocambrian (Gukit and Lower and Upper Padrokan Formations). Also, in dolomite beds in the upper part of the Bodaybo Subgroup (Iligir Formation) katagraphs of the fourth complex have been identified (*Vesicularites concretus* Z. Zhur.). Eocambrian strata overlie granite that cuts the Bodaybo Subgroup. The K—Ar age of similar granites in adjacent parts of the Baikal fold belt is 680 m.y.

The Izyk Group of the Uda—Derba zone (Mana trough) of the *East Sayan fold belt* is composed of two formations with a total thickness of 3,400 m. Each consists of a lower terrigenous sequence (phyllitic slate and sandstone) followed by limestones (Dodin et al., 1968). The group lies unconformably between the Neoprotozoic Kuvay Group and the Eocambrian—Lower Cambrian Anastas'ina Formation. The presence of *Osagia tenuilamellata* Reitl. and *O. undosa* Reitl. (oncolites) in limestones in the lower part of the Izyk Group apparently indicates that it is not Epiprotozoic. However, the stratigraphic position of the group and its close relationships with Eocambrian—Lower Cambrian rocks suggests that assignation of an older age is improbable (formerly these rocks were considered by many to be Cambrian). The oncolites mentioned above have also been found in fossiliferous Cambrian strata in the Eastern Baikal region, the Khankai region of the Soviet Far East and the Shantar Islands of the Okhotsk Sea (data of T.M. Tetyayeva, A.N. Yefimov, M.I. Lipkina, O.A. Stepanov, E.L. Shkol'nik).

There are discrepancies between data on phytolites and geological observations concerning the Epiprotozoic formations of the Irkut—Oka zone of Eastern Sayan. The rocks in question are the Sarkhoy and Gorkhon Formations. The Sarkhoy Formation consists of variegated coarse clastic rocks with subordinate red acid volcanics. It lies with angular unconformity on the Neoprotozoic Orlik Group, largely on the Mongosha Formation which contains stromatolites and oncolites of the second complex, and on the overlying Upper Neoprotozoic Dibin Formation. The Gorkhon Formation generally lies conformably on, but locally has erosional contacts with, the underlying Sarkhoy Formation. It consists of dolomite with stromatolites of the fourth complex (*Sacculia ovata* Korol. in particular) together with stromatolites typical of the second complex (e.g. *Conophyton metula* Kirič. and *C. lituus* Masl.). The Gorkhon Formation is overlain by the Eocambrian—Lower Cambrian Bokson Formation which consists of dolomite, with stromatolites in the lower part. These stromatolites are typical of the fourth complex. Archaeocyathids are present in the upper part. The Bokson Formation lies locally on an eroded and karsted surface of the Gorkhon dolomite, with bauxite at the base. Locally it appears to be conformable (Dodin et al., 1968). The Gorkhon and Bokson Formations may not be greatly separated in time so that the Sarkhoy and Gorkhon Formations are probably not Neoprotozoic.

In the Uda—Derba zone of the Oka River basin there are small intrusive bodies of alkaline and subalkaline granitoids which have given age values up to 620 m.y. by K—Ar analysis (Mitrofanov and Kol'tsova, 1965). These granitoids are considered to be a part of the so-called Paleozoic Ognitsk complex, but it is probable that in Eastern Sayan there are two alkaline granitoid complexes of different age, for one cuts Paleozoic rocks up to the Devonian (Ognitsk complex, s.str.) and the other occurs as clasts in conglomerate (Grebeshkov Mt.) of the Eocambrian Anastas'ina Formation (according to A.Z. Konikov, personal communication, 1972).

In addition to the Siberian occurrences, Epiprotozoic strata are extensive in the northeastern part of the U.S.S.R., in Kazakhstan, Southern China, North Korea and probably in the northern part of the Hindustan Peninsula. Glaciogenic rocks are common in these rocks in the areas listed above.

In the northeastern part of the *U.S.S.R.* Epiprotozoic carbonate-rich and terrigenous rocks including tillites cap the Precambrian section in the Stolbovskoye uplift of the Kolyma area. They consist of a lower division of dolomite and limestone with phytolites of the fourth complex and an unconformable (?) overlying sequence of red terrigenous rocks (sandstone, shale, conglomerate) with a total thickness of up to 400 m. This sequence includes a tillite (up to 130 m) consisting of unsorted pebble—boulder conglomerate with striated boulders (Furduy, 1969).

In certain localities of the Kolyma uplift, tillites are replaced by conglomerate and quartzitic sandstone interbedded with dolomite, named the Tumuss Formation (300—400 m). This is overlain by the Korkodon Formation (550 m) which consists of dolomite with phytolites of the fourth complex. It is considered to be Eocambrian, for it is conformably overlain by variegated limestones and marl of the Kirpichnikov Formation which has a Lower Cambrian fauna (V.P. Rabotnov et al., 1970).

In *Kazakhstan*, Epiprotozoic rocks are present at Bolshoy Karatau and Ulutau. The Ulutau Group unconformably overlies Neoprotozoic strata (Kaynar Formation) which are cut by granite of the Kumystin Massif that yielded 670±70 m.y. by K—Ar analysis of micas and 720 m.y. by the α-Pb method on zircon (Korolev, 1971). Both of these values are probably too young. The Ulutau Group (up to 1,500 m) consists of various terrigenous rocks: sandstone, siltstone and shale with two tillite horizons, one 300 m above the base, the other near the top. Lower Cambrian strata rest on the eroded surface of the Ulutau Group.

In Ulutau, the most complete Epiprotozoic section is situated in the area of the Baykonur synclinorium (Zaytsev and Filatova, 1971). In that region the Upper Neoprotozoic Koksuy Group is unconformably overlain by the Akbulak Group (up to 1,500 m) which is made up of conglomerate and sandstone (partly tuffaceous). The Ulutau Group unconformably overlies these rocks. It is subdivided into: (1) Zhaltau Formation — quartz sandstone and granule conglomerate that passes upwards into phosphate-bearing shale interbedded with limestone; (2) Satan Formation — tillite, sandstone and shale (the lower tillite horizon); (3) Bozigen Formation — dolomite; (4) Kurayli Formation — variegated shale and limestone; and (5) Baykonur Formation — tillite and conglomerate (the upper tillite horizon). As at Bolshoy Karatau the Ulutau Group is, in this region, overlain by Cambrian strata.

According to Zaytsev and others (Zaytsev and Filatova, 1971; Zaytsev et al., 1972) the Akbulak Group is younger than the alkaline granitoid of the Aktas Massif, dated by the Pb-isotope method on zircon. These authors accept the 650±50 m.y. age for these granites, but the age was calculated on an isochron for several samples of zircon yielding discordant values. This value is not reliable for many samples of zircon contain abundant common lead (1—1.4%), and the necessary correction essentially changes the results. Only one sample with a low common lead content (0.325%) yielded concordant values on major isotope ratios and it can be considered reliable. The $^{207}Pb/^{206}Pb$ and $^{206}Pb/^{238}U$ ratios revealed an age of 790±60 m.y. The geological relationships of the Aktas granitoid and the Akbulak Group are not certain, for the contacts are not exposed. The supposed older age of the granitoid is based on the presence, in the Akbulak conglomerate, of granite pebbles similar to the Aktas granite. The identification of granitoid may be erroneous.

Precambrian strata that include tillites are reported from a number of areas in *Tien-Shan*. These strata have been studied in detail by many workers. According to Zubtsov (1972) four units can be identified within this sequence. These are (in upward sequence): (1) Kichitaldysuy, which underlies tillite (100—200 m) and is made up of conglomerate, arkose and siltstone with dolomite interbeds in many areas; (2) Dzhetym or lower tillite (400—

2,500 m) made up of terrigenous rocks with tillite associated with finely bedded shale. In the Dzhetymtau Ridge, hematitic and magnetitic rocks are associated with glacial deposits; (3) Dzhakbolot or intratillitic sequence (200—500 m) consisting of terrigenous and carbonate rocks; and (4) Baykonur or upper tillite (15—100 m) and largely composed of tillite. These four units were found by E.I. Zubtsov, from Kazakhstan to the Kuruktag Ridge in China (Sinkiang), where similar rocks were described by Norin (1937).

In the Karatau—Talas zone of Tien-Shan the Malokaroy Group is assigned to the lower (pre-tillite) part of the Epiprotozoic. It is unconformable on the Upper Neoprotozoic Karagrain Group and is overlain unconformably by Eocambrian (?) and Lower Cambrian strata. The Malokaroy Group in the Talas Ridge area includes the Kizylbel Formation (30—200 m) which is made up of variegated siltstone and sandstone, the Chichkan Formation (50—100 m) of grey siltstone and shale interbedded with sandstones and silicified dolomite with the stromatolites *Patomia* and *Linella* (fourth complex), and the Kurgan Formation (300—700 m) of variegated tuff, porphyry, siltstone and shale (Maksumova, 1967). The Malokaroy Group is here considered to be lowermost Epiprotozoic, but it may be the uppermost part of the Upper Neoprotozoic which is usually characterized by phytolites of the fourth complex.

In Southern China, in Hupeh and Yunnan provinces, platform strata of the Liangtou, Nantou and Toushantou Formations which the Chinese geologists (Lü Hung-yün and Sha Tzu-an, 1965) attribute to the Sinian "system" may be Epiprotozoic. The Liangtou Formation (50—500 m) consists of red arkosic conglomerate and sandstone interbedded with argillite and siltstone. These rocks lie subhorizontally on Neoprotozoic (?) metamorphic rocks, cut by granite dated at 900 m.y. (K—Ar method). The Nantou Formation (up to 200 m) is composed of tillite. It unconformably overlies the Liangtou sandstone or older metamorphic rocks. The younger Toushantou Formation (80—420 m) is dark argillite interbedded with limestone, dolomite and sandstone. Glauconite from this formation is dated at 620—670 m.y. old, but this age is probably slightly "rejuvenated". The Toushantou Formation is overlain (apparently conformably) by the Taning Formation (dolomite and limestone with phosphorite) which contains organic remains resembling hyolites in its upper part. Higher up the section there are Lower Cambrian strata with the trilobite *Redlichia*. The Taning Formation may be Eocambrian and/or lowermost Lower Cambrian.

In *North Korea* the Kuhen Group (Formation) is Epiprotozoic. It is developed in the Phennam trough where it rests unconformably on various Sinian formations of the Middle—Upper Neoprotozoic (the Sanvon Group) and is transgressively overlain by fossiliferous Lower Cambrian rocks (Yandok Formation). This group consists of phyllites or slates with rare beds of calcareous sandstone. It includes one or two tillite units in its lower part. These are preserved as mudstones with disrupted or contact-framework polymictic conglomerate. The group varies greatly in thickness mainly due to erosion just before the Cambrian. It is normally a few hundred metres thick, but may reach 800—1,200 m (Masaytis, 1964).

In the *Hindustan Peninsula*, the Epiprotozoic interval is probably represented by the Melany Group in Western Rajasthan. It is made up of subaerial acid volcanics, including rhyolite (porphyry) and felsite, interbedded with tuff and pyroclastic breccia. The lavas are associated with subvolcanic intrusions of granite porphyry and granite. The volcanic rocks subhorizontally overlie metamorphic rocks of the Aravalli Group. The subvolcanic intrusions cut the post-Delhi (Erinpurian) granite. However, the relationships between the volcanics and the Vindhyan Supergroup are not certain. Many Indian geologists think that the Melany rhyolite is older than Vindhyan, but even they sometimes state that the Melany rhyolite and granite porphyry cut the Vindhyan rocks, but do not give detailed descriptions (Krishnan, 1960a, p. 134). Probably the second concept is correct, for the Rb—Sr isochron age determination on six samples of unaltered rhyolite and tuff yielded a value of 745±10 m.y. (Crawford and Compston, 1970).

North America

In North America Epiprotozoic strata form a single stratigraphic sequence, as in Eurasia. They are separated from the overlying and underlying strata by unconformities. As in the case of the Neoprotozoic, these rocks are represented in the North American platform by flat-lying platform deposits and, in the surrounding Phanerozoic fold belts, by miogeosynclinal units.

The Upper Keweenawan Group is suggested as a stratotype for the platform facies. It is developed in the area of Lake Superior, in the southern part of the *Canadian—Greenland shield*. The Copper Harbour conglomerate occurs at its base. These rocks have been considered by some to be tillites (Murray, 1955). Above the conglomerate is the Nonesuch shale with an age of 1,075 m.y. by Rb—Sr isochron analysis (S. Chandhuri and Faure, 1964). Still higher there are red, cross-bedded arkosic sandstones, siltstones, shales and the Freda (Oronto) conglomerate. These make up the major part of the group with a total thickness of up to 7,000 m. The Upper Keweenawan rests unconformably on the Lower Keweenawan Group and on gabbro—diabase and gabbro of the Duluth complex which cuts the latter and has an age of 1,150—1,250 m.y. The Upper Keweenawan rocks are unconformably overlain by the Eocambrian (?) Bayfield sandstone.

In Minnesota the Fond du Lac sandstone is correlated with the Freda (Oronto) sandstone. It unconformably overlies the Lower Keweenawan rocks and gabbro—diabase, and is in turn unconformably overlain by the Eocambrian (?) Hinkley sandstone.

In the Bear province of northwestern Canada Epiprotozoic sandstones, slates and dolomites unconformably overlie the Upper Neoprotozoic Coppermine River Group (older than 1,200 m.y.), and are cut by diabase with a K—Ar age of 635—735 (600—700) m.y. In the adjacent Slave province and on Victoria Island these strata are known as the Shaler Group. In the Brock Inlier (Balkwill and Yorath, 1970) they consist of grey shale, argillite and siltstone (>1,000 m) overlain by red dolomite with stromatolites including gypsiferous rocks and siltstone in the upper part (up to 800 m). In Victoria Island the Shaler Group is much thicker (3,500—3,900 m). It contains abundant carbonate rocks and evaporites (gypsum and anhydrite). The dolomite and limestone contain abundant columnar stromatolites which form banks and reefs (Thorsteinsson and Tozer, 1962; Young, 1974). The group rests subhorizontally on the Archean (?) gneiss and granite and is unconformably overlain by basalt and agglomerate. These rocks are intruded by diabase sills which do not cut the Cambrian strata. The K—Ar age of these diabases is 670 (640) m.y.

Epiprotozoic strata are most common in the northern part of Greenland. Platform strata of this age are developed in the northwestern part of Greenland in the Thule region (Cape York) and in the northeast in Peary Land and Danmarksfjord. The most complete section of these strata is reported from Southern Peary Land (Berthelsen and Noe-Nygaard, 1965). In this region the base of the Epiprotozoic section is tillite (5—50 m), which is unconformable on the Neoprotozoic Midsummer sandstone. These sandstones are cut by diabase with an age of 1,045 (1,000) m.y. Above the tillite there are sandstones interbedded with shale and dolomite (100 m), followed by dolomite (200 m), topped by sandstones interbedded with quartzite, shale and limestone (420 m). The dolomite is stromatolitic and oncolitic and contains microscopic algal remains similar to those reported from the Epiprotozoic Bitter Spring Formation of Northern Australia (Pedersen, 1970). The Lower Cambrian Brønlund dolomite lies unconformably on the Epiprotozoic strata.

The Upper Thule Group, also known as the Narsarsuk Formation (Group) is correlated with the sequence at Peary Land. It lies unconformably on the Upper Neoprotozoic Lower Thule Group. Red dolomite and siltstone are present in the lower part. The middle part is grey dolomite (locally with oncolites), and the upper part is red sandstone. Salt casts are reported from the lower and middle parts of this group (Berthelsen and Noe-Nygaard, 1965; Pedersen, 1970). The K—Ar age of a diabase dike cutting the group is 710 (676) m.y. Other diabase dykes and sills emplaced in the rocks underlying the Lower Thule Group are 640 (610) m.y. old (Dawes et al., 1973).

In the Danmarks Fjord area the Campanuladale Formation is Epiprotozoic. It is composed of sandstone (lower part) and stromatolitic limestone (upper part) and has a thickness of about 250 m. The overlying Fune Lake Formation (also Epiprotozoic) consists of stromatolitic dolomite (320 m). The Campanuladale sandstone lies unconformably on the Norsemandale sandstone which is cut by diabase. The Fune Lake dolomite is unconformably overlain by Lower Cambrian quartz sandstone. Both the Epiprotozoic strata of Danmarks Fjord and the Upper Thule Group are probably somewhat younger than the tillite horizon in Peary Land (Table III).

In the southeastern part of the *North American plate*, in the Arbuckle and Wichita Mountains (Oklahoma) sedimentary and volcanic rocks of subplatform aulacogen type are developed. They are probably Epiprotozoic and are subdivided into two conformable groups (or formations) — the Tillman, composed of terrigenous rocks, and the Navajo, of basic volcanics (Ham et al., 1964). The Tillman Group nonconformably overlies granite with a Rb—Sr age of 1,050 m.y., and the Eocambrian (?) Carleton acid volcanics unconformably overlie the Navajo Group.

Epiprotozoic miogeosynclinal—subplatform type strata are present in all the fold belts surrounding the North American platform. The Windermere Group may be taken as a stratotype. It is present in the northern part of the *Cordilleran fold belt*, in the mountains of British Columbia. This group lies unconformably on the Neoprotozoic Purcell Supergroup which is cut by diabase dated at 1,200 m.y., and is also unconformably overlain by the Eocambrian Three Sisters Formation. The Windermere Group includes, in ascending sequence, the following formations: Toby (20—600 m) — tillite interbedded with sandstone and siltstone; Irene (0—2,500 m) — basic volcanics interbedded with argillite, conglomerate and sandstone, with local interbeds of limestone; and the Monk or Horsethief Creek Formation (up to 2,000 m) that lies locally with erosional discordance on the underlying rocks and which has a tillite-like conglomerate near its base, but is largely composed of argillite, siltstone and sandstone with interbeds of conglomerate and limestone (Salmo-map, 1965). K—Ar dating of only slightly altered volcanics of the Irene Formation yielded values in the 865—890 (825—850) m.y. range.

The Rapitan Group may be correlated with the Windermere Group. It extends from the Yukon Territory into adjacent parts of the District of Mackenzie. This group consists largely of argillite and mixtite but the upper part is composed of shale with subordinate dolomitic limestones. In the lower part of the group, among argillites and mixtites, there are extensive iron formations. The mixtites contain striated and faceted clasts of various sedimentary and igneous rocks. These mixtites have been regarded by some as tillite, or as formations related to glacial processes. The total thickness of the group is about 1,800 m (Blusson, 1971; Gabrielse et al., 1973). The Rapitan Group lies with an angular unconformity on a thick sequence of terrigenous and carbonate strata with stromatolites. These strata are compared with the Purcell Supergroup of the Middle—Upper Neoprotozoic. Some 3,000 m above the top of the Rapitan Group Olennelid trilobites and Archaeocyathids of the Lower Cambrian are reported. It is quite probable that the sequence above the Rapitan Group includes both Eocambrian and Lower Cambrian rocks. Tillites in the Bonnet Plume River basin (Yukon Territory, Canada) may be correlated with the Rapitan Group. The base of the Bonnet Plume mixtites is not exposed. They are unconformably overlain by Lower Cambrian rocks (Ziegler, 1959).

In southern areas of the North American Cordillera Epiprotozoic strata are present in Utah, in Northern California and in Nevada. In Utah (Great Salt Lake and Wasatch Ridge area) the Mineral Fork tillite lies unconformably on the Neoprotozoic Big Cottonwood Group and is transgressively overlain by quartzite of the Eocambrian Mutual Formation (K.C. Condie, 1967). In California (Death Valley) the Kingston Peak, Noonday and Johnnie Formations are Epiprotozoic. They are all separated by stratigraphic breaks (Wright and Troxel, 1966; Stewart, 1970). The Kingston Peak Formation (300—600 m) lies unconformably on the Middle Neoprotozoic Pahrump Group and consists of sandstone interbedded with siltstone and argillite, grading upwards into tillite with striated flat-iron-shaped boulders of various rock types. The sandstone contains disseminated pebbles of quartz and granite. The overlying Noonday Formation (400—660 m) is composed of grey massive dolomite, and the uppermost Johnnie Formation (600—1,300 m) is sandstone, siltstone and dolomite with stromatolites of the fourth complex (*Linella ukka* Kryl., *Boxonia gracilis* Korol.) (Cloud and Semikhatov, 1969). These strata are overlain by the Eocambrian (?) Stirling quartzite which is overlain by rocks with Lower Cambrian fossils.

In the *East Greenland Caledonides* the upper part of the Hagen Fjord Group in Kronprins Christians Land (Berthelsen and Noe-Nygaard, 1965) and the major part of the Merkebjerg Group in the King Oskars Fjord area are Epiprotozoic (C. Poulsen, 1956). The lower part of these two units is composed of tillite intercalated with sandstone. These are overlain by shale—limestone—dolomite strata (Table III). Tillite of the Hagen Fjord Group (Ulvaberg Formation) rests on the lower (Neoprotozoic) part of the group with a sharp unconformity. The whole group is unconformably overlain by the Lower Cambrian (?) Cape Holbek quartz sandstone. The miogeosynclinal strata of the upper part of the Hagen Fjord Group are divided in the same way, and bear the same names as the platform strata at Danmarks Fjord, with the exception of the lower tillite formation. The Merkebjerg Group tillite lies unconformably on the Neoprotozoic Eleonore Bay Group. Carbonate strata at the top of the Merkebjerg Group are overlain unconformably by red terrigenous—carbonate rocks with salt casts. These red beds are usually included as the upper unit of the Merkebjerg Group. They are probably Eocambrian.

The Ella Bay dolomite in the Innuitian fold belt of Ellesmere Island is correlated with the upper carbonate part of the Epiprotozoic section in East and North Greenland. It lies between the Neoprotozoic Kennedy Channel Group (Formation) and the Eocambrian Ellesmere Group and is separated from both by unconformities (Kerr, 1967).

In the *Appalachian fold belt* sedimentary and volcanic strata, in almost every region of Precambrian outcrops, are Epiprotozoic. Among these is the Mount Rodgers Group (also known as "Grandfather" and "Ashne"). It is present in northwest North Carolina. According to Rankin (1969, 1970) it consists of metamorphosed sedimentary rocks (greywacke, arkose, slate, etc.), interbedded with basic and acid metavolcanics. In the lower part of the group there are arkosic sandstones and unsorted boulder—pebble conglomerates which have some of the characteristics of tillites. The group overlies the Cranberry gneiss complex with a sharp unconformity and is transgressively overlain by Lower Cambrian rocks. The acid volcanics are dated at 820 m.y. by the Pb-isotope method on zircon. This dating, and also the presence of tillite, support its assignment to the Epiprotozoic.

In the northern part of the Blue Ridge (Virginia) a close analogue of the Mount Rodgers Group is present in the metamorphic rocks of the Glenarm complex. This complex includes the Lynchburg metasedimentary formation and the overlying Catoctine Formation, composed of basic metavolcanics (Espendhade, 1970). To the south in the Valley and Ridge area the lower part of the Ocoee Supergroup appears to be Epiprotozoic. One part of it is called the Snowbird Group. American geologists (Espendhade, 1970; Rankin, 1970) correlate this group with the Mount Rodgers Group; it is divided into four formations made up of various terrigenous rocks (Table III). It lies unconformably on an Archean (?) gneiss complex and is overlain unconformably by the Great Smoky Group (upper part of the Ocoee Supergroup). The Great Smoky Group may be Eocambrian.

In the northeast side of the Appalachian fold belt the major part of the Conception Group (2,400 m) of the Avalon Peninsula (Newfoundland) is probably Epiprotozoic (Hutchinson, 1953; McCartney, 1967). This group consists of various terrigenous rocks, largely greywacke, siltstone and argillite, with subordinate chert and conglomerate. In the lower part there are two units of finely bedded siltstone and sandstone with disseminated clasts (up to 30 cm across) which in some areas are striated and faceted. This is one of the reasons for regarding these rocks as glacial marine (McCartney, 1967; Brueckner and M.M. Anderson, 1971; M.M. Anderson, 1972).

In the middle part of the group, among terrigenous rocks, there are andesitic lavas, and in the uppermost part (Cape Kovy Formation) which is mainly shale and chert, there are numerous casts of non-skeletal animals of the Ediacara type, mostly medusoids (Anderson and Misra, 1968; Misra, 1971). The thick terrigenous Hodgewater Group (>4,500 m) unconformably overlies the Conception Group, which is in turn overlain unconformably by the Random quartzite. The quartzite contains a fauna that is lowest Lower Cambrian. Thus, the glacial units are stratigraphically separated from Lower Cambrian rocks by about 7,000—8,000 m of section. The Cape Kovy Formation (according to the organic remains present) seems to be Eocambrian, but it may in fact be Upper Epiprotozoic. The nature of its contact with the major underlying part of the Conception Group is not certain, so that the possibility that these two units are separated by a small break cannot be ruled out.

Isotopic age data apparently contradict the Epiprotozoic age of the Conception Group. Rb—Sr isochron analysis yielded 600 (568) m.y. for the Harbour Main volcanics which are overlain by the group, and 607 (574) m.y. for the Holyrood granite which is probably comagmatic with the volcanics (McCartney, 1967). Considering the enormous thickness of Precambrian strata present and the structural complexity of the area, both values may in fact be strongly "rejuvenated". M.M. Anderson (1972) reached the same conclusion. He also considered the Harbour Main volcanics to be in the lower part of the Conception Group, and that sedimentation (and glaciation) and volcanism were essentially contemporaneous. This idea is supported by the intimate relationship of volcanic and glacial rocks in other Epiprotozoic successions of the Appalachians. The tillite from the Conception Group is probably close in age to volcanics of the Mount Rodgers Group in Virginia and of the Coldbrook Group in New Brunswick. The former gave an age of 820 m.y., and the latter 795 m.y., by the Rb—Sr isochron method (M.M. Anderson, 1972).

General Characteristics

Throughout all the continents Epiprotozoic strata form a well-defined complex that, in almost every case, is separated from the underlying and overlying strata by distinct unconformities. The lower boundary is fixed by the end of the Grenville orogeny (1,000 m.y.) and its upper boundary by the time of the end of the Katangan orogeny (650—680 m.y.). The length of the Epiprotozoic Era is about 320—350 m.y. This is similar to the length of the Paleozoic Era (330 m.y.) but is longer than the Mesozoic or Cenozoic Eras (170 and 70 m.y., respectively).

The most characteristic feature of the Epiprotozoic is the widespread occurrence of glaciogenic rocks. These deposits appear to have been laid down over a very wide range of latitudes, whether recent or ancient latitudes are considered (Fig.33). Some areas, for example in Eastern Siberia, lack tillite although Epiprotozoic strata are present. This may be due to the existence there, of local conditions that were unfavourable for glaciation, or possibly glacial units were simply not preserved. Recent petrographic studies by O.N. Sennov (personal communication, 1973) on Epiprotozoic strata of the Western Baikal region (Ushakov Formation) and of the Sayan region (Mara Formation of the Olkha Group) suggest that these rocks formed under cold-climatic, possibly glacial, conditions.

Studies of Epiprotozoic sequences in both Northern and Southern Hemisphere continents show that tillites occur at two distinct stratigraphic positions, usually separated by thick deposits showing no evidence of glacial conditions. The lower tillite may lie directly on an eroded surface of Neoprotozoic rocks, or may have been deposited on a thick pre-glacial sequence of Early Epiprotozoic age. The upper tillite is always close to the top of the

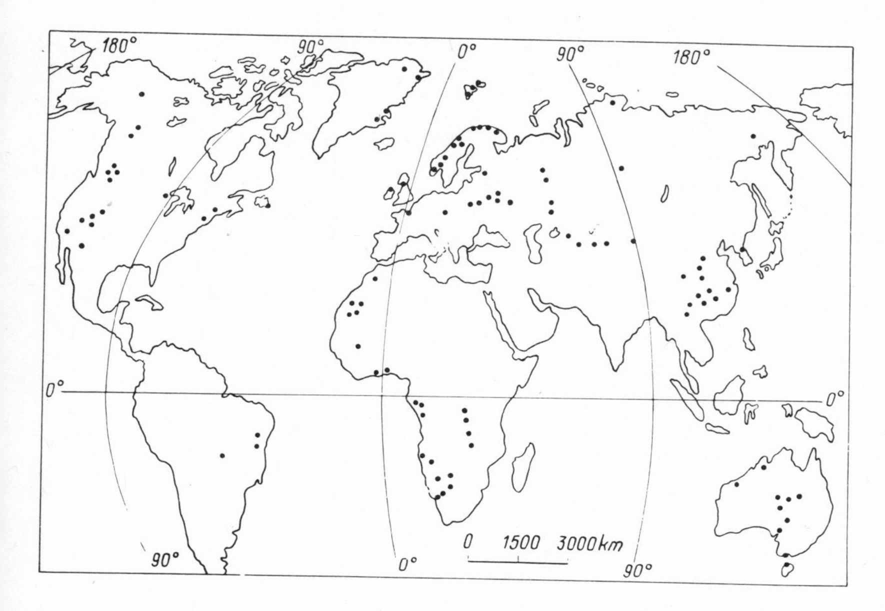

Fig. 33. Distribution of Epiprotozoic glacial deposits (tillites).

Epiprotozoic sequence. The stratigraphy of the glaciogenic units is generally complex, for the tillites are interbedded with terrigenous, carbonate and some volcanic rocks together with either interglacial deposits or various fluvioglacial and glacial-marine units.

The age of the lower tillite can be roughly estimated from Pb-isotope data (on zircon) from volcanics that overlie the tillite in the Appalachian fold belt. This value is 820 m.y. The K—Ar age of volcanics overlying tillite of the Windermere Group in Western Canada is 860—890 m.y. The tillites in Equatorial Africa appear to be of similar age. The age of the lower tillite in Northern Australia (Moonlight Valley tillite) is, on the basis of Rb—Sr isochron analysis on phyllitic slate, 740 m.y. (Compston and Arriens, 1968; Perry and Roberts, 1968), but this value may represent the time of the initial metamorphism. The age of the upper tillite, including the Churochinskaya Formation (Polyudov Ridge of the Urals) can be estimated from two K—Ar dates on glauconite. One gave an age of 680 m.y. from glauconite in rocks of the Ust'Churochin Formation which underlies tillite. The other is 640 m.y., for glauconite from the rocks of the Eocambrian Il'yavozh Formation which unconformably underlies tillite. Both of these values (especially the first one) are probably too young because of argon loss. Pringle (1973) recently published results of Rb—Sr isochron dating of shale from the Nyborg Formation in the Varanger Fjord area of Northern Norway. He obtained two parallel isochrons corresponding to 665±25 m.y. and 680±55 m.y., respectively.

This shale was deposited during the Late Epiprotozoic (Varangerian) glacial period. The second value is probably closer to the time of the Varangerian glaciation, but both values are probably slightly too young, for shale from the Eocambrian Stappuggiedde Formation gave a value of 530—550 m.y. The Stappuggiedde Formation lies on the upper tillite and is overlain by the Lower Cambrian strata. The age obtained may reflect Early Caledonian "rejuvenation" for the true age of the Stappuggiedde shale must be more than 570 m.y. — the lower boundary of the Cambrian.

The upper tillite of North Australia is slightly older than 685 m.y. according to Rb—Sr isochron dating of overlying shale (Compston and Arriens, 1968; Perry and Roberts, 1968). Shale from the upper tillite (the Gbelia tillite) in Western Africa is of the same age according to Rb—Sr isochron dating (N.M. Chumakov, personal communication, 1973). Since the upper age limit of the Epiprotozoic is 650—680 m.y., and since the upper tillite is close to the top of most sections, its age is probably about 690—670 m.y. The lower and upper tillite horizons appear to belong to two separate glacial epochs separated by a time interval of about 150—200 m.y.

Some workers (e.g. M.A. Semikhatov) assign the lower tillite to the Upper Riphean, and the upper tillite to Vendian (terminal Riphean) which is defined by the presence of phytolites of the fourth complex. However, this complex of phytolites has a great vertical range (from the upper part of the Upper Neoprotozoic to the Cambrian), and both glaciogenic units occur in a single stratigraphic complex.

Formation of the Epiprotozoic glacial deposits may have been a global phenomenon. If this was the case, then the tillites would provide useful marker horizons for intercontinental correlation. The tillite horizons may also be used for subdivision of the Epiprotozoic into four parts; pre-Glacial, Lower Glacial, Interglacial and Upper Glacial. However, in some areas, there are difficulties, for it is not certain whether two tillitic horizons were formed as part of one glaciation (say the older one) or whether they represent both.

Evidence of Epiprotozoic glaciation varies from place to place. In Europe, according to Chumakov (1971), the Late Epiprotozoic (Varangerian or Laplandian) glaciation was of continental type. A huge ice cap covered most of Eastern Europe. It was centred on the head waters of the Volga River. Continental glacial deposits in the periphery of the ice-sheet was interpreted as grading out into fluvio-glacial and glacial-marine strata. K.E. Jacobson suggests that a transition from glacial-marine facies occurs in the central part of the East European platform (Pachelma trough). In other areas of the Northern Hemisphere the data available do not permit detailed definition of the paleogeography during deposition of the glacial units. However, certain regions are characterized by continental glacial deposits, and others by marine and fluvio-glacial deposits. In many cases the glacial sediments were redeposited by rivers, mudflows, subaqueous slumps and turbidity currents. Some regions were probably sites of mountain glaciation, but the probability of preservation of such deposits is very low. There is abundant evidence in favour of the primary glacial origin of many such formations, but more detailed studies are required.

In many regions pre-Glacial and Interglacial deposits are mainly terrigenous, in some cases they are carbonate-rich terrigenous rocks and rarely volcanics. Red beds, salt and gypsum casts, stromatolitic limestone and dolomite are present. These are generally interpreted as deposits formed in a region of hot climate. The climatic change leading to glaciation appears to have been relatively sudden (see Chapter 11).

In many respects the Epiprotozoic strata resemble those of the Neoprotozoic, with the exception of the glacial formations. Red beds are abundant in platform and miogeosynclinal sequences of the Epiprotozoic as is the case in similar facies of the Neoprotozoic. However, bedded hematite and siderite ores are relatively rare in the Epiprotozoic, as also are epigenetic magnesite deposits related to chemogenic dolomite, although similar host strata are present in many Epiprotozoic sequences. Many of the Epiprotozoic glacial deposits are associated with bedded hematite—magnetite iron formations. In the northern continents iron formations of this kind occur in the Dzhetym Formation in Central Asia and in the Rapitan Group of Northwestern Canada. There are many similar occurrences in the southern continents (the Tsumeb, Chuos and Holgat (Hilda) Formations of Southwestern Africa, the Buem Formation of Western Africa and a number of formations in Australia). In most cases this association occurs in the lower glacial level of the Epiprotozoic. Unfortunately, there is no adequate explanation of this association at present.

Phytolites of the fourth complex are most common in the Epiprotozoic: the most characteristic stromatolites are *Boxonia* Korol., *Linella* Kryl. and *Sacculia* Korol. The oncolites include *Osagia minuta* Z.Zhur., and katagraphs, *Vesicularites bothrydioformis* (Krasnop.), *V. lobatus* Reitl., *V. concretus* Z. Zhur., *Vermicularites tortuosus* Reitl. Casts of medusoids are reported from some Epiprotozoic units in several continents. Casts of other coelenterates are reported from the Khattyspyt Formation of the Olenek uplift (Northern Siberia) and in the Charnian Group (England). Thus, the Epiprotozoic strata contain the first traces of non-skeletal metazoans of Ediacara type.

The tectonic pattern of the Epiprotozoic and Neoprotozoic is also similar, for the major structural elements (platforms and geosynclines) which appeared in the Neoprotozoic, also existed during the Epiprotozoic. However, there was a distinct decrease in platform sedimentation, probably related to a decrease in size and depth of submergence of the intracratonic troughs. An abrupt decrease in amount, or even complete absence, of platform-type volcanism, is typical. During the Epiprotozoic the older platform underwent further stabilization and consolidation. The Katangan orogenic cycle that terminated the Epiprotozoic, caused tectonic movements and associated intrusive magmatism. The most intensive orogeny occurred in the

Baikal fold belt. In the Baikal geosynclinal system very intense folding and emplacement of huge granite plutons (among the greatest in the world) were associated with the Katangan (Early Baikalian) orogeny. This orogeny terminated the Baikal geosynclinal systems and caused development of an orogenic phase in the Eocambrian. In the platforms, small intrusions of alkaline rocks and diabase occurred during the Katangan orogeny.

The isotopic dating of intrusive rocks of the Katangan cycle shows that the upper age limit of the Epiprotozoic lies in the 650—680 m.y. range. Pb-isotopic and K—Ar dating of ultrabasic and alkaline rocks of the Ingili pluton in the Aldan shield gave an age of 650—680 m.y. These data are used in establishing the boundary. The Ingili pluton cuts Epiprotozoic strata of the Uy Group and is unconformably overlain by the Udoma Formation which is an Eocambrian stratotype. Dating of the granite in folded areas (K—Ar method) has given slightly "rejuvenated" values (550—630 m.y.).

The above upper age boundary is in good agreement with glauconite dates from the youngest Epiprotozoic and Eocambrian strata. The age (glauconite) from the uppermost part of the Starorechenskaya Formation in the Anabar uplift is 675 m.y., from the upper part of the Turkut Formation in the Olenek uplift is 670 m.y., from the Ust'-Kirba Formation in the Yudoma—Maya area of the Aldan shield is 680 m.y., from the Suvorov Formation of the Yenisei Ridge is 688 m.y., etc. A similar age is reported from shale of the upper tillite horizon by the Rb—Sr method. The glauconite age of the oldest Eocambrian beds, which unconformably overlie Epiprotozoic strata, is 650 m.y. (Yudoma Formation), 610 m.y. from the Zherba and Moty Formations of Siberia, 640 m.y. from the Il'-yavozh Formation of the Polyudov Ridge and 607 m.y. from the Redkino Formation (Valday Group) in the central areas of the European part of the U.S.S.R. All these rocks form part of the Upper Precambrian platform cover and were never deeply submerged. The glauconite ages therefore probably approximate the time of sedimentation, at the beginning of the Eocambrian.

In the Northern Hemisphere there are few important syngenetic economic deposits that are known to be related to Epiprotozoic strata. There are some occurrences of iron and iron—manganese ores, principally as matrix material in clastic rocks of platform complexes. In other regions, particularly in the Southern Hemisphere (e.g. in Equatorial Africa) large sedimentary copper deposits occur in the lower (pre-Tillite) part of the Epiprotozoic. Endogenic mineralization is associated with plutonic granites. Mica and rare-metal pegmatites of great economic importance typify this kind of mineralization. Such deposits occur in several regions of Siberia.

CHAPTER 9

THE EOCAMBRIAN

The Eocambrian complex formed between the end of the Katangan orogenic cycle and prior to deposition of sediments containing the oldest skeletal fossils of the Lower Cambrian. The Eocambrian complex can only be differentiated in areas where it underlies the oldest fossiliferous Lower Cambrian beds and/or contains casts of non-skeletal animals of the Ediacara type (medusoids of the Ediacara fauna are also found in older Epiprotozoic strata), and overlies Epiprotozoic rocks. In many cases, strata which do not fall within this definition may be correlated reliably with well defined stratotypes.

Regional Review and Principal Rock Sequences

Europe

In Europe the Eocambrian is fully represented in the *East European platform* and surrounding fold belts. Everywhere the Eocambrian rocks are unconformable on older formations, and are composed of interbedded sandy and muddy rocks which are locally tuffaceous. These sequences may be divided into platform and orogenic facies. Certain strata that are locally developed on the platform margins may be described as miogeosynclinal. Their Eocambrian age is not certain in all cases.

The Valday Group in the central part of the Moscow syneclise is the European stratotype for Eocambrian platform facies. This group comprises the Redkino, Lyubim and Reshma Formations. The Redkino Formation consists of interbedded brown and greenish-grey sandy mudstones with thin beds of tuffaceous mudstone. The Lyubim Formation consists mainly of greenish-grey sandy mudstones with films of organic residue on bedding surfaces. The Reshma Formation is principally composed of red sandstone (Solontsov and Aksenov, 1970). The maximum thickness of the Valday Group is more than 1,100 m in the most complete sections of the Moscow syneclise (Kotlas). The Valday Group extends, practically without any change, from the Moscow syneclise to the Upper Kama trough of the Urals region where three formations are distinguished. These are the Kirs, Velva and Krasnokamsk Formations, corresponding to the three that are present in the central regions. The Valday Group of Bashkir is divided into the Kairovo and Shkapovo Formations.

Many authors think that the peculiar brown tuffaceous clays are of special importance in correlating the Valday strata, assuming that they may represent a single unit that was contemporaneously deposited throughout

(Klevtsova, 1963; Kirsanov, 1968; Aksenov et al., 1967). However, the clays appear to be at several stratigraphic levels. In the central regions (Redkino, Pereslavl', Lyubim) they occur at the base of the Valday Group, and in Smolensk they form the upper parts of two cyclic units and are separated by sandstones. Thus it is impossible to correlate the Smolensk and Redkino sections on the basis of these tuffaceous clays. Even greater complications exist in the lowermost Eocambrian section in northern parts of the Russian plate (Nenoksa, Obozerskaya, Ust'-Pinega) where there are three brown mudstone horizons of the Redkino type (Aksenov and Igolkina, 1969). These brown mudstones appear to be typical of the lower part of the Valday Group in general, but the number of beds present varies from place to place.

The sequence of the Valday Group in Byelorussia is different. On the margins of the older Byelorussia uplift the group is composed of coarse clastics, sandstones and granule conglomerate. The stratigraphy is quite complex. In Southern Byelorussia (Stolin, Ozernitsa, Minsk) and in certain sections of the Northwestern Ukraine (Ratno) the Valday Group includes the Svisloch Formation which is sandy mudstones with some admixture of volcanic material, grading into tuffs. Formerly the Svisloch Formation was compared with volcanic rocks of the Volyn' Group (Bruns, 1957; Makhnach and Veretennikov, 1970), but recent data suggest correlation with the Valday Group (Jacobson, 1971b).

The sequence in the Svisloch sections is highly variable. In the Pripyat trough the formation consists of 155 m of interbedded sandy and silty tuff and terrigenous clastic rocks (Makhnach and Veretennikov, 1970). To the north, in Minsk, it comprises two cyclic units with a lower tuffaceous sandstone and tuff unit overlain by siltstone with some volcanic material (Bruns, 1957). Still farther north, in Smolensk, two cycles are also present, but in this area the upper part of the cycle is not siltstone, but rather, brown tuffaceous mudstones similar to those of the Redkino Formation. Thus, there appears to be a gradual replacement of the Svisloch tuff by mudstones of the Redkino Formation.

The Valday Group is not rich in organic remains, but contains casts of medusoids, the problematic form *Vendia*, worm tracks and algal remains (*Laminarites*) in its several areas of preservation (Podolia, Moldavia, Yarensk and the eastern part of the Russian plate) (Menner, 1963).

The Valday Group lies on the eroded surface of the older units, including Neoprotozoic and Epiprotozoic strata which are locally developed in the cover of the Russian plate. The base of the group is 590—607 m.y. old on the basis of glauconite dates (Podolia and Kaluga). The group is unconformably overlain by the Lower Cambrian Baltic Group which in some places possibly includes some upper Eocambrian rocks.

In the Russian plate at Volyn', the Rovno Formation is probably Upper Eocambrian (Kir'yanov, 1969). It is made up of glauconitic sandstone and mudstones (up to 53 m) and is separated by breaks from the underlying

Valday Group and the overlying Stokhod Formation, which is considered to be Lower Cambrian. Skeletal faunal remains are not known from either the Rovno or Stokhod Formations. The only forms recorded are *Sabelliditides* (from the Rovno Formation) and *Platisolenites* (from the Stokhod Formation). The Stokhod Formation is assigned to the Cambrian on the basis that a unit considered to be correlative — the Lontova Formation in Estonia ("the blue clays") — contains hyolites and primitive gastropods (*Pleurotomaria*?). This formation is therefore assigned to the Tommotian stage (the pre-trilobite Cambrian). Rocks similar to the Rovno Formation occur in central areas of the Russian plate but in that region it is difficult to distinguish them from "the blue clays". In Byelorussia they cannot be distinguished from the Valday sandy mudstones (Smorgon' bore hole). In Eastern Poland the Rovno Formation corresponds to all or part of the Lyubel' Group. In Estonia correlatives are not known.

Eocambrian strata distributed in the northwestern and northern margins of the Baltic shield are transitional in type between typical platform and miogeosynclinal deposits. In Northern Norway (Eastern Finnmark) shales and sandstone of the Stappugiedde Formation (which includes problematic organic remains) lie unconformably on Epiprotozoic glacial deposits and are conformably overlain by Lower Cambrian rocks (of Tommotian stage?) siltstone of the Breyvik Formation, with *Platysolenites antiquissimus*, followed by quartzite of the Duolbasgaissa Formation with the trilobite *Holmia*. In the western part of Finnmark the lower part of the Dividal terrigenous sequence correlates with the Stappugiedde Formation. These strata contain only casts of medusoids and worm burrows. These strata pass gradationally into beds with skeletal remains of the Cambrian Tommotian type (hyolites, brachiopods, gastropods). In Southern Norway the Vardal sandstone and the overlying Ringsack quartzite which contains worm burrows (*Scolithus*, *Monocraterion* and *Diplocraterion*) appear to be Eocambrian. The sandstone lies unconformably on the Moelv tillite and the Epiprotozoic Äkre shale, and the quartzite is unconformably overlain by Lower Cambrian strata with *Holmia*.

In Southern Sweden (Bornholm Island) the Nekse sandstone and the overlying Balka quartzite which contains worm burrows (*Scolithus* etc.) are assigned to the Eocambrian. These rocks are overlain by siltstone of the lowermost Cambrian, with hyolites, brachiopods and gastropods (C. Poulsen, 1967). These strata are typical platform deposits. They are not very thick and lie subhorizontally on a peneplaned surface of various older rocks.

Eocambrian orogenic molasse-like deposits are widespread in the *Urals*. The Asha Group, studied in detail by Bekker (1966) in the Southern Urals, is the formational stratotype of these strata. In upward succession, the Asha Group consists of the Uryuk Formation (200—350 m) of polymict conglomerate and conglomerate—breccia: the Basa Formation (up to 1,000 m) which is sandstone with interbedded siltstones in the upper part; the Kurkarauk

Formation (240 m) which comprises lithic sandstone with interbeds of granule conglomerate; and the Zigan Formation (500 m) which is sandstone, passing gradationally into conglomerate near the top. Glauconite from the Basa Formation is dated between 570 and 625 m.y. The group rests unconformably on various Neoprotozoic and Epiprotozoic rocks and is unconformably overlain by Devonian strata.

The Sylvitsa Group of the Middle Urals (Mladshikh and Ablizin, 1967) and the Il'-yavozh and Koshechor Formations of Polyudov Ridge (Borovko et al., 1964) have close stratigraphic and lithological similarity to the Asha Group. The age of glauconite from the Il'yavozh sandstone is 620—640 m.y., and from the Koshechor Formation is 550—590 m.y. These terrigenous strata may be traced in drill holes from the western slopes of the Urals to the easternmost sections of the Valday Group of the Russian plate. They are undoubtedly Eocambrian.

Miogeosynclinal strata of presumed Eocambrian age are developed in the central part of the Urals, in Mugodzhary, in the Great Caucasus and probably in the Northern Caucasus area. These strata are better defined in Southern Moldavia and adjacent areas of Roumania.

In the central part of the Urals, in the Uraltau zone, the Suvanyak Group (s.str.) is probably Eocambrian. It consists of the Ukshuk—Arvyak, Akbiik and Belekey Formations composed of sericitic and chloritic slates, phyllite, quartzite (or quartzitic sandstone) with local basic lavas interbedded with tuff. These strata overlie the Epiprotozoic (?) Mazarino Formation and are unconformably overlain by Ordovician rocks. These are tentatively assigned to the Eocambrian on the basis of rather poor correlation with the Asha Formation.

The age of the Shebekta Group in Mugodzhary, which is composed of terrigenous rocks of variable metamorphic grade, and lacks organic remains, is even less reliable. Middle Cambrian limestone (with trilobites) occurs higher in the section. Some consider the Middle Cambrian rocks as the upper part of the Shebekta Group but others think that the upper strata are separated by a stratigraphic break from the underlying rocks. Thus, the Shebekta Group may in fact be Eocambrian though it cannot be proved.

The Urlesh Formation in the Elbrus area of the *Great Caucasus* is also problematic. This formation has a basal conglomerate and lies unconformably on the Neoprotozoic (?) Khasaut Formation. The Urlesh Formation consists of quartzo-feldspathic and quartz sandstone with red siltstone interbeds. Clasts of Middle Cambrian limestone and of sandstone similar to the Urlesh sandstone have been found in Upper Silurian rocks that unconformably overlie this formation (Potapenko and Momot, 1966). These results point to a Cambrian age for the formation, but it may be older (Eocambrian).

In *Southern Moldavia*, drill holes revealed thick, highly deformed dark siltstones with organic remains (*Vendotaenia*). These rocks are lithologically similar to the Eocambrian platform strata of the Dniester area and probably present a miogeosynclinal equivalent of them. Similar rocks are present in drill holes in the Lower Danube River in Roumania (A.Ya. Dubinsky personal communication, 1973).

In Europe, outside the limits of the East European platforms and surrounding fold belts, it is only in *Scotland* that Eocambrian rocks can be distinguished with some certainty. There, the upper part of the Lower Dalradian Group (with a thickness of up to 2,000 m) appears to be Eocambrian. The group lies unconformably on the Epiprotozoic tillite. Quartzite is present in its lower part, locally interbedded with conglomerate. Above these rocks there is a unit of rhythmically bedded black bituminous shale. This unit is conformably overlain by the Upper Dalradian which contains Middle Cambrian fossils in its upper part. The Eocambrian—Cambrian boundary in the Dalradian section of the Scottish Highlands is not quite certain. It is possible that the lower part of the Upper Dalradian is Eocambrian.

As stated earlier (Chapter 8) the upper carbonate and terrigenous rocks of the Polarisbreen and Gothian Groups in Spitsbergen, occur above a tillite-bearing sequence and may also be assigned to the Eocambrian.

Asia

In Asia Eocambrian strata are widely distributed in Siberia and to a lesser extent in the Far East. In the fold belts of Kazakhstan, Middle and Central Asia (with the exception of the Maly Karatau area) Eocambrian rocks are missing or cannot be reliably identified. Formerly it was supposed that the Sinian rocks of China were Eocambrian, but now there is evidence to suggest that they are as old as Upper Neoprotozoic, for in North Korea Epiprotozoic strata (containing tillites) unconformably overlie rocks that are very similar to the Sinian. These rocks are called the Sanvon Group. The Epiprotozoic rocks are overlain by a sequence with Lower Cambrian fossils. Isotopic dating of glauconite from the upper formation of the Sinian sequence also favours this idea (see Chapter 7).

The Eocambrian strata in *Central Siberia* are principally represented by platform deposits and "orogenic" facies typical of the last stages of geosynclinal development. Geosynclinal units in the presumed Eocambrian succession of the southern part of the Baikal fold belt and in the eastern part of Eastern Sayan provide exceptions to the rule.

In the Khantai—Rybninsk uplift the Izluchinsk and Sukharikhinsk Formations and the Platonovo Formation and correlatives are examples of Eocambrian platform deposits. In the Yenisei anteclise similar rocks are represented by the Uglovskaya Formation and its correlatives — the Redkolesnaya and Ostrov Formations, and the lower part of the Yartsevo Formation. Other examples are provided by the Yudoma Formation of the Aldan shield, in the Zherba, Tinnaya and Moty Formations close to the platform margin of the Baikal fold belt and the lower parts of the Chabura, Manykay and probably also the Kessyusa and Tyuser Formations in the northern part of the Siberian platform.

The Yudoma Formation (150—250 m), which is extensively developed in the Aldan shield, possesses the attributes of a good stratotype. Its lower part is made up of quartz sandstone or granule conglomerate and the upper (major) part is dolomite and dolomitic limestone. The formation unconformably overlies various Precambrian formations ranging in age from Archean (gneisses) up to the Epiprotozoic Uy Group, and is conformably overlain by variegated carbonates and terrigenous strata of the Yuedey Formation. This unit has various skeletal fossils (hyolites, gastropods, kamenids, etc.) (Rozanov et al., 1969). The age of the formation was determined by K—Ar dating of glauconite. The Yudoma Formation has been studied in detail and is well defined on the basis of different microphytolites and stromatolites of the fourth complex (Semikhatov et al., 1970). The following medusoids are also reported from it; *Cyclomedusa* sp., *Suvorovella aldanica* Volog. et Masl. (Sokolov, 1972). The strata of the lower part of the formation were dated (glauconite) at about 650 m.y. old (several determinations in the 635—650 m.y. range). Recent results indicate that at the base of the variegated unit there are erosional features that must detract from the value of the Yudoma Formation as a stratotype (Cowie and Rozanov, 1973). Similar erosional features are also reported in a number of places within the Yudoma Formation. The break between the Yudoma and the variegated (Yuedey) formations may not be significant.

Zhuravleva et al. (1959) demonstrated that the Yudoma Formation could be traced to the northern margin of the *Patom Highlands* (Baikal fold belt), where it passes into the Zherba Formation of quartzitic sandstone and the overlying Tinnaya Formation (dolomite and limestone with microphytolites of the fourth complex). The Tinnaya Formation is also overlain by a variegated unit of Cambrian age as is the Yudoma Formation. The age of glauconite from the base of the Zherba Formation is 610 m.y.

The Zherba and Tinnaya Formations extend along the whole northern platform margin of the Patom Highlands and along the northwestern margin of the North Baikal Highlands. On the margin of the *Western Baikal region* these formations are replaced by red carbonate and terrigenous strata of the lower and middle part of the Moty Formation. In the middle part of this formation Sokolov (1972) identified casts of *Pteridinium* sp. and traces of the burrowing organism *Cylindrichnus*. The upper part of the formation contains remains of the oldest skeletal fauna and corresponds to the Lower Cambrian variegated formation.

From the western part of the Baikal area the Moty Formation may be traced into the northern margin of the *Sayan region*. It is continuous along the whole Sayan uplift on the Siberian platform. It lies transgressively on Archean rocks, on the Neoprotozoic Olkha Formation and on the Epiprotozoic Oselochnaya Group.

The Eocambrian strata of the *Yenisei anteclise* comprise the Uglov Formation in the Teya trough, the Redkolesnaya and overlying Ostrov Formations in the Angara—Kansk trough and the lower part of the Yartsevo Formation in the Yenisei trough. All these formations are so similar to the Moty Formation that they are commonly given the same name. In some cases even the upper red-coloured rocks of the Epiprotozoic Chingasan Group were erroneously assigned to the Moty Formation.

In the western part of the *Khantai-Rybninsk uplift* the Izluchinsk (Polban) and the overlying Sukharikha Formations are very similar to the Moty Formation. They consist of red and greenish-grey carbonate and terrigenous rocks. Eocambrian strata in the eastern part of the uplift (Platonovo Formation) are mainly grey and greenish-grey dolomites with minor variegated dolomites. The Sukharikha and Platonovo Formations contain abundant microphytolites of the fourth complex, and the Polban and Platonovo Formations also contain well preserved *Sabelliditides*. The Eocambrian strata are everywhere conformably overlain by Lower Cambrian fossiliferous rocks (Dragunov, 1967).

In the *northern part of the Siberian platform*, in the Anabar anteclise and the Udzhin uplift only the upper part of the Eocambrian succession is present. Some workers (Savitsky, 1962; Rozanov et al., 1969; Semikhatov et al., 1970) designate it as the Nemakit—Daldin horizon with worm burrows or *Anabarites trisulcatus* Miss. These hyolithes (?) are also known from the uppermost part of the stratotype of the Yudoma Formation. This horizon is regarded by many as the lower part of the Chabura and Manykay Formations with the oldest Cambrian faunas at a higher stratigraphic level. The Manykay and Chabura Formations (the Nemakit—Daldin horizon) are unconformable on the Epiprotozoic Starorechenskaya Formation (Fig. 34).

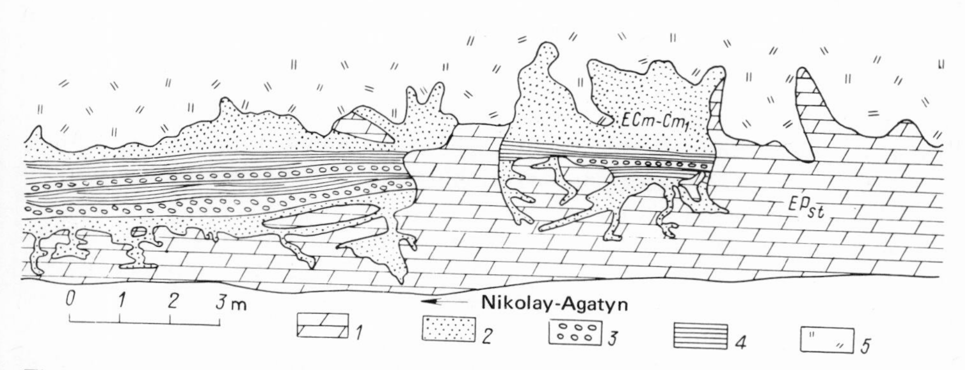

Fig. 34. Mode of occurrence of the Eocambrian—Lower Cambrian Manykay Formation on the eroded surface of dolomites of the Epiprotozoic Starorechenskaya Formation on the Nikolay—Agatyn River, eastern slope of the Anabar uplift. Illustration of V.A. Kaban'kov after a photograph by K.S. Zaburdin.
1 = dolomite; *2* = sandstone; *3* = conglomerate; *4* = argillite; *5* = soil layer.

The lower parts of the Kessyuse Formation in the Olenek Uplift and the Tyuser Formation of Northern Kharaulakh are probably uppermost Eocambrian (they are mainly of Cambrian age). The lowermost part of the Kessyuse Formation has remains of *Anabarites trisulcatus* Miss., and the Tyuser Formation has microphytolites typical of the fourth complex.

Eocambrian strata of orogenic aspect are distributed in the *East Sayan fold belt* and in the internal parts of the *Baikal mountain area*. The Anastas'ina Formation and the correlative Angul Formation of the Uda—Derba zone of East Sayan, the major part of the Mamakan Subgroup of Middle Vitim, and the Kholodnaya Formation of the North Baikal Highlands are also Eocambrian of this facies. All these units are represented by coarse-grained molasse-type strata that are commonly red in colour. These rocks formed in intermontane troughs following strong tectonic deformation of the geosynclinal systems. The Mamakan Group is a standard unit for such strata. It includes the Gukit, Lower Padrokan and Upper Padrokan Formations. These formations are confined to narrow, deep depressions (troughs or grabens) bounded by deep, long-lived faults. These rocks are red and greenish-grey polymictic conglomerates and sandstones with a thickness ranging from several tens of metres (bordering uplifts) up to 4,500 m in the central parts of the troughs. These strata consistently lack any organic remains and are conformably overlain by carbonate—terrigenous rocks largely composed of quartzitic sandstone, marl and dolomite with rare small inarticulate brachiopod remains. These rocks are known as the Sidel'te Formation (upper formation of the Mamakan Group) and should be considered Lower Cambrian. This formation is overlain by thick (up to 3,500 m) carbonates (the Yangud Subgroup) containing archaeocyathids and trilobites of the Lower Cambrian Lena stage, trilobites of the Middle Cambrian Amga stage are recorded (Salop, 1964—1967) from its upper part.

In the Verkhne—Angara Ridge the thick Kholodnaya Formation (up to 3,500 m), which consists of conglomerate and sandstone, is comparable to the three Eocambrian formations of the Mamakan Subgroup. The conformably overlying Tukalomi Formation is similar to the Lower Cambrian Sidel'te Formation; it contains worm burrows. The Tukalomi Formation is overlain by Lower and Middle Cambrian carbonates similar to those of the Middle Vitim area (Salop, 1964—1967).

The age of metamorphosed conglomerate and tuff—conglomerate of the Tataurovo Formation in the Selenga River basin in the southernmost Baikal mountain area is problematic. This formation appears to be at the base of Lower Cambrian sedimentary and volcanic successions of the Transbaikalian zone of the early Caledonides in southern Siberia (Salop, 1964—1967). It is probably Eocambrian, but direct relationships with Cambrian strata are not seen.

The lower and middle members of the Bokson Formation in the Bokson—Sarkhoy synclinorium, in the eastern part of *East Sayan* (Dodin et al., 1968) are Eocambrian rocks of miogeosynclinal facies. This formation overlies the Epiprotozoic Gorkhon Formation, either unconformably or, in rare cases, conformably. It consists largely of carbonate rocks: dolomite, limestone and rare marl. A bauxite sheet and a unit of variegated carbonate and terrigenous rocks is present near its base. Phosphate is also present in the dolomites. The lower part of the middle member contains stromatolites of the fourth complex (*Boxonia*, etc.), and the upper ("calcareous") unit contains archaeocyathids. Thus, the Eocambrian boundary lies somewhere in the middle member or at the base of the upper one.

The Maly Karatau Ridge is the only region of *Middle Asia* where Eocambrian strata are reliably defined. The Kyrshibakta and Berkuta Formations, which underlie the Tamda Group and unconformably overlie the Epiprotozoic Malokaroy Group (Keller et al., 1974), are Eocambrian. The Kyrshibakta Formation (0—170 m) is formed of variegated carbonate rocks with some glauconite-bearing sandstone. The overlying Berkuta Formation (up to 30 m) consists principally of dolomite with remains of fourth-complex microphytolites and also of protoconodonts (*Protohertzina*) and hyolithes (?): *Anabarites ex. gr. trisulcatus* Miss. The dolomite is overlain by the Chulaktaus Formation with a slight (?) stratigraphic break. This formation is made up of shale with rich phosphorite deposits, abundant remains of various hyolithes and inarticulate brachiopods, gastropods and kamenids similar to those of the oldest Lower Cambrian strata (the Tommotian stage). These are overlain by beds with Lower Cambrian trilobites.

In the *Far East of the U.S.S.R.* the Rudonosnaya (ore-bearing) Formation and the overlying Londoko Formation appear to be Eocambrian miogeosynclinal strata. These formations are overlain by the Lower Cambrian Chergilen Formation which contains archaeocyathids. The Rudonosnaya Formation (400 m) lies unconformably on the Neoprotozoic Murondava dolomite. Its lower part is sedimentary breccias which pass upwards into bituminous phyllitic slate, followed by manganese-rich and ferruginous quartzite, and finally by bituminous shale interbedded with limestone and dolomite. The Londoko Formation (up to 800 m) is made up of dark limestone interbedded with slate. The limestone contains remains of *Chlorellopsis* algae. There are reports of pelecypod moulds resembling *Modioloides priscus* Walc. in the Rudonosnaya Formation, but these are thought to be erroneous, and for the moment there is insufficient evidence to support the claim of a Lower Cambrian age for these strata.

In the *Northeastern part of the U.S.S.R.*, in the Kolyma region, the Korkodon Formation mentioned above (Chapter 8), may be assigned to the Eocambrian. It consists of dolomite with phytolites of the fourth complex. These rocks lie above Epiprotozoic strata (with tillite) and are conformably overlain by Lower Cambrian rocks (V.P. Rabotnov et al., 1970).

North America

As in Europe and Asia the Eocambrian strata of North America are of two major types, platform and miogeosynclinal. They are commonly transitional to orogenic facies.

It is difficult to select a good stratotype of platform rocks for this region. With some reservations the sedimentary strata in the Nevada—California border region in the *Cordilleran fold belt* may be regarded as a standard section (Stirling and Wood Canyon Formations). The Stirling Formation is composed of red quartzitic sandstone (up to 1,000 m thick) and lies transgressively on the Epiprotozoic Johnnie Formation which has stromatolites of the fourth complex. The second is more variable and consists of quartzitic sandstone, siltstone, shale, with some limestones (up to 700 m); it is overlain conformably or with a slight stratigraphic break, by red quartzitic sandstone of the Zabriskie Formation (70 m). These rocks are overlain conformably by the Carrara Formation which contains fossils in its lower part typical of the Lower Cambrian, and in its upper part has a lowermost Middle Cambrian fauna. The first organic remains (*Scolithus* worm burrows) are encountered about 100—300 m below the top of the Wood Canyon Formation, and in the uppermost part of this formation the Lower Cambrian archaeocyathids and trilobites are recorded. Fossils typical of the lowermost Cambrian have not been found as yet, so that the position of the Eocambrian—Lower Cambrian boundary remains problematic. Barnes and Christiansen (1967) suggested that it is within the Wood Canyon Formation on the basis of the first appearance of *Scolithus*. However, such structures are also reported in the upper part of many Eocambrian successions in Eurasia below the oldest Cambrian beds. The lower boundary of the Eocambrian in this region is taken as the bottom of the Stirling Formation because the underlying Johnnie Formation is stratigraphically more closely related to the Epiprotozoic Noonday dolomite and Kingston Peak tillite. It is everywhere separated from the quartzitic sandstone of the Stirling Formation by a break.

Red quartzites of the Mutual Formation may be correlated with the Stirling Formation. This unit is developed in Utah. It lies unconformably on an Epiprotozoic tillite and is overlain unconformably by the Lower Cambrian Tintic quartzite (K.C. Condie, 1966). Farther east in Utah, in the Uinta Mountains which form part of the North American plate, the Eocambrian is represented by the Uinta Group which is a thick sequence (several thousand metres) of terrigenous strata (Williams, 1953; K.C. Condie, 1966). Its lower (or middle) part (grey and red quartzite, arkose and siltstone) may be correlated with the Mutual Formation and is sometimes known by that name. The upper part of the group, the Red Pine Formation, is composed of red shale and siltstone and unconformably underlies the Tintic quartzite.

In Central Arizona the Bolsa quartzite with *Scolithus* seems to be Eocambrian. It lies unconformably on Neoprotozoic rocks but is overlain with an erosional break by the Middle and Upper Cambrian Abrigo strata.

In the *Midcontinent* region Eocambrian rocks are probably present only in the mountains of Southern Oklahoma. In that region the Eocambrian may be represented by the Carleton subaerial red porphyry and tuffs which rest unconformably on Epiprotozoic (?) sedimentary—volcanic rocks and granites that cut them. The Eocambrian rocks are in turn unconformably overlain by the Upper Cambrian Rigan Formation. The porphyry age is 525±25 m.y. by Rb—Sr isochron and Pb-isotope methods (Ham et al., 1964), but this value seems to be slightly young.

In the *Canadian—Greenland shield* the Bayfield Group (and the correlative Jacobsville Formation) may be Eocambrian. These rocks are in the Lake Superior area. They consist of quartz sandstone and shale, and rest on the Epiprotozoic Upper Keweenawan Group. The Upper Cambrian St. Croix sandstone appears to overlie them unconformably. In Minnesota the lower part of this group is comparable to the Hinkley sandstone.

The stratotype for the Eocambrian miogeosynclinal rocks, partly transitional to the "orogenic" facies, is present in Newfoundland, in the northern part of the *Appalachian fold belt*. These strata are the upper part of the Conception Group and the overlying Hodgewater Group (McCartney, 1967). The Cape Kovy Formation (up to 700 m) is in the upper part of the Conception Group (the major part of which is Epiprotozoic). It consists of shale and siliceous slate, together with siltstones with abundant casts of soft-bodied animals of the Ediacara type (mostly medusoid casts; Anderson and Misra, 1968; Misra, 1971). The Hodgewater Group (>4,500 m) and the probably correlative Musgravetown Group (up to 7,000 m) developed in western parts of the Avalon Peninsula, are of orogenic association. These groups are red arkose, siltstones and conglomerate. Basic and acid lavas are present in the lower part of the Musgravetown Group. In the upper part of the Hodgewater Group, just below the unconformable contact with the Random quartzite, some fossils are present as follows: *Acrothella* and *Lunnarssonia* (inarticulate brachiopods), *Volborthella* (cephalopods), hyolithes and *Epiphyton* sp. (algae). All of these are typical of the lower (pre-trilobite) horizons of the Lower Cambrian (Greene and Williams, 1974). The Random quartzite is overlain by shale and limestone of the Bonavista Formation which contains hyolithes. Above this unit there are limestones and argillites of the Smith Formation (with inarticulate brachiopods) followed by the Brigus Shale which carries remains of brachiopods and trilobites typical of higher parts of the Lower Cambrian. Thus, the oldest Cambrian rocks appear to be present in Newfoundland so that the underlying rocks can be safely attributed to the Eocambrian.

In the southern part of the Appalachian fold belt (in Eastern Tennessee) the Great Smoky Group of the Ocoee Supergroup is an example of an Eocambrian miogeosynclinal sequence. It consists of various terrigenous rocks which are coarse-grained in the lower part and fine-grained in the upper part.

Sections of this group are somewhat different in the west and in the east and local units have been established. Correlation of these units after Colton (1970) and others is shown in Table III. The Great Smoky Group lies unconformably on the Epiprotozoic Snowbird Group (lower part of the Ocoee Supergroup) and it is overlain, also unconformably, by the Lower Cambrian Chilhowee Group that seems to include all of the Lower Cambrian.

The Ellesmere Group also belongs to the miogeosynclinal Eocambrian. It is developed on *Ellesmere Island* in the Canadian Arctic (the Innuitian fold belt). This group is composed of conglomerates, sandstones and phyllites. It lies between the Epiprotozoic Ella Bay dolomite and the Lower Cambrian Scoresby dolomite and is bounded by an angular unconformity below and by a disconformity above. Worm burrows (*Scolithus*) are reported from one of the formations of this group (the Rawlings Bay Formation) (Kerr, 1967).

The Eocambrian Three Sisters Formation in the *British Columbia* Mountains is of miogeosynclinal or orogenic aspect. It consists of granule conglomerate, sandstone, quartzite and phyllitic slate. These strata lie unconformably on the Epiprotozoic Windermere Group (with tillite) and are separated by a break from the overlying Cambrian Range quartzite. Thick miogeosynclinal terrigenous and carbonate strata (about 2,000—3,000 m) in the *Yukon Territory* and the *Mackenzie district*, are probably Eocambrian. These strata unconformably overlie the Epiprotozoic Rapitan Group and underlie (without any marked break) Lower Cambrian strata with Olenellids (Blusson, 1971; Gabrielse et al., 1973).

General Characteristics

As defined here, the Eocambrian consists of all rocks formed after the Katangan orogeny (680—650 m.y. ago) but before the beginning of the Cambrian. The lower boundary of the complex is almost everywhere sharply defined and is commonly marked by an unconformity which separates it from older units (including the Epiprotozoic). The upper boundary is fixed by the first appearance of skeletal remains, typical of the Lower Cambrian (the pre-trilobite fauna of the Tommotian stage). In many localities the Eocambrian strata are similar to those of the Cambrian and the boundary commonly lies within a homogeneous sequence.

The upper age limit of the Eocambrian may be estimated from dates on glauconite from the overlying Lower Cambrian rocks. Dates have been obtained from carbonates in the northern part of the Siberian platform and in the Aldan shield. Glauconite from the Lower Cambrian Kessyuse Formation (which also includes some Eocambrian rocks) in the Olenek uplift gave an age of 558 m.y., and glauconite from the overlying Erkeket Formation (which contains faunal remains of the Aldan stage) yielded 527 and 556 m.y. The second value may give a better indication of the age of the Erkeket Formation, for glauconite from various Lower Cambrian formations of the Olenek uplift gives similar ages. Thus the ages obtained from the Kessyuse glauconites are "rejuvenated". Glauconite from the Tyuser Formation of Northern Kharaulakh gave an age of 545 m.y. (the Tyuser Formation is considered to correspond to both the Kessyuse and Erkeket Formations). These datings are not useful in establishing the Eocambrian—Cambrian boundary, for they only give the age of strata that are unconformable on the Epiprotozoic, and it is not certain in all cases that the topmost Eocambrian beds are present. For solving the problem we are more interested in the isotopic age data on glauconite from Cambrian strata of the Aldan shield, overlying with apparent conformity, the Eocambrian stratotype Yudoma Formation. These data provided several age values ranging from 527—575 m.y. The greatest value is probably closest to the true age of the boundary for it corresponds to the age (570 m.y.) accepted by the majority of Soviet and other geologists. Thus, the duration of the Eocambrian appears to have been relatively short (80 m.y.) and is similar in length to the other Paleozoic periods.

The Eocambrian is more closely related to the Paleozoic (Cambrian) than to the Epiprotozoic from the point of view of stratigraphic history. However, it is clearly differentiated from the Cambrian on the basis of its contained organic remains, and in this respect is more closely related to the Epiprotozoic. It contains phytolites of the fourth complex, as also does the Epiprotozoic. Medusoids and rangeids, which form a typical element of the Eocambrian Ediacara non-skeletal fauna, are also present in the Epiprotozoic. The Eocambrian Ediacara fauna is, however, more abundant and varied than the Epiprotozoic fauna. Eocambrian sequences also contain worm burrows, films of brown algae (*Laminarites*), typical *Sabellidites* and the lime-secreting algae (*Epiphyton*, *Renalcis*), and even hyolithes (?) *Anibarites trisulcatus* Miss. This fauna is quite different from the more abundant, more varied and complex skeletal fauna of the Lower Cambrian, which embraces thousands of species belonging to different taxonomic groups. The Eocambrian—Cambrian boundary is unique in the whole history of the earth in this strikingly sharp change in the composition of organic remains.

The rank of the Eocambrian and where it should be placed in a stratigraphic scheme are debatable problems. On the basis of stratigraphic history, the Eocambrian should be considered Paleozoic, but if biostratigraphic criteria are used as in the case of subdivision of the Phanerozoic, then this unit should surely be considered Protozoic. This dual classification is the result of application of different principles in subdivision of Precambrian and Phanerozoic sequences. It is also due to the fact that the Eocambrian lies at the boundary of two eons. To avoid fruitless debate, the Eocambrian is described as a single unit, transitional in character, between the Protozoic (Epiprotozoic) and Phanerozoic (Paleozoic). The Eocambrian is lower in rank than an erathem and in duration it corresponds to the Phanerozoic systems (periods). The systems are largely based on biostratigraphic criteria, but paleontological data from the Eocambrian are not sufficiently good to permit establishment of a new system. For the present the Eocambrian is considered under the informal term "complex" until such time as new and more definitive data are accumulated.

In the Soviet Union Sokolov (1952a, b; 1958) was the first to delimit the strata we now attribute to the Eocambrian. He worked on a sequence that covers the platform in the northwestern part of the Russian plate. For these rocks he used the name "Vendian complex". Initially Sokolov considered only the Valday Group to be Vendian, but because of the work of E.P. Bruns, B.M. Keller and others, he added some older formations of the Volyn' Group. It was, however, suggested in recent publications by Jacobson (1966), Solontsov and Aksenov (1970) that the Volyn' Group and correlatives in the Russian plate are of quite different age, and are separated by a great unconformity from the overlying Valday Group. Thus, there is no reason why these two separate complexes should be combined into the Vendian. Many workers (Keller, 1968; Keller et al., 1967) define the Vendian complex (or Terminal Riphean) on the basis of presence of phytolites of the fourth complex. If this criterion is accepted, then many formations older than 1,000 m.y. and even older than 1,400 m.y. (Nadezhdinsk and Shtandinsk Formations of the Russian plate) should be considered Vendian. These workers also accept the lower age boundary of the Vendian at 650 m.y. so that there is an obvious contradiction in definition of the Vendian.

Thus, the widely used term — "Vendian complex" — is no longer accepted, for formations of widely different age have been included in it. In particular, the units considered as its stratotype — the Valday and Volyn' Groups — belong to different parts of the Precambrian succession. The former is Eocambrian and the latter is Upper Neoprotozoic. If this term is used, then it is necessary to state which way it is used in every case (s.str. and s.lato). The term "Yudoma complex" has been proposed to replace "Vendian complex", for its stratotype (the Yudoma Formation of Siberia) falls within the age limits of the Eocambrian (680—570 m.y.). However, those in favour of this term (Semikhatov et al., 1970) have used as a definitive criterion, the presence of phytolites of the fourth complex, so that when they attempt correlation with Upper Precambrian sequences of Siberia, they erroneously correlate formations of different age with the stratotype. In view of these difficulties it is suggested that the old term "Eocambrian", which is known throughout the world, is the best to designate the strata in question. The term "Eocambrian" is considered a poor one by some people because it was proposed by W.C. Brögger at the beginning of the 20th century for the sparagmite complex in Scandinavia. These rocks include the Precambrian strata of various ages (from Middle Neoprotozoic to Cambrian). However, the meaning of this term has changed through time. Recently Scandinavian geologists defined the boundaries of the Eocambrian by the underlying Varanger (Lapland) tillite, and the sub-Holmia beds above. The Eocambrian, so defined, would correspond to the "Vendian" (s.str.). Now the term "Eocambrian" is widely applied by many European and American geologists to designate a complex of different strata directly underlying the Lower Cambrian, but definitions differ from author to author.

Regional review of the rocks shows that the Eocambrian of the Northern Hemisphere consists largely of platform, orogenic and miogeosynclinal units. Eugeosynclinal strata may exist in certain Phanerozoic fold belts, but it is difficult to distinguish these strata from the Lower Cambrian sedimentary—volcanic sequences.

The platform formations are largely shallow-marine strata. In the East European platform and in most of the North American platform they are principally terrigenous rocks, whereas in the Siberian and Hindustan platforms, well-sorted "mature" terrigenous rocks (quartzitic sandstones) are present only in the lower part of the complex, while the major (upper) part consists of carbonate or terrigenous—carbonate strata. Some Eocambrian units in Siberia consist almost entirely of carbonate rocks (for example, the Yudoma Formation). Red or variegated rocks are common, but considerably less abundant than in the Epiprotozoic or Neoprotozoic sequences. Certain red beds contain inclusions (or casts) of gypsum and rock salt that indicate evaporitic conditions.

A peculiar feature of almost all of these platform strata is the absence of volcanics. Only some terrigenous rocks of the Valday Group in the Russian plate contain some tuffaceous material, probably derived from adjacent geosynclinal areas to the west. There is a distinct lack of basaltic (trappean) volcanics like those of the Middle and Upper Neoprotozoic. Acid terrestrial volcanics of taphrogenic type are only reported from parts of the American Midcontinent (Arbuckle and Wichita Mountains).

Many fold belts of the Northern Hemisphere include Eocambrian terrigenous strata of orogenic facies. These terminated the geosynclinal phase of development of many mobile belts. The thick conglomerates and sandstones of the Baikal fold belt, Eastern Sayan, western slopes of the Urals, British Columbia and the Appalachians are of this type. In the Appalachians basic and acid volcanics are widespread among these rocks. Some of the rocks listed above, for example, red terrigenous strata of the Baikal fold belt (Kholodnaya Formation and correlatives), and also the Asha Group of the Western Urals, are similar to molasse-type deposits. However, they differ from typical Phanerozoic molasse in some respects (lesser thickness, predominance of sandstone over conglomerate, etc.; Salop, 1964—1967; Bekker, 1968). The Eocambrian includes strata that are in many respects comparable to molassic sequences; these were the first in geological history.

The Eocambrian miogeosynclinal strata have no specific peculiarities. They are largely composed of terrigenous rocks which in some regions (especially in North America) are very similar to orogenic assemblages.

The Eocambrian saw initiation of a great marine transgression onto the older platforms. This reached its maximum in the Early—Middle Cambrian. It is for this reason that the Eocambrian strata, together with the overlying Cambrian and younger rocks, form a vast platform cover. These rocks either lie unconformably on older platform units, or directly on older basement. In the Russian plate Eocambrian rocks cover the locally developed Neoprotozoic and Epiprotozoic platform strata (see Fig.30) showing that there were significant tectonic and paleogeographic changes in the configuration of the East European platform at the beginning of the Eocambrian.

Eocambrian mineral deposits are not numerous and are not commonly related to sedimentary sequences. The following deposits are known: bauxite and phosphorite in dolomite of the Bokson Formation of Eastern Sayan and adjacent areas of the Mongolian People's Republic (Ugakhol and Tabayanur phosphorite deposits), manganese and iron-rich quartzites in the Londoko Formation of Maly Khingan (Southern Khingan manganese and Kimkan iron-ore deposits). Oil is also present in the Eocambrian Moty Formation of the Irkutsk amphitheatre (Markovo deposit) where it is highly probable that the bitumen is secondary.

PART III

MAJOR FEATURES OF GEOLOGICAL EVOLUTION DURING THE PRECAMBRIAN

CHAPTER 10

COMMON FEATURES OF THE PHYSICAL AND CHEMICAL EVOLUTION OF THE OUTER ZONES OF THE EARTH DURING THE PRECAMBRIAN

Analysis of the major characteristics of supracrustal assemblages of the different subdivisions of the Precambrian throws light on many important aspects of the evolution of the physical and chemical environment of the earth's lithosphere, biosphere, hydrosphere and atmosphere. Certain generalities can already be made, but new data, particularly in regard to stratigraphic and geochronological succession, now permit a more specific treatment of these problems.

Archean

The supracrustal rocks of the Archean are perhaps the most peculiar. They differ not only in having high-grade metamorphism, but also in certain aspects which are probably related to the conditions under which they accumulated. Volcanics form a significant part of all these sequences. This probably indicates a higher temperature in the interior of the earth and greater mobility of the ancient crust. Striking similarity of sedimentary rocks in widely separated areas (probably of global extent) indicates that the environment of deposition was the same in different areas. Sedimentation may have taken place in a huge ocean ("Pantalassa") covering most of the planet. This is suggested by the absence of clearly expressed facies zonation (at least of linear type) and of any evidence of extensive land areas. The proto-ocean was probably not very deep so that in some areas the sea-bottom was exposed as banks and low islands. This is suggested by a general lack of conglomerates among the Archean strata. Many layered rocks containing alumino-silicates and closely intercalated with volcanics, are probably metamorphosed tuffs of varied composition. True terrigenous strata appear to be present only in the uppermost parts of Archean sequences. These rocks were pelitic sediments before metamorphism.

The problem of quartzite genesis in the Archean is particularly interesting. Quartzites are common in some Archean complexes, usually in the lower units. For example, in the Aldan shield, the lower part of the Archean complex (the Aldan Group) is composed of quartzite interbedded with metabasite. The oldest unit among these units is the Kurumkan Formation which consists of quartzite with rare interbeds of sillimanite—quartz schist (or quartzitic gneiss). The apparent thickness of this unit is more than 1,000 m, and the total thickness of three major quartzite units of the lower (Iyengra) subgroup of the Aldan Group reaches 2,800 m. Quartz sandstones and quartzites are normally considered as the end product of extreme sedimentary differentiation. They typify platform and miogeosynclinal assemblages. They are generally absent from or uncommon in younger eugeosynclinal assemblages and are not normally associated with submarine (spilitic) lavas.

What then was the origin of the Archean quartzite and in what tectonic environment did they form?

The Archean quartzites are always completely recrystallized and are commonly microclinized. No relict clastic textures are preserved. The fact that they are distinctly layered and are commonly interbedded with high-alumina rocks indicates that they are definitely of sedimentary origin. The fact that these quartzites are commonly associated with volcanic rocks suggests that the silica was a product of volcanic activity and was formed as a gel. The Archean quartzite—amphibolite association could thus be considered analogous to the chert—volcanic association of younger geosynclinal areas. This interpretation is improbable because of the exceptionally great thickness of the rocks and also because of the presence, in some quartzites, of rare, small (0.1—0.01 mm), well-rounded grains of clastic zircon. The source material, still older (Katarchean) silica-rich rocks, must have been of granitic or gneissic composition. However, such formations are not reported as yet anywhere. The so-called "Katarchean blocks", defined by some authors on the basis of erroneously interpreted isotopic dates, appear to consist of the same rocks as the surrounding Archean supracrustal and plutonic formations and belong to a single complex. All the granites associated with Archean supracrustals appear to be younger than them. However, the occurrence of thick quartzites in the lower part of many Archean complexes suggests that either the Katarchean granites were not removed by erosion or were strongly reworked by later ultrametamorphic processes. The latter is more likely.

It might be suggested that the silica for these quartzites could have been derived by chemical weathering of basic rocks, but the amount of material that could result in this way is much less than is actually present in the Archean metasedimentary complexes. For example, could chemical weathering of basic rocks result in the formation of quartzites as thick (2,800 m) as those in the lowermost Aldan Group? Great volumes of rock would have to undergo chemical weathering and complementary deposits such as iron formations, magnesite and other carbonates are not recorded in sufficient volume in the oldest complexes. Also the presence of (clastic) grains of zircon and monazite (?) in the quartzites contradicts this idea. Certain aspects of the chemistry of the quartzites and associated alumina-rich rocks also provide evidence against this concept. Thus, it is necessary to presume the existence of pre-Archean granite and gneiss, a pre-Archean crust already rich in sialic matter.

If, previous to metamorphism, the quartzite had been a typical clastic rock (quartz sand), then why are there no obvious facies changes? In spite of detailed studies in some regions there are no data to indicate the source areas for this clastic material.

Perhaps the Archean quartz sediments were largely formed as a result of precipitation of silica derived by chemical weathering of sialic rocks. However, for this process to be efficient, very specific physical and chemical conditions must have existed during the Archean.

Most workers acknowledge that the Archean atmosphere was very thick, dense and consisted largely of water vapour, CO_2, HCl, HF, H_2S, NH_3, CH_4, S and other minor components (Vinogradov, 1964, 1967). Intensive volcanic activity in the Archean probably supplied these products. Under these conditions the so-called "greenhouse effect" would have been very important so that temperatures in the lower atmosphere and in the surface waters would have been higher than at present by some tens of degrees (at the very beginning of the Archean it could even have been more than 100°C). High atmospheric pressure (some tens of atmospheres) would have prevented seawater from boiling. Under this hot atmosphere including many acid gases, chemical weathering would have been extremely active. The high temperature of seawater and its composition would have favoured quick dissolution of decomposition products and of silica in particular.

It is well known that silica is easily dissolved in warm water, particularly hot mineralized water. Experiments done by Balitsky et al. (1971) showed that the solubility of quartz greatly increases as temperature increases, especially when the water contains dissolved sodium sulphide. Sodium sulphide was probably abundant in the Archean seas, for there was virtually no free oxygen in the ancient atmosphere and compounds of Na_2S-type probably took the place of sodium sulphate. This may be indicated also by the common occurrence of lazurite (containing Na_2S) deposits in Archean marbles. These were probably sulphurous limestones or dolomites before metamorphism (see Chapter 4).

Silica from paleosols was largely dissolved in hot mineralized waters of the older ocean and was later deposited as a chemical precipitate. This is a possible explanation of the absence of an association between the Archean quartzitic strata and areas showing evidence of erosion. In some cases the Archean quartz rocks were of mixed chemical—clastic origin. Some small clastic particles of quartz and insoluble heavy minerals such as zircon and monazite, could have been transported for rather long distances in suspension in colloidal or true solutions.

The idea that Archean quartzite formed as a result of redeposition of material derived from subaerial (and subaqueous?) chemical weathering is supported by the association of these quartzites with high-alumina and locally with ferruginous rocks. In some cases sillimanite and less commonly magnetite is present in the quartzite.

Of particular interest is the first appearance of carbonate rocks at about the middle of the Archean section (metabasite—carbonate complex, A^{III}). These rocks are of great importance in correlation of Archean strata. As this feature is present in Archean rocks of many regions of the world it probably has some common cause. Possibly precipitation of carbonates was related to a decrease of carbon dioxide content and especially of strong acid in the atmosphere and hydrosphere. This may have been the result of a reduction in volvanic activity and/or loss of CO_2 due to other processes. Once the critical level was achieved, precipitation of carbonates became possible, and the content of carbon dioxyde in the Archean atmosphere would have decreased abruptly as a result of its being fixed in carbonate sediments. Probably carbon dioxide was being constantly supplied from volcanic activity, and it is only at the end of the Archean that its content significantly decreased due to a general reduction in volcanism (during sedimentation of the AIV complex). Thus., the evolution of the older atmosphere may provide an explanation for the strikingly similar sequences found in many Archean complexes.

Many peculiarities of older rocks, of Archean, Paleoprotozoic and partly of Mesoprotozoic age, suggest that Early Precambrian sedimentation took place in the absence of free oxygen in the hydrosphere or atmosphere. Examples are the common association of lazurite deposits with Archean carbonate rocks, composition of the Paleoprotozoic iron formations and the environment under which gold—uranium conglomerates formed at the beginning of the Mesoprotozoic. Isotopic analysis of sulphur in Archean and Paleoprotozoic sedimentary rocks indicates that sulphur in the older sulphates was only slightly different from sulphur in sulphides of the same age. The isotopic composition of the sulphates is similar to that of sulphur of meteorites. The oxidation process leads to formation of "heavier" sulphur rich in ^{34}S (relative to ^{32}S), so that the sulphur of modern sulphates is notably heavier than the sulphur of sulphides (E. Perry et al., 1971). Some relatively high values of $\delta^{34}S$ (Vinogradov et al., 1969) obtained from sulphur-bearing minerals from Archean rocks are probably related to later hydrothermal processes which accompanied oxidation reactions.

Low Fe_2O_3/FeO ratios, typical of Archean and Paleoprotozoic metavolcanics, may also indicate a lack of free oxygen in the atmosphere at that time. An absence or deficiency of oxygen in the Early Precambrian seems to have been typical, not only of the atmosphere and hydrosphere, but also in relatively deep parts of the lithosphere which had some connection with shallower zones.

In the absence of biological activity free oxygen in the Archean atmosphere could be formed only as a result of photochemical processes. Even this could have taken place only in the uppermost layers of the dense atmosphere. Any oxygen so formed would have been used up in partial oxidation of ammonia, sulphur, methane, etc.

These speculations on the composition of the oldest atmosphere are also based on observation of the geological record. Together with L.V. Travin a preliminary attempt was made to determine the composition of the Archean atmosphere by analysis of gas inclusions (bubbles) from quartzites of the

Iyengra Subgroup (Aldan River). It was assumed that during the chemical precipitation of silica there was some gas (and liquid) captured from the surrounding solutions. During later metamorphism the bubbles could have been redistributed throughout the rock, but it was assumed that, on the whole, the system was relatively closed. Analysis of the composition of gas inclusions is shown in Table 3. These analyses were done in the laboratory of the Institute of Geology and Geophysics, Siberian Department of the Academy of Science of the U.S.S.R. by N.A. Shugarova and V.K. Katayeva, with Yu.P. Kazansky.

TABLE 3

Composition of gas bubbles in the Iyengra quartzite

Specimen No.	Temperature of homogenization (°C)	Diameter of bubbles (mm)	$V_{bub.}/V_{incl.}$	Concentration (vol. %)			
				O_2	CO_2	N_2+rare gases	H_2S, SO_2, NH_3, HCl, HF
1819g-1	157	0.31	276	—	57.5	8.0	34.5
1819g-2	156	0.023	563	—	61.2	2.0	36.8
1819g-3	156	0.038	112	—	56.2	5.1	38.7
1823a-1	308	0.04	745	—	30.2	5.0	15.0
1823a-2	199	0.006	152	—	65.8	0.8	33.4
1823a-3	197	0.023	196	—	61.0	1.0	35.0
1823a-4	170	0.015	42	—	63.0	4.0	33.0
1823a-5	167	0.004	40	—	60.0	5.0	35.0

Note: hydrocarbons, CO, and H_2 not observed.

If the above assumptions are accepted, then the table indicates that free oxygen was absent from the older atmosphere (hydrosphere), which consisted largely of carbon dioxide (about 60%), H_2S, SO_2, and acid gases — HCl, HF (about 35%), small amounts of nitrogen and rare gases were also present. Thus, the analytical data confirm in a general way the concept of an anoxygenic primitive atmosphere. Hydrocarbons were absent from the rocks analysed, in spite of the fact that many workers (Calvin, 1969) have suggested that methane was present in the primitive atmosphere, and was the material from which organic molecules were made. From analysis of the isotopic composition of carbon from Precambrian carbonates, it has been suggested (Galimov et al., 1968) that the Precambrian atmosphere was made up essentially of carbon dioxide and that methane was not essential to the origin of life. The analytical data are obviously not sufficient to suggest that the problem has been solved. However, if these analytical data do indicate the approximate composition of the older atmosphere (or the gas composition of the hydrosphere, which is almost the same), then they support the chemical origin of the Archean quartzites, for had they originated as quartz sand, gas inclusions of this kind would not have been present.

Kazansky et al. (1973) showed that gas inclusions in younger siliceous chemical rocks showed a regular increase in oxygen content from 5.5% in the Paleoprotozoic to 12% in Upper Precambrian—Lower Paleozoic rocks. They obtained a value as high as 18% in Middle Paleozoic—Cenozoic rocks. A corresponding decrease in carbon dioxide content was also reported. Values range from 42% in Paleoprotozoic rocks down to a value close to the modern one in Middle Paleozoic—Cenozoic rocks. This regular change suggests that gas inclusions may give some idea of the composition of the atmosphere and how it changed with time.

There are no direct data on the salinity of the Archean ocean. Analyses of liquid inclusions from Archean rocks (quartzites of the Aldan shield and from the Slyudyanka area in the Baikal region) yielded highly varied results which can probably be explained by the presence of hydrothermal (metamorphic) solutions in the inclusions. A.P. Vinogradov and others considered that the total concentration of salts in the older ocean "was slightly different from the modern one, for the juvenile water was always accompanied by volatiles approximately in the same proportions as it is now" (Vinogradov, 1967, p. 30). Taking into account the higher temperature and different composition of the Archean atmosphere, however, it is possible that the saline components of "Pantalassa" were very different in comparison to those of the modern ocean. It can also be presumed that ocean water in the Early Archean was principally chloride-rich and lacked sulphates because of the absence of free oxygen necessary for oxidation of sulphur.

The presence (in rare cases) of sulphate (anhydrite) in carbonate rocks, as for example, in the Aldan and Grenville Groups, may indicate local oxidizing environments in parts of the older ocean in Late Archean time. Such occurrences may have been related to local biogenic photosynthetic activity. However, sulphur from the analysed Aldan sulphates has abundant light isotopes ($\delta^{34}S$ = 6‰), perhaps indicative if oxidation processes of low intensity and possibly relatively inefficient photosynthesis in an essentially oxygen-free atmosphere.

Organic remains appear not to be preserved in Archean rocks. However, the presence of thick bands of graphitic gneiss or marble might favour the idea of the existence of primitive organisms at that stage in the earth's evolution. This is also suggested by the isotopic composition of carbon from Archean rocks (A.V. Sidorenko and S.A. Sidorenko, 1971; S.A. Sidorenko, 1971). Certain phosphatic Archean rocks may also be related to biogenic activity as was the case for younger strata of this type in the Precambrian and Phanerozoic.

The existence, at the very beginning of the Paleoprotozoic, of photosynthetic blue-green algae suggests that such plants already existed in the Archean. The proposed dense, thick atmosphere of Archean time would have effectively prevented penetration of sunlight so that evolution of the photosynthetic plants (algae) could have begun only in the second half of the Archean when more carbon dioxide had been extracted from the atmosphere in carbonate sediments. At the very beginning of the Archean, the atmosphere was probably very dense. High temperature of the older ocean water would not have prevented evolution of the blue-green algae, for these primitive plants appear to be able to adapt to extreme environmental changes. Some modern blue-green algae are able to live in temperatures up to 85°C and are common in the water of hot springs. These blue-green algae are also able to survive in waters of different salinity and composition (they are reported from sulphate-, chlorine- and sodium-rich lakes), and in water with highly varied gas content. They live in water almost completely saturated with carbon dioxide. Some forms of blue-green algae are anaerobic and others develop where there is abundant oxygen. Blue-green algae are also reported from sulphurous waters associated with volcanoes (Goryunova et al., 1969).

Highly metamorphosed rocks of amphibolite and granulite facies are particularly widespread in Archean terrains. This, together with abundant granitization, indicates a very high thermal regime in the interior of the earth during the oldest Precambrian. Metamorphic rocks of a distinctive type occur only in these complexes — these are cordierite granulites that indicate high temperature—low pressure (or shallow depth) during metamorphism. This suggests that there were high heat flow rates and a steep temperature gradient in the Archean compared to later periods.

Paleoprotozoic

At the beginning of the Paleoprotozoic there was a significant change in physical and chemical conditions of the outer parts of the earth. Firstly, radioactive heat became much less important in the total energy budget of the earth because of a great decrease in activity of certain widely distributed radioactive elements with relatively short half-lives (mainly of ^{40}K and ^{235}U). According to calculations made by V.G. Khlopin (and more exact calculation by Lyubimova (1960) the greatest decrease in radioactive heat production took place approximately 3—4 b.y. ago, that is, at the boundary between the Archean and Protozoic (Fig.35). At the same time fixation of carbon dioxide in carbonate rocks and its decomposition by photochemical processes caused a decrease in density of the atmosphere and a concomitant reduction in importance of the greenhouse effect. The temperature near the earth's surface was probably not greater than 60—70°C, and the atmospheric pressure was no more than a few tens of atmospheres. The atmosphere at that time continued to prevent excessive loss of heat derived from the sun (or of internal heat) from the earth's surface, so that a hot and humid climate may have prevailed throughout the whole planet.

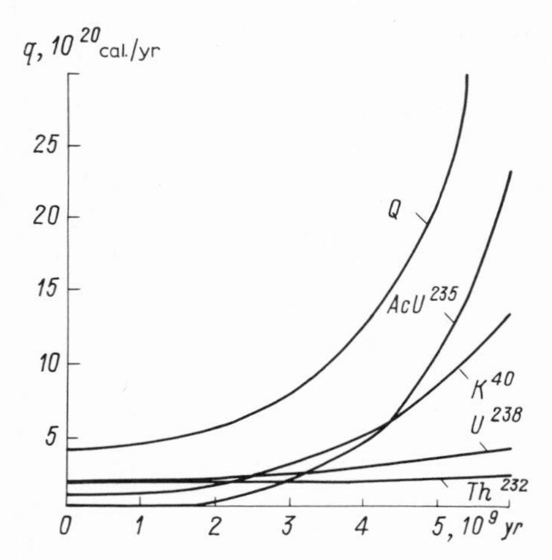

Fig. 35. Heat emanation by major radioactive elements during geological time (after Lyubimova, 1960).
Q = curve of total heat emanation by all major radioactive elements.

The Paleoprotozoic Era, as noted above, was typified by strong volcanism of eugeosynclinal type. The most intensive volcanism occurred in the middle of the era. Volcanic gases restored to some extent, the reserves of CO_2 and acid gases in the atmosphere, which had been partly depleted in the second half of the Archean Era and at the beginning of the Paleoprotozoic. This may be the reason for the dearth of carbonate rocks among the Paleoprotozoic volcanogenic strata. Carbonates are found mainly in the lower and upper parts of the erathem where there are only subordinate volcanics.

With increasing atmospheric density, there was increased transparency which brought about more intensive cosmic photosynthesis in the whole atmosphere and provided favourable conditions for biogenic photosynthesis in the hydrosphere. Thus, there is some reason to believe that free oxygen could have appeared in the Paleoprotozoic, but as it was immediately fixed in oxidation processes, the atmosphere still lacked significant free oxygen.

Very low concentrations of oxygen in the ancient atmosphere and also in seawater (which was in equilibrium with it), according to Strakhov (1963) and others, would have resulted in high mobility of protoxide compounds of iron (and manganese). It may be for this reason that Precambrian jaspilitic iron formations are widely distributed in the Paleoprotozoic. According to Lepp (1968) the geochemical behaviour of iron and manganese in an oxygenated atmosphere are quite different, but in an atmosphere deficient in oxygen their behaviour is very similar. The fact that sedimentary iron—manganese ores are common in the Paleoprotozoic is an additional piece of evidence suggesting low concentration of oxygen in the older atmosphere.

In the Paleoprotozoic, as in the second half of the Archean, seawater was rich in chlorides and carbonates and poor in sulphates.

Organic remains are present in Paleoprotozoic sedimentary rocks, including those in the lower Paleoprotozoic (Steep Rock Group and the Soudan Formation of North America, Onverwacht and Fig Tree Groups of the Swaziland Supergroup, and the Bulawayan Group in South Africa). These stromatolites, probably formed by blue-green algae and bacteria, indicate that the atmosphere at that time was sufficiently transparent for the penetration of sunlight and for biological photosynthesis to occur. However, such organic remains are relatively rare and the photosynthetic processes were of minor importance at that time.

Mesoprotozoic

In the lowermost part of the Mesoprotozoic, gold- and uranium-bearing conglomerates occur. Their formation may have been possible only under relatively anoxygenic conditions (see Chapter 6). In carbonate strata at the top of the lower part of the Mesoprotozoic, above the strata with the gold- and uranium-bearing conglomerates, there are abundant and varied phytolites that, for the first time, from large bioherms and even make up thick stratigraphic units. In chemical cherts associated with jaspilites, there are abundant remains of blue-green algae. The great increase of biomass and concomitant biogenic photosynthesis must have greatly augmented the free-oxygen content of the atmosphere. Above the phytolitic limestones and dolomites syngenetic red beds appear (for the first time) in many Mesoprotozoic sequences. This indicates that there was sufficient free oxygen in the atmosphere to oxidize at least part of the iron present. The concentration of oxygen was probably no more than a few tenths of a percent of the present level, for the "first critical level" constituting 1% of the modern value was not reached till much later (see below).

Thus, the end of the first half of the Mesoprotozoic was a most important period. For the first time there were significant amounts of free oxygen in the atmosphere. This had profound effects on many geochemical processes. This event is considered to have taken place about 2,400—2,500 m.y. ago on the basis of Pb-isochron datings of the time of sedimentation of the first stromatolitic dolomites (unpublished work of A.D. Iskanderova and S.L. Mirkina, laboratory of V.S.E.G.E.I., on material supplied by the author).

The increased oxygen content in the atmosphere led to an increase of oxidation—reduction potential (Eh) in the hydrosphere and resulted in a lowered geochemical mobility of iron in seawater. This explains why there are so many Mesoprotozoic jaspilitic iron formations in miogeosynclinal belts of this age. Accumulation of iron in shallow-water areas was probably caused by prolific development of oxygen-producing blue-green algae. The process consisted of oxidation of iron compounds to produce colloidal oxides which are commonly precipitated together with silica. Many Mesoprotozoic sedimentary iron ores, including jaspilite, have an oolitic structure. This aspect of Mesoprotozoic ores, which clearly differentiates them from similar Paleoprotozoic ores, is related to the accumulation of these ores in shallow-water regions, rich in algal activity. The ore particles were coated with algal mucus that reduced their density and coherence, and hence contributed to their mechanical mobility. The algal mucus also served as a medium for precipitation of colloidal iron particles. It is possible that certain blue-green algae affected the iron precipitation. Abundant microscopic remains of algae are present in Mesoprotozoic jaspilite in Canada (Gunflint Formation) and in Australia (Barghoorn and Tyler, 1965).

The products of volcanic activity in eugeosynclinal regions were probably the main source of iron in both Mesoprotozoic and Paleoprotozoic iron formations. Fossil soil formation was probably also an important process in putting iron into solution. Iron compounds so formed could not be carried far from the source area into the open sea, and had to precipitate in the shallow coastal areas in the near-platform zone. This was the origin of iron formations of Superior or Urals type.

In the Mesoprotozoic biochemical (stromatolitic) dolomites predominate among the carbonate rocks. According to Strakhov (1963) and Vinogradov (1967), the abundance of dolomite in Precambrian basins was related to a greater concentration of CO_2 in the atmosphere, providing a high alkaline reserve in the seawater. Under these conditions dolomitic material became oversaturated more readily than calcite, so that with evaporation, it readily precipitated.

As in Mesoprotozoic times much of the carbon dioxide was fixed in carbonate sediments, and its concentration did not increase appreciably due to volcanism. Thus there was an overall reduction in the carbon dioxide content of the atmosphere and of the greenhouse effect. This was the reason why, in the second half of the Mesoprotozoic, there was a change from a hot azonal climate to one that was less hot and had a poorly expressed zonation. It has been argued (Roberts, 1971) that the reduction of the greenhouse effect due to decreased carbon dioxide content in the atmosphere, could even have led to glaciation. Such an anti-greenhouse effect was not sufficiently great to cause global glaciation but it could have contributed to the Mesoprotozoic glaciation that is recorded in some countries of the world.

In the upper part of the Mesoprotozoic some evaporites are recorded for the first time in the stratigraphic record. These probably indicate arid climatic conditions during the second half of the Mesoprotozoic. The climate may have become more arid generally due to the appearance of relatively large platform regions at that time. An increase in free-oxygen content in the atmosphere in the second half of the era must have led to more intensive oxidation of sulphur and hydrogen sulphide in seawater, and to a change in seawater composition from being chloride- and carbonate-rich, to chloride—carbonate—sulphate rich waters (Strakhov, 1962).

Neoprotozoic

The geochemical phenomena that were initiated in Mesoprotozoic times became more pronounced in the Neoprotozoic. An abrupt increase in the abundance of phytoplankton resulted in higher concentration of oxygen in the atmosphere. Accumulation of thick sequences of carbonate rocks caused extraction of much carbon dioxide from the atmosphere, with corresponding reduction of the greenhouse effect. The temperature at the surface of the earth must have been somewhat lowered. According to paleoclimatological studies (Sinitsyn, 1967), the earth's surface temperature in the Early Paleozoic was about 30—32°C (and was falling). Thus, the average annual temperature in the Neoprotozoic may have been about 40°C. The lowering of temperature was accompanied by the establishment of latitudinal and vertical climatic zonation similar to that of the present time. The preserved Neoprotozoic strata suggest that it was less distinct.

The change in atmospheric composition resulted in strong development of certain sedimentation processes. The higher oxygen content of the atmosphere caused the mobility of iron in seawater to be sharply reduced, and formation of finely bedded ferruginous formations of jaspilite type ceased completely. Accumulation of sheet iron ores of the Urals type (related to semi-closed lagoonal-marine basins) decreased somewhat. In the coastal zones of shallow-water basins it became possible for chamosite and glauconite — highly oxidized iron minerals — to form. Sedimentary iron ores were less common and confined to coastal-marine and continental deposits. Thus, deposition of iron ore migrated from the eugeosynclinal regions (during the Paleoprotozoic) to miogeosynclinal and platform zones. Strakhov (1963) suggested that in the Phanerozoic there was a continued migration of iron-ore formation from miogeosynclines to platforms. There was a distinct separation of iron and manganese deposition. In the Neoprotozoic these metals generally form separate deposits and occur in quite different facies. Iron occurs only in miogeosynclinal and sub-platform deposits, while manganese is largely in eugeosynclinal deposits.

The lowering of CO_2 concentration in the atmosphere resulted in reduction of the alkaline reserve of seawater and a corresponding increase in calcite precipitation relative to dolomite. The alkaline reserve was also decreased by influx of fresh river water from the continental massifs. The chloride- and carbonate-dominated seawater with little admixture of sulphate was converted to chloride- and sulphate-rich water similar to that of the present time.

In the Neoprotozoic, the largest older platforms were formed. They were partly covered with epicontinental seas, but were most emergent. Red clastics accumulated in alluvial plains, lakes, deltas and in coastal regions. Eolian sands were prominent among these deposits. The absence of vegetation favoured accumulation of such sands. This was also part of the reason for preservation of abundant primary red sedimentary beds; the ferric iron was not reduced to the ferrous state because of a lack of organic matter. For this reason there are widespread red beds in the Late Precambrian, even more than in the Phanerozoic, in spite of the fact that later atmosphere contained much more oxygen.

With the appearance of extensive land areas, aridity was prevalent in many areas. This explains the relatively widespread development of rocks with gypsum, anhydrite and rock salt among Neoprotozoic platform and partly miogeosynclinal sequences.

The ozone layer was probably not sufficiently developed to afford protection to organisms from ultraviolet rays, but under a relatively thin layer of water, about 10—15 m deep (Berkner and Marshall, 1965), life could have successfully developed. The increase in free oxygen, lowering of temperature and other changes in the physical and chemical environment, probably favoured the appearance of the first Metazoa (?) determined as pogonophores in the Neoprotozoic. New forms of algae with a nucleus, and capable of sexual reproduction (eukaryotes) also appeared in the Neoprotozoic.

Epiprotozoic

The earlier major evolutionary trends of the earth's outer shells continued into the Epiprotozoic. However, this time period saw the widespread development of glaciations that affected almost every region of the world. The cause (or causes) of these glaciations is not certain, as is also the case for the younger glaciations, including those of the Quaternary. The available data on physical and chemical conditions in Precambrian times do not suggest that these glaciations resulted from normal evolutionary development of the atmosphere and hydrosphere. Red beds, dolomites, limestones with stromatolites and microphytolites, and evaporites are common in the pre-glacial, inter-glacial and post-glacial Epiprotozoic strata. These rock types suggest that the climate was typically hot. The global distribution of the Epiprotozoic glaciations (Fig.33) suggests that they were not related to latitudinal climatic zonation. The idea of "the anti-greenhouse effect" (Roberts, 1971), suggested to explain these glaciations, fails to explain the scale of the phenomena. Also there are no exceptionally thick carbonate accumulations before either Epiprotozoic glaciation. Such accumulations would be required to produce the proposed anti-greenhouse effect. It is presumed that cosmic phenomena were responsible for the Eocambrian glaciations. It is possible that they were due to the solar system passing through a dust cloud which absorbed part of the sun's energy. Whatever the cause of these phenomena the same thing happened twice in the Epiprotozoic. The two glaciations appear to have been separated by an interval of about 200 m.y.

Terrestrial factors were perhaps also favourable for the development of glaciations. The Epiprotozoic atmosphere was probably similar to the modern atmosphere. It consisted principally of nitrogen, oxygen, argon, water vapour and carbon dioxide with perhaps less oxygen and more carbon dioxide and water vapour than the present atmosphere. The high concentration of CO_2 is shown by the presence of thick chemical dolomites. Atmospheric pressure may still have been slightly higher than at present. The greenhouse effect ceased to have an important influence on the earth's climate so that a decrease in intensity of the sun's heat due to cosmic phenomena could easily give rise to glaciation.

Organic evolution in the Epiprotozoic is characterized by the first obvious forms of Metazoa, belonging to the Ediacara non-skeletal fauna. These are mainly medusoids, and quite rare rangeids.

Eocambrian

In Eocambrian time biological photosynthesis may have caused oxygen concentration in the atmosphere to have reached the "first critical level" or "Pasteur point", when the partial pressure of oxygen was about 1% of the modern value. This led to the development of the ozone layer and to a change from anaerobic life processes to more efficient aerobic ones. Berkner and Marshall (1965, 1967) and others have suggested that this change involved the evolution of multicellular organisms of Ediacara type and "the evolutionary outburst" at the Eocambrian—Cambrian boundary.

However, the increase in oxygen content up to the Pasteur point could have created conditions necessary for development of animals with aerobic breathing (fauna of Ediacara type) rather than conditions that led to the sudden appearance of several thousands of animal species with outer and inner skeleton at the beginning of the Cambrian. The greatest problem in explaining "the evolutionary explosion" was that the oldest skeletal animals were represented by complex and highly specialized forms, whose ancestors were not known. In 1914 Wallcott argued that the predecessors of the Cambrian fauna were unknown because they had lived at a time which is not represented in the geologic record — the so-called "Lipalian interval". However, Upper Cambrian strata are now widely studied and the faunal ancestors have not been found. Thus, the theoretical concept of a "Lipalian interval" is no longer acceptable. Many ideas have been suggested to explain the accelerated biological processes at the Precambrian—Cambrian boundary ("Pasteur point", radiation mutations, "cooperative" biological processes, etc.), but none of these provides a satisfactory explanation of the absence of ancestors. An adequate explanation of this dramatic fact would be a major contribution to natural science.

CHAPTER 11

TECTONIC EVOLUTION IN THE PRECAMBRIAN

Evolution of Rock Types

Lithology may reveal many important features of the tectonic evolution of the planet. Table IV (inset plate) is a series of correlation charts showing the distribution of various rock types in Protozoic and Eocambrian sequences in the principal regions of the Northern Hemisphere continents. Archean formations are not included in the table because they are considered separately. The vertical succession of Archean units is related not only to the tectonic evolution, but also to changes in the chemical (and partly physical) environment of the atmosphere and hydrosphere at an early stage in the geological evolution of the planet.

The table graphically shows the distribution of different rock types in time and space. Such a table permits reconstruction of the geological evolution of individual regions. However, in this book only the general trends will be emphasized.

From Table IV it can be seen that, in the Paleoprotozoic, rocks of eugeosynclinal aspect were the dominant rocks developed in the northern continents. True miogeosynclinal formations are not recorded, but some eugeosynclinal complexes near older cratons (protoplatforms) are typified by abundant terrigenous rocks, and thus, represent a transitional facies. Clastic quartzites are common and indicate a subdued topography in the hinterland. True platform sequences of this age are also missing. They were probably destroyed by later erosion.

It is possible that a number of rather large protoplatform blocks (Fig.18) existed in the Paleoprotozoic. This is suggested because of the composition and sequence of the geosynclinal assemblages, presence of facies changes and also because of the orientation of fold structures and zonal distribution of regional metamorphism of the rocks. In certain areas of Eastern Siberia (mainly in the Aldan shield) some sedimentary and volcanic assemblages accumulated in narrow fault-bounded grabens on the older cratons, near the margins of eugeosynclinal belts. The composition and metamorphic grade of these strata are similar to those of eugeosynclinal successions and are quite different from those of aulacogens and younger taphrogenic rocks, although their tectonic situation is closer to that of the latter. The peculiarities of the Early Protozoic thermal and tectonic regime probably resulted in the unique nature of these rocks. The strata in question are of a very specific proto-taphrogenic facies.

In Mesoprotozoic times the miogeosynclinal and platform facies appeared for the first time, and were of wide distribution. Rocks of miogeosynclinal facies invariably border geosynclinal areas, so that the eugeosynclinal belts are only rarely contiguous with protoplatforms. The rocks of eugeosynclinal facies differ from those of the Paleoprotozoic in having a greater proportion of sedimentary rocks relative to volcanics. Clastic quartzites are only reported near miogeosynclinal belts. The clastics in eugeosynclines were not sedimented directly from platforms, but after transport across the miogeosynclinal zones.

Mature clastic rocks are present in both platform sequences and in miogesynclinal areas. They are typical of many Mesoprotozoic sequences and seem to have been related to wide distribution of peneplaned platforms and median massifs.

Mesoprotozoic platform successions differ from analogous younger ones in the marked metamorphism of the rocks, more intensive tectonic deformation (especially close to faults), by the presence of locally abundant volcanics (mainly basic) and by the presence of many stratigraphic breaks. All of these indicate a relatively high tectonic mobility of the older platforms, and it is for this reason that they are designated as a special geohistorical category — protoplatforms. The strata developed there are regarded as a special type — protoplatform deposits. Some Mesoprotozoic strata, especially those in the upper parts of the protoplatform complexes, are hardly distinguishable from typical platform deposits.

The Lower Neoprotozoic taphrogenic formations are very characteristic. These rocks are widely distributed in all continents and almost everywhere are represented by subaerial red clastics and volcanics of porphyritic (rhyolitic) or andesite—trachyte—porphyrite composition. These rocks are thought to have been formed in response to epeirogenic movements of the earth's crust accompanied by formation of deep faults. In the Middle and Late Neoprotozoic there are extensive platform deposits that are only slightly different from those of the Phanerozoic. Strata deposited in aulacogens are recognized in many platforms. The geosynclinal formations are also similar to younger ones. Tectonic differentiation of geosynclinal areas was highly developed. Coarse-grained terrigenous rocks are common in all facies. In geosynclinal facies volcanics are less abundant than in Mesoprotozoic sequences and, in general, the eugeosynclinal facies was of lesser importance.

All of these trends are even more pronounced in Epiprotozoic sequences. In some geosynclinal systems true orogenic sediments appear and are quite widespread, indicating a change in tectonic regime.

In the Eocambrian, geosynclinal formations are only locally important and are replaced by platform and orogenic strata. Thus, at the end of Precambrian time there was an appreciable stabilization of the earth's crust. Some regions underwent later disturbance during the Baikalian and Caledonian orogenies.

Evolution of Magmatism and Metamorphism

Magmatism shows many changes with time. Only those directly related to the questions dealt with in this study will be considered here.

Geosynclinal volcanism decreased with time as is obvious in comparing the areas occupied by eugeosynclinal belts of different periods. The content of volcanic rocks in corresponding complexes of different periods, was also much less with time. This trend is poorly expressed in regard to platform magmatism where quantitative measurements are needed to evaluate it. Nevertheless, it appears that volcanics were more abundant in the oldest platform (protoplatform) deposits of the Mesoprotozoic than in analogous younger strata. Two major periods of platform volcanism are recognized. The first one is in the Early Neoprotozoic and comprises taphrogenic andesite—porphyrite volcanics. The second was at the end of the Middle Neoprotozoic and continued into the Late Neoprotozoic. It was typified by abundant basic lavas.

The granitoid plutonism was different at different times. In particular there was a steady decrease through time in the abundance of autochthonous syntectonic metasomatic granites, and a relative increase in the abundance of allochthonous magmatic syntectonic and late-tectonic granitoids. A particularly striking change in the development of these granites occurred after the Saamian orogeny at the Archean—Protozoic boundary. This change may be related to a reduction of the earth's internal heat production at that time because of radioactive decay (see Fig. 35). The Archean was characterized by widespread metasomatic granitization, whereas at the beginning of the Paleoprotozoic it became localized into narrower zones. It is not certain that during the Protozoic there was a further general reduction in volume of granite plutonism, for in some mobile belts (Baikal geosynclinal area, for example), unusually large syntectonic and late-tectonic granitic plutons were formed, even at the end of the Epiprotozoic, during the Katangan orogeny.

In geosynclinal belts there was a successive diminution in volume of basic and ultrabasic intrusive rocks in each tectonic cycle. However, platform intrusions of the same rock types were increasing both in size and abundance with time. The same is true of platform bodies of alkaline and nepheline syenite, alkaline gabbroids and hyperbasites. The decrease in basic intrusions in geosynclines is related to a general decrease in basic volcanism in mobile belts through time, which was, at least partly, due to "stopping up" by granite bodies, of faults that, within geosynclinal areas, had formerly been passageways for the magma. On the other hand, the increasing abundance of basic platform intrusions was probably related to a general "rigidification" of cratonic regions so that deep faults were more common.

The reasons for the occurrence of all large anorthosite intrusions in the Precambrian, principally in the Early Precambrian, and of all rapakivi granites at the end of the Early Neoprotozoic (during the Vyborgian tectono-plutonic cycle) are not understood.

The idea that anorthosites form as a result of a filter pressure mechanism seems to be the most reasonable. This process involves separation of plagioclase crystals from a slow-flowing magma that had already undergone some crystallization and gravitational differentiation at depth. For plagioclase crystals to grow large and for the process of fractionation to continue, required high magma temperature and an absence of intense tectonic movements, particularly for the formation of anorthosites. The only possible way of forming such large masses of these rocks is by quiet crystallization of magma during a long period of transport by laminar flow. Tectonic movements would have disturbed the fractionation process and produced homogenization of the magma. Many features of the earth's mobile belts suggest that initially they were poorly differentiated and also that the intensity of Precambrian tectonics gradually increased with time. This may help to explain why the major development of anorthosites occurred in the Early Precambrian and why they are localized in relatively quiet tectonic regions on platforms, and in marginal zones of older geosynclinal systems.

The same requirements seem to be necessary for the formation of rapakivi granites which are commonly associated with anorthosites. The fact that they occur almost exclusively in the Early Neoprotozoic is noteworthy. Rapakivi granites are found associated, not only with anorthosites, gabbroids and alkaline rocks, but also with taphrogenic andesite—trachyte porphyries. These intrusions may be related to a period of post-Karelian epeirogenic tectonics with block faulting and fractures which permitted extrusion and intrusion of magma. The epeirogenic movements of the Early Neoprotozoic may be considered as a distant effect of the Karelian orogeny. The younger Vyborgian tectono-plutonic cycle was of lesser importance.

During the Vyborgian orogeny the energy sources appeared to be somewhat less than those involved in the major global tectono-plutonic cycles. The source of energy for the weaker orogenic cycles may have been in the simatic layer of the crust, rather than in the mantle. This idea is supported by the absence (or scarcity), in the Early Neoprotozoic, of ultramafic intrusions, the poor expression of geosynclinal processes, by the fact that tectonic movements and magmatism were largely confined to platforms, and by the abundance of intermediate and acid volcanics.

The available facts on petrographic and petrochemical characters of rapakivi granites, their association with basic rocks, the presence of composite, multiphase plutonic bodies, their situation near the Conrad discontinuity (see Chapter 7) all suggest that rocks of the gabbro—rapakivi—alkaline—syenite association were derived from magma chambers close to the boundary between the basaltic and granitic layers of the earth's crust. Initially the material of the basaltic layer was melted due to the ascent of a thermal "plume" from the interior of the earth. This melting created intrusions of basic magma, and when the heat began to affect the granite layer, intrusions and extrusions of acid magma were formed. If the thermal front oscillated near the Conrad boundary, then both basic and acid magma intrusions would have alternated with time. Magmas of mixed composition in which syntectic (hybrid) rocks of diorite, syenite and monzonite type were formed, could also have originated in this zone. In the presence of deep-seated alkaline solutions, the magma could also have been a source of alkaline and subalkaline rocks. Later diabase dike swarms could form in the following way: after the granite melt had crystallized, contraction joints developed in plutonic bodies and basic magma could thus rise from the deeper parts of magma chambers rooted in the basalt layer.

The high primary $^{87}Sr/^{86}Sr$ ratio in rapakivi granites suggests that some radiogenic strontium (^{87}Sr) was derived from an older granitic crust that was remelted. The presence in rapakivi granites, of inequilibrium mineral assemblages (armoured relict minerals) and of inclusions of basic composition are probably related to magmatic phenomena such as syntexis and hybridism, and the presence of "melted" ovoids of potassium feldspar with plagioclase coatings could be related to pulsations of a thermal front and changes in composition of residual melts.

The change in thermal regime of the interior of the earth with time is also reflected in the evolution of metamorphic rocks. The Archean is typified throughout by development of unzoned (or weakly zoned) metamorphism of granulite and amphibolite facies, and by exceptionally intensive, almost universal processes of ultrametamorphism (granitization and migmatization). In Protozoic times the metamorphism is more irregularly distributed and has a distinct linear-zonal character. Ultrametamorphic processes are localized in certain zones which decreased in size with time. Rocks of granulite facies are relatively rare. Starting with the Neoprotozoic, unaltered rocks are widely distributed on the platforms, and in geosynclinal areas, apart from zones of plutonic metamorphism resulting from the effects of large granitic intrusions, the dominant rocks are in the greenschist facies.

Evolution of Tectonic Processes

Worldwide Precambrian orogenic cycles appear to be well established. These cycles occur in the following time intervals: 3,700—3,500 m.y. — the Saamian cycle; 2,800—2,600 m.y. — the Kenoran cycle; 2,000 (2,100?)—1,900 m.y. — the Karelian cycle; 1,700—1,600 m.y. — the Vyborgian (Sanerutian) cycle; 1,400—1,300 m.y. — the Prikamian (Kibaran, Avzyan) cycle; 1,100—1,000 m.y. the Grenville cycle; 680—650 m.y. — the Katangan cycle. In the Phanerozoic, orogenic cycles occurred in the following time intervals: 540—500 m.y. — the Baikalian cycle; 450—400 m.y. — the Caledonian cycle; 340—240 m.y. — the Hercynian cycle; 190—110 m.y. — the Kimmerian cycle; 85—12 m.y. — the Alpine cycle. It appears that orogenic periods were more frequent with time, and that periods of tectonic stability were correspondingly shortened. The duration of such anorogenic epochs (in m.y.) changed in the following way: Paleoprotozoic — 700; Mesoprotozoic — 600; Early Neoprotozoic — 200; Middle Neoprotozoic — 200; Late Neoprotozoic — 200; Epiprotozoic — 300; Eocambrian—Cambrian — 100; Ordovician — 60; Devonian — 40; Triassic — 40; Cretaceous — <30 (Table V, inset plate).

The large Precambrian units: eras and sub-eras (erathems and sub-erathems) are divided by orogenic cycles so that the younger units are, in general, shorter than the older ones (the Neoprotozoic Era is regarded as being made up of three sub-eras). On the basis of this regular progression it is possible to estimate the length of the Archean Era. The Paleoprotozoic includes an interval of about 900 m.y., and the Mesoprotozoic Era was shorter by 200 m.y. The Archean may have been correspondingly longer than the Paleoprotozoic Era by about the same period of time. If this is the case, then the pre-Archean or Katarchean orogeny should have taken place about 4,500—4,700 m.y. ago.

At the moment geophysicists and astronomers estimate the age of the earth as a cosmic body at about 5000 m.y. ago. This would allow some 300—500 m.y. for the "pregeological" stage of evolution of the planet. Prior to the formation of the Archean supracrustals most of the granitic crust was probably already formed so that the oldest quartzites resulted from its disintegration. If 5,000 m.y. is accepted as the age of the earth, then it suggests that geochemical and physical processes responsible for differentiation of planetary matter were very efficient during the early stages of the earth's evolution when it was hot and still retained much of its volatile content. This conclusion is also supported by many reliable geochronological and geological data. For example, the oldest granites appear to be as much as 3,500—3,750 m.y. old according to various radiometric methods and these granites are not Katarchean but rather Archean.

Many modern geophysicists, geochemists and astronomers think that the original differentiation of the planet into spheres of different composition (including formation of the primary sialic crust) must have occurred some millions or tens of millions of years after the original condensation of the protoplanet.

The age of the oldest tectono-plutonic cycle on earth (the presumed Katarchean and the established Saamian cycles) agrees with the age of two widely distributed types of moon rocks, anorthosite and basalt, which are dated at 4,500 and 3,500 m.y. respectively. This may be quite fortuitous but it may show that at a definite stage of the earth's and moon's evolution, thermal events on both planets were synchronous and related to a certain common cause. This may also indicate that both planets were derived from a single protoplanetary nebula, but tends to contradict the idea that the moon was torn from the earth or captured at a later time. Widespread development of anorthosites on the moon — rocks formed as a result of differentiation of basic magma — indicates that segregation of the interior part of the planet took place very early in its history (about 4,500 m.y. ago). Intensive processes of differentiation of the interior of the moon appeared to end after formation of the anorthosite crust. This may be due to the smaller size of the moon, and the consequent lack of internal thermal energy. The granitic crust of the earth may have originated at about the same time.

Phanerozoic orogenic cycles are subdivided into a number of phases, separated by rather quiet tectonic periods. In the Precambrian this kind of subdivision into phases has not been possible. This may be due to the lack of precise (paleontological) methods of correlation of minor stratigraphic units. During the Precambrian, tectonic movements appear to have continued throughout the entire period of the orogenic cycle, though they may have taken place in different places at different times by migration of zones of tectonic activity (Chapter 2). The same conclusion can be reached from the fact that nowhere in the Precambrian are there different syntectonic granitic intrusions belonging to a single orogenic cycle, the emplacement of which was separated by a period of sedimentation (excluding hypabyssal granitic massifs, related to volcanics). In Phanerozoic orogenic complexes such intrusions are numerous. The number of important tectonic phases in Phanerozoic orogenic cycles appears to increase with time. According to Stille (1949) there were two such phases in the Caledonian cycle, 4—5 in the Variscan cycle and 10 in the Alpine cycle. Thus the general tendency is clear in the Phanerozoic and can be extrapolated back into the Precambrian.

Another aspect of tectonic evolution is a gradual shortening of Precambrian orogenic cycles. If we consider radiometric dates on granitoids and metamorphic rocks related to orogenies, then the Kenoran cycle lasted about 200 m.y., the Karelian cycle 100—200 m.y., the Vyborgian, Prikamian and Grenville 100 m.y., and the Katangan about 30—50 m.y. The time boundaries of Phanerozoic cycles such as the Kimmerian and Alpine in particular, are quite arbitrary, for the nonorogenic periods separating them are quite comparable in scale with the quiet (interphase) periods within the cycles. It is for this reason that certain tectonic phases are attributed to one or another cycle. Considering the fact that tectonic events appear to be more frequent with time, it might be better to regard the major Phanerozoic tectonic phases (the Taconic, Late Caledonian, Sudet-Asturian, Indo-Sinian, Andian, Laramide, etc.) as separate orogenic cycles, in which case the shortening of orogenic cycles through time would be much clearer (Table V).

In previous publications (Salop, 1964, 1964—1967) it has been stated that not only did the frequency of tectonic movements increase during the evolution of the earth, but also the vertical amplitude of these movements. For example there is a marked increase in the importance of coarse-grained terrigenous rocks with time. In the Early Precambrian such sediments were relatively insignificant (in the Archean they were almost completely absent). Molasse-type sediments comparable to those of the Phanerozoic, first appear in the Epiprotozoic, but their thickness (<4 km) is generally much less than that of younger deposits of similar type. According to Kay (1955) the maximum thickness of the Caledonian molasse is 6 km, the Hercynian is 12.5 km and the Cenozoic is 20 km (the last value is probably too great). These data suggest that mountains and associated marginal and intermontane troughs that originated during the orogenic stage of mobile belt evolution had a much greater relief in younger fold belts.

The gradual increase through time, of the amplitude of relative vertical movements in the earth's crust, is also indicated by the apparent increase in sedimentation rate, or to be more exact, by the increase in the maximum thickness of sediments accumulated during an identical period of time in different geological epochs. If the thickest complexes of the various Precambrian erathems are considered and, for each, the total thickness accumulated during a million years is calculated, maximum sedimentation rates are obtained as follows: Paleoprotozoic — 20 m; Mesoprotozoic — 25 m; Neoprotozoic — 30 m; Epiprotozoic — 50 m; Eocambrian — 88 m in a million years. According to Gilluly (1949) the maximum sedimentation rate in geosynclinal areas of the Cambrian had already reached 150 m in 1 m.y. and there was a continued increase up to 500 m in the Pliocene. As is shown in Fig. 36 sedimentation rates regularly increased through time, and appear to have been most rapid at the beginning of the Paleozoic and at the beginning of the Cenozoic.

Some authors have presented arguments against these conclusions (Raaben, 1966). The expression "rate of sedimentation" should not be taken literally.

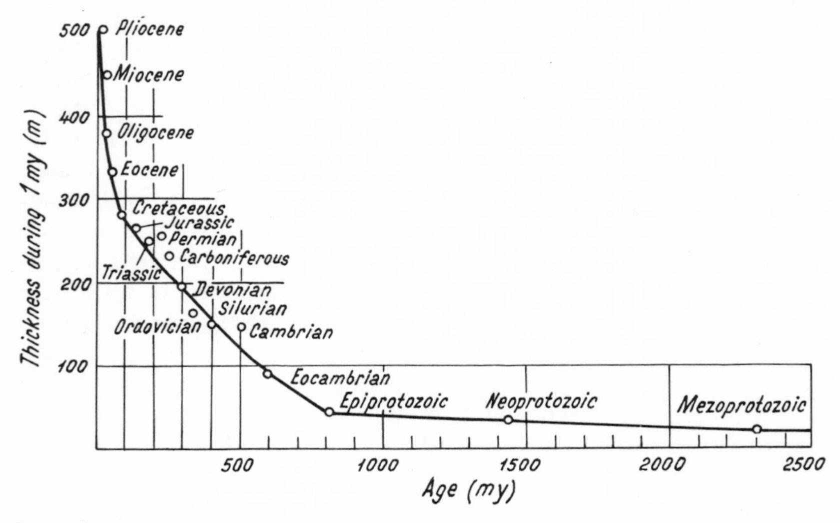

Fig. 36. Change in the rate of downwarping of geosynclines with time and the maximum thickness of strata accumulated during 1 m.y. in different geological epochs (revised after Salop, 1964). Data of Gilluly (1949) on Phanerozoic were also used in the plot.

What is involved is not the sedimentation rate proper, but rather the maximum rate of basin subsidence as reflected in the thickness of strata accumulated in a certain period of time. The measurements are difficult, for accurate measurement of the time involved in deposition of the measured Precambrian sedimentary complexes is virtually impossible. Evaluation of the time involved in sedimentation on the basis of glauconite dating is not necessarily very accurate. The older complexes selected for measurement, however, are not only the thickest, but also have a particularly continuous stratigraphy. These complexes correspond to a complete (or almost complete) transgressive—regressive megacycle, and correlation of these complexes with sections in adjacent areas shows that the stratigraphic sequence is fairly complete. Regardless of the age of the rocks, the maximum thickness of strata formed during one complete sedimentary cycle (between two orogenic cycles) is not greater than 15 km. This maximum thickness corresponds to the maximum possible depth of basin subsidence and is governed by isostatic processes. Since atectonic sedimentation phases appear to become shorter with time, the maximum "sedimentation rate" (subsidence rate) is automatically different.

The Precambrian geosynclinal systems are characterized by an exceptionally long period of evolution (Salop, 1964). Many of the systems existed throughout several tectonic cycles and, in spite of minor complications, a single structural plan was preserved. The types of sedimentation evolved in a unidirectional manner as is shown by the evolution of the rock types concerned (Table IV). In the Phanerozoic there was a marked acceleration of geosynclinal development. In Precambrian time geosynclinal development continued for several hundreds of millions of years in the same region. For example the Baikal region appears to have been the site of polycyclic geosynclinal accumulation for about 3,000 m.y. (Salop, 1964—1967). In the Phanerozoic, mobile belts appear to have existed for only a few tens of millions of years.

Thus, there appear to be general trends in the geological evolution of the earth as shown by development of more rapid and greater tectonic movements in younger orogenic belts. These ideas were established for the Phanerozoic in the works of Stille (1948), Von Bubnov (1954), Rukhin (1959) and Khain (1958). They are even clearer in the much longer Precambrian part of the geological record. Recent data on the tectonic evolution of the earth in Precambrian times support these general tectonic principles.

The reasons for speeding up of the tectonic pulse of the earth are not clear. Jeffreys (1959) stated that, according to contraction theory, shortening of the earth's crust should be greater with time. Thickening of the earth's crust and consequent increased strength would have favoured increased elevation of uplifted regions. Greater heterogeneity of the crust would favour its division into smaller blocks, emphasize contrasts and cause greater differentiation of structural patterns.

The most probable cause of acceleration of tectonic movements is the shift of radioactive elements to higher levels in the interior of the planet through time. Radioactivity is the major source of heat and is accordingly responsible for orogenic processes. Approximately 300 m.y. is needed for conductive heat transmission through 100 km thick strata (Verhoogen et al., 1970). It is suggested that the length of an anorogenic epoch is determined by the time necessary for heat transfer from the area of heat generation (the layer rich in radioactive elements) to the surface of the earth where the heat is used up in deformation and metamorphism of rocks, in magma formation and is partly dissipated into outer space. The length of the Paleoprotozoic anorogenic period is about 900 m.y., so that we can infer that at the beginning of this era the heat-generating layer was at a depth of about 300 km. The length of the anorogenic period of the Mesoprotozoic Era is about 700 m.y., so that at the beginning of the era (after the Kenoran folding), the generating layer ascended to a level about 230 km below the surface. The duration of the later anorogenic epochs in the Precambrian is about 300—350 m.y., indicating that the generating layer was about 100 km from the surface. From this point to the end of the Mesoprotozoic, the heat-generating layer was moved up about 100—130 km in relation to its former position. The proposed great elevation may have been related to intensive migration of radioactive elements during the exceptionally strong Karelian orogeny which was accompanied by formation of huge granite masses. Little change in the position of the generating layer in the Late Precambrian may be correlated with a relatively weak orogeny in the Neoprotozoic. Judging from the short duration of Phanerozoic anorogenic epochs (see above), the generating layer moved from a depth of 30 km at the beginning of the Paleozoic up to about 10 km at the end of the Mesozoic. In the Cenozoic the major part of the radioactive elements appeared to be concentrated in the granitic layer of the earth's crust, so that heat generation now mainly occurs at a shallow depth (practically in the realm of tectonic and magmatic activity).

Periodicity of orogeny with continuous emission of radioactive energy is probably due to heat accumulation up to a critical value (sufficient for triggering tectonic and plutonic processes) under a blanket of sedimentary or volcanogenic—sedimentary rocks, with much lower heat-conductivity in comparison to the underlying crystalline rocks (Smyslov, 1969). Heat loss during orogenic phenomena together with loss to outer space results in a temporary "freezing" of the heat system.

Through time the growing thickness of the sedimentary cover on the earth may have provided better thermal insulation of the interior and resulted in greater tectonic effects (increase in rate and amplitude of movements). Upward movement of the generating layer led to a reduction of the area involved in active tectogenesis and metamorphism, or in other words, narrowing of mobile belts. The deeper the "disturbing source" the broader the surface structures caused by it. The upward movement of the heat-generating layer would have caused a shortening of orogenic cycles with time, for the closer the layer is to the earth's surface the more rapidly the radiogenic heat is consumed. Because of the decay of radioactive elements, the overall heat flow is also gradually reduced.

The above reasoning is in agreement with geochemical data which record the migration of radioactive elements (K,U,Th) from the interior parts of the earth both during differentiation processes and also during metamorphism and granite formation. The estimates of depth of the generating layer are no more than that, no allowance was made for transmission of heat by emission and convection. Many geophysicists and petrologists consider conductive heat transmission to be of major importance. This was the basis of the calculations reported above and the results appear to be in good agreement with the periodicity of orogenic cycles.

Evolution of Major Tectonic Elements and Structures

In Chapter 4 it was stated that the tectonic pattern of the Archean Erathem was one of its most unique features. It was quite different from that of later periods. Archean tectonic processes and related metamorphic, ultrametamorphic and magmatic events probably took place under conditions of total mobility of the earth's crust. There was no differentiation into elements showing lesser and greater degrees of mobility so that the name "permobile stage" (Salop, 1969; Salop and Scheinmann, 1969) was proposed for this period in the evolution of the earth. Pavlovsky previously proposed the term "nuclear stage" for this period. However, it is based on the concept of Archean greenstone nuclei with older platforms being "crystallized" around them. This idea does not appear to be valid so that the term is not considered to be a good one. The beginning of the Proterozoic Eon saw the introduction of a new stage in the earth's evolution — the platform—geosynclinal stage. A distinct linear-structural, formational and metamorphic zonation of Paleoprotozoic sequences, and also the belt-like arrangement of fold structures, indicate the differentiation of tectonic elements. Greenstone-type strata of eugeosynclinal facies, and abundant clastic rocks, occur around areas of Archean rocks where the Paleoprotozoic sequences are absent or only locally developed and rather thin. These massifs are surrounded by greenstone belts which are approximately parallel to their boundaries. The massifs are interpreted as ancient cratonic blocks, either platforms or median masses (see Fig.18).

Small fault-bounded median masses (blocks) of Archean rocks that are present in such fold belts, display a specific tectonic pattern. In East Siberia (Baikal fold belt, Eastern Sayan and the Stanovoy Range) such blocks have been shown to be made up of Archean rocks similar to those of the large cratonic blocks. The tectonic pattern in such fragments is that of gneiss fold ovals similar to those that are widely developed in adjacent platform regions (Salop, 1964—1967). This suggests that the Paleoprotozoic geosynclines developed not on oceanic crust, but on a faulted Archean basement of granite gneiss that underwent local thermal and tectonic mobilization to produce geosynclinal belts. This is also suggested by the fact that tectonically reworked Archean gneisses form a basement to greenstone assemblages in some deeply eroded geosynclinal systems (in the Stanovoy belt, for example), and in marginal parts of geosynclines where there was less mobilization, there are well preserved remnants of the Archean tectonic pattern in the Archean basement (see Fig. 12).

Paleoprotozoic platforms may be differentiated from younger platforms by their higher degree of mobility, related to the existence of a thinner crust, and a higher thermal gradient. In many places, particularly on the periphery, the platforms were broken up by faults into separate blocks. Specific volcano-sedimentary assemblages accumulated in the tectonic depressions of graben or aulacogen type. These strata are considered as a special proto-taphrogenic category. Heat flow in platform areas was lower than in geosynclinal belts, but was still sufficiently great to produce relatively high-grade metamorphism in taphrogenic strata that were never buried to a very deep level. Argon loss from the basement rocks was related to thermal mobilization causing "rejuvenation" of the K—Ar ages from Archean gneisses to values corresponding to the time of the Kenoran orogeny. Thermal mobilization in the platform areas was not sufficient to cause strong deformation or destroy the older tectonic pattern.

After strong tectonic and plutonic activity at the end of the Paleoprotozoic there was significant consolidation of older cratonic blocks during the Kenoran orogeny, and many of the blocks combined into larger platforms. Mesoprotozoic platform strata formed in these regions were, however, different from younger ones in that they appear to have been more mobile and were characterized by intense volcanic activity. It is for this reason that the older platforms of the Paleoprotozoic and Mesoprotozoic are designated as a separate type — here called protoplatforms.

In the Mesoprotozoic, geosynclinal systems became more differentiated tectonically than they were previously. Whereas in the Paleoprotozoic the eugeosynclinal regime dominated virtually all mobile areas, in the Mesoprotozoic, miogeosynclinal belts appeared on the borders of geosynclinal systems for the first time. These oldest miogeosynclines also display a number of peculiar features, for their contained strata differ from younger formations of similar facies by the presence of volcanics (locally rather abundant), by high-grade metamorphism and by intensive plutonism (largely in the internal zone, adjoining the eugeosynclinal belt). All of these peculiarities are not so great as to merit designation of a special kind of miogeosynclinal belt.

The Neoprotozoic Era is typified by development of very large platforms and of development of a complex structure in geosynclinal systems. The major modern platforms were essentially formed at that time. In some cases they were even greater than at present. Neoprotozoic platform-cover strata and their tectonic preservation indicate that these platforms were characterized by higher stability than the older protoplatforms. Neoprotozoic geosynclines were sharply differentiated into zones characterized by different tectonic regimes. Many internal troughs and uplifts (intrageosynclines and intrageoanticlines) appeared within the limits of these geosynclines, and marginal troughs were, for the first time in geological history (see Figs. 29 and 31), developed peripherally.

Studies of the older fold belts (e.g. the Baikal fold belt) showed that the idea that marginal troughs (and foredeeps) developed only since the Late Paleozoic, was wrong (Salop, 1964—1967). Development of marginal troughs, however, evidently took place. The Precambrian (and Caledonian) peripheral troughs differ from the Hercynian and younger ones in that they underwent less subsidence and have correspondingly thinner sedimentary successions and lesser abundance of coarse-grained detrital sediments.

Termination of the long Precambrian stage of evolution involved a development of platform facies in many formerly geosynclinal regions. Certain geosynclinal systems completely terminated and in others an orogenic regime was established. In some cases there was a short period of platform development which was followed by the Baikalian and Caledonian movements. A few geosynclines of the Northern Hemisphere continued as before (Table IV).

Tectonic analysis shows that most Phanerozoic geosynclines (together with eugeosynclinal zones) were not new developments, but rather, tended to be confined to wide belts of long-lived older geosynclinal systems, so that they were to a large extent "inherited" from older mobile regions.

There were important changes in the nature of fold belts at the boundary between the Cryptozoic (Archean) and Protozoic Eons, signifying transition from a permobile to platform—geosynclinal stage in the geological evolution of the earth. This involves a change from gneiss fold ovals, typical of the Archean, to the linear or arcuate fold belts, typical of the Protozoic and Phanerozoic. The basic cause of this change in tectonic style probably lies in an abrupt decrease in intensity of the thermal regime of the earth's crust (see Fig. 35). Mobility and plutonic activity were markedly decreased.

The linear and arcuate shape of later fold systems was due to the development of deep faults at the beginning of the Paleoprotozoic. They are indicated by a great subsidence in the zones of sedimentation, by the large scale of the faults and by their protracted development (Salop, 1964—1967). During the Archean, deep faults could not exist on account of high mobility of the crust. Conditions favourable for the development of such fractures developed later with thickening of the earth's crust. Thus, deep faults are also very important elements which governed later evolution of geosynclines. The location of many Phanerozoic geosynclinal systems was to a large extent controlled by older major fault zones which were later reactivated.

Gneiss domes are a striking feature in the tectonic evolution of Precambrian regions (Salop, 1971b). They are abundant in the Archean and in some Paleoprotozoic terrains, but are much less important in the Mesoprotozoic, and quite rare in still younger structural stages. They are virtually absent since the end of the Mesozoic. This evolution is related to a decrease in the total energy reserves of the earth.

The most abrupt changes in geological evolution occurred at the boundary of the Archean and Paleoprotozoic Eras, that is, soon after the Saamian orogeny about 3,700—3,500 m.y. ago. This critical turning point in the earth's evolution permits a twofold subdivision of the Precambrian into two eons: Cryptozoic (Archean) and Protozoic. The major differences in these eons are shown in Table 4.

TABLE 4

Diagnostic features of the Cryptozoic and Protozoic Eons

Major aspects	Cryptozoic (Archean) Eon	Protozoic Eon
Major structural elements	earth's crust not differentiated into platforms and geosynclines; total mobility of the earth's crust (permobile stage)	platforms and geosynclines (platform—geosynclinal stage)
Tectonic pattern (types of fold belts)	isometric and irregularly shaped gneiss fold ovals	linear fold belts and arcs; deep faults
Metamorphism of rocks	rocks are everywhere metamorphosed in granulite and amphibolite facies. Metamorphic zonation is absent or weakly expressed and displays no linear aspects; granitization is ubiquitous, but irregular	rocks are of variable metamorphic grade; linear metamorphic zonation is common. Granulite facies of regional metamorphism absent-to-rare; granitization local
Structural facial zonation	absent or weakly expressed, no linear aspect	obvious and of linear character
Sedimentation	chemical sedimentation predominant; coarse clastic rocks absent	different types of sedimentation; chemical sediments subordinate
Organic remains	determinable organic remains are absent	various organic remains; phytolites, algae; Metazoa appear at the end of the Protozoic

In different regions certain other geological boundaries can be easily fixed, especially those that separate intensely folded basement rocks and platform-cover strata which are generally only gently folded. For instance, in the Superior tectonic province of the Canadian—Greenland shield the most striking boundary is the unconformity between the Keewatin greenstone belts and the Huronian sedimentary rocks, that is, between the Paleoprotozoic and Mesoprotozoic. It was for this reason that Canadian and American geologists assigned the Keewatin to the Archean and the Huronian to the Proterozoic. In other regions a similar boundary may occur at different stratigraphic levels. All such boundaries obviously cannot be taken as the boundary of the two eons. The differences between the Archean and Protozoic complexes are fundamental, although in certain regions the boundary may be masked by the effects of superimposed metamorphism and deformation.

The major aspects of geological evolution dealt with in this book are graphically represented in synoptic form in Table V. Data from the Precambrian of the southern continents are also included in this table.

Conclusions

In this book an analysis has been made of the sequence of older stratigraphic successions in three continents of the Northern Hemisphere. A new scheme for subdivision of the Precambrian has been proposed and various correlations were proposed on the basis of the material studied. In the majority of cases these ancient strata may be correlated with some degree of certainty, though sometimes they are separated by many thousands of kilometres. Not long ago such correlations could not have been contemplated. The situation has changed, however, because of significant advances in the field of Precambrian regional stratigraphy, and because of development of new correlation techniques such as isotopic dating and biostratigraphic methods. We are now at the starting point of a new stage in understanding Precambrian geology. This is indicated by the current interest in studies of older strata and by the many recent publications on Precambrian geology. Each new study makes new data available so that present concepts will have to be modified. We should not, however, be too pessimistic. We should keep in mind the words of Sederholm: that Precambrian stratigraphy is like Penelope's weaving — what has been woven by day, is unravelled every night. These words, however, were pronounced at the very beginning of the study of ancient rocks. It is hoped that only some threads of the proposed correlation scheme will be unravelled, but that the framework will remain.

Further work on the fundamental concepts and details of the proposed stratigraphic subdivision of the Precambrian will be continued. An analysis of the Precambrian sequences of the continents of the Southern Hemisphere will be made together with a compilation of proposed correlations according to the scheme used for the Northern Hemisphere. At the moment, with the preliminary material available, it appears that the major features established for the Northern Hemisphere, are identical throughout the planet.

In conclusion the following is a listing of the major results of this study.

(1) A single stratigraphic (and geochronological) scheme for subdivision of the Precambrian was presented; it has been shown to be applicable to the stratigraphic sequences of all the continents of the Northern Hemisphere.

(2) Correlation charts for correlation of Precambrian rocks of the major areas were compiled. These include the East European, Siberian and North American platforms and surrounding fold belts. Correlation of Precambrian strata was carried out for other areas of Europe, Asia and North America.

(3) It has been shown that coeval Precambrian strata in different continents display remarkable similarities in rock sequence, in lithology and in the nature of subsequent metamorphism and tectonics.

(4) Orogenic cycles that separate the Precambrian erathems and which are related to both basic and granitoid plutonism took place in different continents of the Northern Hemisphere at approximately the same time.

(5) Many Precambrian units consist of quite specific sedimentary formations in an irreversible sequence. Certain exogenic mineral deposits are associated with some of these formations. The evolution of iron formations was considered. Synchroneity of deposits of gold—uranium conglomerates in the lower part of the Mesoprotozoic was demonstrated (Salop, 1972b).

(6) It was shown that in every continent there was synchronous (Early Neoprotozoic) development of peculiar volcano-sedimentary assemblages. These were accompanied by intrusions of rapakivi granite, anorthosite, alkaline syenite and gabbroic plutons.

(7) Red beds appeared for the first time in earth's history on all the continents at the end of the Mesoprotozoic, after development of the first stromatolitic limestones and dolomites. The presence of free oxygen in the atmosphere and oceans was shown to be related to intense photosynthetic activity of algae in the first half of the Mesoprotozoic.

(8) A global twofold glaciation in the Epiprotozoic is discussed. The possibility of using these tillites as important stratigraphic markers was considered.

(9) A discussion was given of definition of age limits of Precambrian phytolite complexes.

(10) Major stages of the physical and chemical evolution of the outer layers of the earth in the Precambrian were dated, and an attempt was made to outline some aspects of geochemical evolution.

(11) The basic similarity in the tectonic evolution of the continents of the Northern Hemisphere during the Precambrian was established.

(12) Certain general trends in tectonic evolution during the Precambrian were presented.

Many of the established trends in geological evolution during the Precambrian provide clues to a better understanding of the evolution of the earth during the Phanerozoic.

REFERENCES

Abdulin, A.A. and Smirnov, Yu.D., 1971. On the age of the oldest strata in the Urals and Mugodzhary. *Izv. Acad. Sci. Kaz. SSR*, 1971(6): 1—15.

Afanas'ev, G.D., Bagdasar'yan, G.P. and Borovikov, L.I., 1964. Geochronological scale in absolute chronology based on data of the USSR. In: *Absol. Vozrast Geol. Formatsyi.* Nauka, Moscow, pp. 287—324.

Ahmad, F., 1954. An ancient tillite in Central India. *Q.J. Geol. Min. Metall. Soc. India*, 26: 15—20.

Ahmad, F., 1965. Glaciation in the Vindhyan System. *Curr. Sci.*, 34(7): 8—12.

Ahrens, L.H., 1955. Oldest rocks exposed. In: *Crust of the Earth (A Symposium). Geol. Soc. Am. Spec. Pap.*, 62: 155—168.

Aksenov, E.M., 1967. The Vendian complex in the east of the Russian platform. *Izv. Acad. Sci. USSR, Geol. Ser.*, 1967(9): 81—91.

Aksenov, E.M. and Igolkina, N.S., 1969. Subdivision of the Redkino formation within the Valdai group in the north of the Russian platform. *Izv. Vuzov Geol. Razved.*, 1969(8): 22—25.

Aksenov, E.M., Lyashenko, A.I. and Solontsov, L.F., 1967. On the lower limit of the Vendian in the central and eastern areas of the Russian plate. *Byull. MOIP, Otd. Geol.*, 42(3): 35—40.

Alcock, F.J., 1934. Report of the National Committee on Stratigraphical Nomenclature. *Trans. R. Soc. Can., Ser. 3*, 28 (Sect. 4): 113—121.

Alcock, F.J., 1938. Geology of the Saint John region, N.B. *Geol. Surv. Can. Mem.*, 216: 58 pp.

Aldrich, L.T., Davis, G.L. and James, H.L., 1965. Ages of minerals from metamorphic and igneous rocks near Iron Mountain, Michigan. *J. Petrol.*, 6(3): 445—472.

Allsopp, H.L., Ulrych, T.J. and Nicolaysen, L.O., 1968. Dating some significant events in the history of the Swaziland System by the Rb—Sr isochron method. *Can. J. Earth Sci.*, 5(3): 605—619.

Ambrose, J.W. and Burns, C.A., 1956. Structures in the Clare River syncline: a demonstration of granitization. In: *The Grenville Problem. R. Soc. Can. Spec. Publ.*, 1: 42—53.

American Commission on Stratigraphic Nomenclature, 1955. Report No. 3. *Am. Assoc. Pet. Geol. Bull.*, 39(8).

Anderson, C.A., 1951. Older Precambrian structure in Arizona. *Geol. Soc. Am. Bull.*, 62(11): 1331—1346.

Anderson, D.H., 1965. Uranium—thorium—lead ages of zircons and model lead ages of feldspars from the Saganaga, Showbank, and Giants Range granites of northeastern Minnesota. *Diss. Abstr.*, 26(1): 308.

Anderson, J.G.C., 1965. The Precambrian of the British Isles. In: K. Rankama (Editor), *The Precambrian, 2.* Interscience, New York, N.Y., pp. 25—111.

Anderson, M.M., 1972. A possible time span for the Late Precambrian of the Avalon Peninsula, Southeastern Newfoundland in the light of worldwide correlation of fossils, tillites, and rock units within the succession. *Can. J. Earth Sci.*, 9(12): 1710—1726.

Anderson, M.M. and Misra, S.B., 1968. Fossils found in the Precambrian Conception Group of South-eastern Newfoundland. *Nature*, 220(5168): 680—681.

Anhaeusser, C.R., 1971. Cyclic volcanicity and sedimentation in the evolutionary development of Archaean greenstone belts of shield areas. *Geol. Soc. Aust. Spec. Publ.*, 3: 57—70.

Appel, P.U., 1974. On an unmetamorphosed iron-formation in the Early Precambrian of South-West Greenland. *Miner. Deposit*, 9(1): 75—82.

Armstrong, R.L., 1968. Mantled gneiss domes in the Albion Range, Southern Idaho. *Geol. Soc. Am. Bull.*, 79(10): 1295—1314.

Aswathanarayana, U., 1964. Isotopic ages from the Eastern Ghats and Cuddapahs of India. *J. Geophys. Res.*, 69 (16): 3479—3486.

Baadsgard, H., 1973. U—Th—Pb dates on zircons from the Early Precambrian Amitsoq gneisses, Godthaab district, West Greenland. *Earth Planet. Sci. Lett.*, 19(1): 22—29.

Balitsky, V.S., 1971. *Solubility of Quartz in Hydrothermal Solutions of Sodium Sulphide, and Fluoride of Potassium, Sodium, and Ammonium.* Nauka, Moscow, 220 pp.

Balkwill, H.R. and Yorath, C.J., 1970. Brock River map-area, district of Mackenzie (97 D). *Geol. Surv. Can. Pap.*, 70-32: 25 pp.

Banks, P.O. and Cain, J.A., 1969. Zircon ages of Precambrian granitic rocks, North-eastern Wisconsin. *J. Geol.*, 77(2): 208—220.

Baragar, W.R., 1966. Geochemistry of the Yellowknife volcanic rocks. *Can. J. Earth Sci.*, 3(1): 9—30.

Baragar, W.R.A., 1969. The geochemistry of Coppermine River basalts. *Geol. Surv. Can. Pap.*, 69-44: 43 pp.

Baragar, W.R.A. and Donaldson, J.A., 1973. Coppermine and Dismal Lakes map-areas. *Geol. Surv. Can. Pap.*, 71-39: 20 pp.

Barghoorn, E.E. and Tyler, S.A., 1965. Microorganisms from Gunflint chert. *Science*, 147 (3658): 567—577.

Barnes, H. and Christiansen, R.L., 1967. Cambrian and Precambrian rocks of the Groom district, Nevada, Southern Great Basin. *US Geol. Surv. Bull.*, 1244-G: 34 pp.

Bart, T.F. and Reitan, P.H., 1963. Precambrian of Norway. In: K. Rankama (Editor), *The Precambrian, 1.* Interscience, New York, N.Y., pp. 27—80.

Bass, M.N., 1961. Regional tectonics of part of the southern Canadian Shield. *J. Geol.*, 60(6): 668—702.

Bekker, Yu.R., 1966. Stratigraphic chart and correlation of the Asha strata in the Urals and Pred-Urals areas. *Dokl. Acad. Sci. USSR*, 169(4): 891—894.

Bekker, Yu.R., 1968. *Late Precambrian Molasse of the Southern Urals.* Nedra, Leningrad, 155 pp.

Bekker, Yu.R., Negrutza, V.Z. and Polevaya, N.I., 1970. Age of the hyperborean glauconite horizons and of its upper boundary in the eastern part of the Baltic shield. *Dokl. Acad. Sci. USSR*, 193(5): 1123—1126.

Belevtsev, Ya.N., Rudnitsky, L.M. and Sukhinin, A.N., 1971. Stratigraphy and structure of the central part of the Ukrainian shield (contribution to the discussion). *Geol. Zh.*, 31(2): 119—134.

Bel'kova, L.N. Ognev, V.N. and Tashchilov, A.F., 1969. *Precambrian in Middle Tien-Shan.* Nedra, Moscow, 42 pp.

Bell, C.K., 1966 (1965). Churchill—Superior province boundary in Northeastern Manitoba. *Geol. Surv. Can. Pap.*, 66-1: 133—137.

Bell, R.T., 1968. Study of the Hurwitz Group, district of Keewatin (55 J, K, L; 65 J, G, H). *Geol. Surv. Can. Pap.*, 68-1 (part A): 28 pp.

Bell, R.T., 1970. The Hurwitz Group — a prototype for deposition on metastable cratons. *Geol. Surv. Can. Pap.*, 70-40: 159—168.

Bell, R.T., 1971. Geology of Henik Lake (east half) and Ferguson Lake (east half) map-area, district of Keewatin. *Geol. Surv. Can. Pap.*, 70-61: 31 pp.

Belokrys, L.S. and Mordovets, L.F., 1968. Precambrian plant remains from the Krivoy Rog area. *Dokl. Acad. Sci. USSR*, 183(1): 196—199.

Belyakova, L.T., 1972. *Riphean Geosynclinal Areas in the Northern Urals.* Thesis, GIN Acad. Sci. USSR, Moscow, 22 pp.

Bennison, G.M. and Wright, A.E., 1969. *The Geological History of the British Isles.* Martin Press, New York, N.Y., 406 pp.

Berdichevskaya, M.E. and Leytes, A.M., 1961. Geologic and petrographic aspects of Proterozoic cupriferous sedimentary strata in the northern part of the Chita area. In: *Ocherki po Metallogenii Osad. Porod.* Acad. Sci. USSR, Moscow, pp. 145—186.
Berkner, L.V. and Marshall, L.C., 1965. Oxygen and evolution. *New Sci.*, 28: 415—419.
Berkner, L.V. and Marshall, L.C., 1967. Rise of oxygen in the Earth's atmosphere. In: *Advances in Geophysics, 12.* Academic Press, New York, N.Y., pp. 309—332.
Bersenev, I.I. (Editor), 1969. *Geology of the USSR, 32. Primorian Area, 1.* Nedra, Moscow, 695 pp.
Berthelsen, A. and Noe-Nygaard, A., 1965. Precambrian of Greenland. In: K. Rankama (Editor), *The Precambrian, 2.* Interscience, New York, N.Y., pp. 113—262.
Bertrand-Sarfati, J. and Caby, R., 1974. Précisions sur l'age précambrien terminal (vendien) de la série carbonatée à stromatolites du groupe d'Eléonore Bay (Groenland oriental). *C. R. Acad. Sci., Ser. D.*, 278(18): 2267—2270.
Bessonova, V.Ya., 1968. Riphean basal strata in the western part of the Russian platform. *Dokl. Acad. Sci. USSR*, 178(5): 1149—1152.
Bessonova, V.Ya. and Chumakov, N.M., 1969. Upper Precambrian glacial strata in western areas of the USSR. *Litol. Polezn. Iskop.*, 1969(2): 73—89.
Bhattacharya, A., 1971. An occurrence of Semri basal conglomerate near Dalchipur, Sagar district, M.P. *Q. J. Geol. Min. Metall. Soc. India*, 43(2): 99—101.
Bibikova, E.V., Tugarinov, A.I., Zykov, S.I. and Melnikova, G.L., 1964. The age of the Karelian formation. *Geokhimiya*, 1964(8): 754—757.
Bibikova, E.V., Tugarinov, A.J. and Gracheva, T.V., 1973. Age of granulites from the Kola Peninsula. *Geokhimiya*, 1973(5): 664—675.
Birck, J.L. and Allègre, C.J., 1973. Rb^{87}—Sr^{87} systematics of Montsche Tundra mafic pruton (Kola Peninsula, USSR). *Earth Planet. Sci. Lett.*, 20(2): 266—274.
Birkenmajer, K., 1964. Course of geological investigation of the Hornsund area, Vestspitzbergen in 1959—1960. *Stud. Geol. Pol.*, 4(11): 7—33.
Biterman, I.M. and Gorshkova, E.P., 1969. Stratigraphy of Sinian formations in the Olenek and Kuoysk—Daldyn Uplifts. *Mater. Geol. Polezn. Iskop. Yakutsk. ASSR*, 1969(13): 163—170.
Björlykke, K., Englund, J.O. and Kirkhusmo, L.A., 1967. Latest Precambrian and Eocambrian stratigraphy of Norway. *Nor. Geol. Unders.*, 251: 5—17.
Blacet, P.M., 1966. Unconformity between gneissic granodiorite and overlying Yavapai series (Older Precambrian), Central Arizona. *US Geol. Surv. Prof. Pap.*, 550-B: 3—15.
Black, L.P., Gale, N.H., Moorbath, S., Pankhurst, R.J. and McGregor, V.R., 1971. Isotopic dating of very Early Precambrian amphibolite facies gneisses from Godthaab district of west Greenland. *Earth Planet. Sci. Lett.*, 12(3): 245—259.
Black, L.P., Moorbath, S., Pankhurst, R.J. and Windley, B.F., 1973. Pb^{207}—Pb^{206} whole rock age of the Archaean granulite facies metamorphic event in West Greenland. *Nature Phys. Sci.*, 244(134): 50—53.
Blackadar, R.G., 1967a. Geological reconnaissance of Southern Baffin Island, district of Franklin. *Geol. Surv. Can. Pap.*, 66-47: 32 pp.
Blackadar, R.G., 1967b. Precambrian geology of Boothia Peninsula, Somerset Island, district of Franklin. *Geol. Surv. Can. Bull.*, 151: 62 pp.
Blackadar, R.G., 1970. Precambrian geology of Northwestern Baffin Island, district of Franklin. *Geol. Surv. Can. Bull.*, 191: 89 pp.
Blackwelder, E., 1932. An ancient glacial formation in Utah. *Geol. Soc. Am. Bull.*, 40(2): 289—304.
Blagoveshchenskaya, M.N., 1959. Chadobets dome-like uplift. *Inf. Sb. VSEGEI*, 8: 71—82.
Blusson, S.L., 1971. Sekwi Mountain map-area, Yukon Territory and district of Mackenzie. *Geol. Surv. Can. Pap.*, 71-22: 17 pp.

Bobkov, Yu.B., Bulayevsky, D.S. and Zaytsev, A.A., 1970. Stratigraphical chart of the Precambrian formations in the Ukrainian shield. *Geol. Zh.*, 30(4): 153.

Bobrinskaya, O.G., Bobrinsky, V.M. and Bukatchuk, L.D., 1964. *Stratigraphy of Sedimentary Formations in Moldavia*. Kartya Moldavenyaske, Kishinev, 131 pp.

Bobrov, A.K., 1964. *Geology of the Pred-Baikal Marginal Trough (Northeastern Part)*. Nedra, Moscow, 228 pp.

Bobrov, A.K. and Moskvitin, I.E., 1970. Stratigraphy and correlation of Proterozoic strata in the southeast part of the Siberian platform. In: *Stratigrafiya i Paleontologiya Proterozoya i Kembriya. Vostoka Sibirskoy Platformy*. Yakutsk. Knizhn. Izdatelstvo, Yakutsk, pp. 5—35.

Bobrov, E.T., Levchenko, S.V. and Chaika, V.M., 1973. Bauxite indications in Archean strata of the KMA. In: *Polezn. Iskop Osad. Tolshchakh*. Nauka, Moscow, pp. 56—61.

Bogdanov, Yu.B., 1971a. On the so-called basal conglomerate of the Kola group in the area of Chuna—Monche—Volch'y tundra in Kola Peninsula. In: *Problemy Geologii Dokembriya. Baltiyskogo Shchita i Pokrova Russkoy Platformy. Tr. VSEGEI*, 175: 199—205.

Bogdanov, Yu.B., 1971b. Types of Lower Proterozoic sections and their correlation. In: *Problemy Geologii Dokembriya. Baltiyskogo Shchita i Pokrova Russkoy Platformy. Tr. VSEGEI*, 175: 106—121.

Bogdanov, Yu.B. and Voinov, A.S., 1968. On Karelide—Belomoride relationships in Eastern Karelia. *Tr. VSEGEI, N. Ser.*, 143: 97—109.

Bogdanov, Yu.B., Negrutza, V.Z. and Suslova, S.N., 1971. Precambrian stratigraphy of the eastern part of the Baltic shield. In: *Stratigrafiya i Izotop. Geokhronologiya Dokembriya Vost. Chasti Baltiyskogo Shchita*. Nauka, Leningrad, pp. 160—170.

Bogdanov, Yu.V., Kochin, G.G. and Kutyrev, E.I., 1966. *Cupriferous strata of the Olekma—Vitim Mountain Land*. Nedra, Leningrad, 385 pp.

Bondarenko, L.P. and Dagelaysky, V.B., 1968. *Geology and Metamorphism of Archean Rocks in the Central Part of the Kola Peninsula*. Nauka, Moscow, 168 pp.

Bondesen, E., 1970. The stratigraphy and deformation of the Precambrian rocks of the Craenseland area, south-west Greenland. *Grönl. Geol. Unders. Bull.*, 86: 210 pp.

Bondesen, E., Pedersen, K.R. and Jorgenson, O., 1967. Precambrian organisms and the isotopic composition of organic remains in the Ketilidian of SW Greenland. *Medd. Grönl.*, 4(164): 20—32.

Borovikov, L.I. and Spizharsky, T.N., 1965. Criteria for subdivision and correlation of the Precambrian. *Geol. Geofiz.*, 1965(1): 21—29.

Borovko, N.G., 1967. *Vendian and Lower Paleozoic of the Polyudov Ridge in the Northern Urals*. Thesis, VSEGEI, Leningrad, 22 pp.

Borovko, N.G., Kell', G.N. and Smirnov, Yu.D., 1964. Stratigraphy, environment and diamond content of the Churochnaya formation (Northern Urals). In: *Materialy po Geologii Urala. Tr. VSEGEI, N. Ser.*, 119: 23—50.

Bostock, H.H., 1967. Itchen Lake map-area. *Geol. Surv. Can. Pap.*, 66-24: 30 pp.

Boutcher, S.M., Edhorn, A.S. and Moorhouse, W.W., 1966. Archean conglomerates and lithic sandstones of Lake Timiskaming, Ontario. *Proc. Geol. Assoc. Can.*, 17: 21—42.

Bridgwater, D., 1965. Isotopic age determinations from South Greenland and their geologic setting. *Grönl. Geol. Unders. Bull.*, 53: 56 pp.

Bridgwater, D., 1973 (1972). General compilation of isotopic work on rocks from Greenland. *Grönl. Geol. Unders. Rep. Act.*, 55: 51—60.

Bridgwater, D., 1974. West Greenland conglomerate. *Geotimes*, 179(1) (photo on journal cover).

Bridgwater, D., Sutton, I. and Watterson, I.S., 1966. The Precambrian rapakivi suite and surrounding gneisses of the Kap Farvel area, South Greenland. *Geol. Surv. Greenl. Rep. Act.*, 11: 52—54.

Bridgwater, D., Watson, J. and Windley, B.F., 1973. The Archaean craton of the North Atlantic region. *Philos. Trans. R. Soc. London, Ser. A*, 273: 493—512.
Bridgwater, D., Collerson, K.D., Hurst, R.W. and Jesseau, C.W., 1975. Field characters of the Early Precambrian rocks from Saglek, coast of Labrador. *Geol. Surv. Can. Pap.*, 75-1: 287—296.
Brookins, D.G., 1968. Rb—Sr and K—Ar age determinations from the Precambrian rocks of the Jardine—Crevice Mountain area, Southwestern Montana. *Earth Sci. Bull.*, 1(2): 5—9.
Brooks, C. and Hart, S., 1972. An extrusive basaltic komatiite from a Canadian Archean metavolcanic belt. *Can. J. Earth Sci.*, 9(10): 1250—1253.
Brotzen, O., 1973. Rb—Sr dating and microclinization in dioritic rocks of the Haparanda group, northern Sweden. *Geol. Fören. Stockh. Förh.*, 95/1(552): 19—24.
Brown, J.S., 1973. Sulfur isotopes of Precambrian sulfides in the Grenville of New York and Ontario. *Econ. Geol.*, 68(3): 362—370.
Brückner, W.D. and Anderson, M.M., 1971. Late Precambrian glacial deposits in Southeastern Newfoundland — a preliminary note. *Proc. Geol. Assoc. Can.*, 24(1): 99—102.
Brummer, J.J. and Mann, E.L., 1961. Geology of the Seal Lake area, Labrador. *Geol. Soc. Am. Bull.*, 72(9): 1361—1381.
Bruns, E.P., 1957. Stratigraphy of the older Pre-Ordovician strata of the western part of the Russian platform. *Sov. Geol.*, 1957(59): 3—24.
Bruns, E.P., 1963. Northwestern, central, and northern areas of the Russian plate. In: *Stratigrafiya SSSR. Verkhniy Dokembriy*. Gosgeoltekhizdat, Moscow, pp. 22—46.
Bubnof, S., von, 1954. *Grund-probleme der Geologie*. Akad. Verlag, Berlin, 234 pp.
Butin, R.V., 1966. Fossil algae of the Proterozoic in Karelia. In: *Fossil Forms and Problematical Organisms of Proterozoic Formations in Karelia*. Karel'sk. Knizn. Izdatel'stvo, Petrozavodsk, pp. 25—45.
Buzikov, I.P., Nikitina, L.P. and Khil'tova, V.Ya., 1964. The Precambrian in Eastern Sayan. *Tr. Lab. Geol. Dokembr. Acad. Sci. USSR*, 18: 328 pp.
Calvin, M., 1969. *Chemical Evolution*. Clarendon, Oxford, 278 pp.
Campbell, F.H., 1974. Paragneisses of the Prince Albert Group. *Geol. Surv. Can. Pap.*, 74-1: 159—160.
Casshyap, S.M., 1969. Petrology of the Bruce and Gowganda formations and its bearing on the evolution of Huronian sedimentation in the Espanola—Willisville area, Ontario (Canada). *Palaeogeogr., Palaeoclimatol., Palaeoecol.*, 6(1): 5—36.
Catanzaro, E.I., 1963. Zircon ages in Southwestern Minnesota. *J. Geophys. Res.*, 68: 2045—2048.
Catanzaro, E.I. and Kulp, J.L., 1964. Discordant zircon from Little Belt (Montana), Beartooth (Montana) and Santa Catalina (Arizona) Mountains. *Geochim. Cosmochim. Acta*, 28(1): 87—124.
Chadwick, B. and Coe, K., 1973. Field work on the Precambrian basement in the Buksefjorden region, Southern West Greenland. *Grönl. Geol. Unders. Rep. Act.*, 55: 32—37.
Charbonneau, B.W., 1973. A Grenville front magnetic anomaly in the Megiscane Lake area, Quebec. *Geol. Surv. Can. Pap.*, 73-29: 20 pp.
Chaudhuri, A., 1970. Precambrian stromatolites in the Pranhita—Godovari Valley. *Palaeogeogr., Palaeoclimatol., Palaeoecol.*, 7(4): 309—340.
Chaudhuri, S. and Faure, G., 1964. The whole rock Rb—Sr age of the Precambrian Nonesuch shale in Michigan. *Mass. Inst. Technol., 12th Annu. Progr. Rep.*, 221: 20—22.
Chaudhuri, S. and Faure, G., 1967. Geochronology of the Keweenawan rocks, White Pine, Michigan. *Econ. Geol.*, 62(8): 1011—1033.
Chen yü-ch'i, Shen Ch'i-han and Wang Tzu-chiu, 1964. The geologic age of metamorphic and magmatic rocks of the Tai Shan group in Sint'an area, Shantung province. *Dichzhi Lunping*, 22(3): 115—128 (translated from the Chinese).

Chernov, G.A., 1948. New data on the geology and tectonics of the western slopes of the Pri-Polar Urals. *Dokl. Acad. Sci. USSR, N. Ser.*, 61(5): 887—890.

Chernov, V.M., Inina, K.A., Gor'kovets, V.Ya. and Rayevskaya, I.B., 1970. *Volcanogenic Ferruginous—Siliceous Formations in Karelia.* Karel'sk. Knizn. Izdatel'stvo, Petrozavodsk, 284 pp.

Chernyshev, N.M., 1972. Precambrian intrusive complexes of basic and ultrabasic rocks of the Voronezh crystalline massif and their common aspects in the field of ore content. *Izv. Acad. Sci. USSR, Geol. Ser.*, 1972(4): 35—47.

Choubert, G.A. and Faure-Muret, A., 1967. The tectonic map of Africa. In: *Tektoniki Karty Kontinentov.* Nauka, Moscow, pp. 83—98.

Chumakov, N.M., 1956. Stratigraphy of the northern margin of Patom Highland. *Dokl. Acad. Sci. USSR*, 111(4): 863—865.

Chumakov, N.M., 1971. The Vendian glaciation in Europe and the North Atlantic (the Upper Precambrian). *Dokl. Acad. Sci. USSR*, 198(2): 419—422.

Chumakov, N.M., 1974. The Laplandian glaciation. In: *Etudy po Stratigrafii.* Nauka, Moscow, pp. 71—96.

Cloud, P.E., 1968. Pre-metazoan evolution and the origin of the metazoa. In: E.T. Drake (Editor), *Evolution and Environment.* Yale Univ. Press, New Haven, Conn., pp. 1—72.

Cloud, P.E. and Semikhatov, M.A., 1969. Proterozoic stromatolite zonation. *Am. J. Sci.*, 267(9): 1017—1061.

Cloud, P.E., Gruner, J.W. and Hagen, H., 1965. Carbonaceous rocks of the Soudan formation (Early Precambrian). *Science*, 148(3678): 1713—1716.

Cogné, J., 1962. Le Briovérien. *Bull. Soc. Géol. Fr., 7ième Sér.*, 4(3): 413—430.

Colton, G.W., 1970. The Appalachian Basin — its depositional sequences and their geologic relationships. In: *Studies of Appalachian Geology, Central and Southern.* Interscience, New York, N.Y., pp. 5—48.

Compston, W. and Arriens, P.A., 1968. The Precambrian geochronology of Australia. *Can. J. Earth Sci.*, 5(3): 561—583.

Condie, K.C., 1966. Late Precambrian rocks of the northeastern Great Basin and vicinity. *J. Geol.*, 74(5): 631—636

Condie, K.C., 1967. Petrology of the Late Precambrian tillite (?) association in Northern Utah. *Geol. Soc. Am. Bull.*, 78(11): 1317—1343.

Condie, K.C., Macke, J. and Reimer, T., 1970. Petrology and geochemistry of Early Precambrian greywackes from the Fig Tree Group, South Africa. *Geol. Soc. Am. Bull.*, 81(9): 2759—2776.

Cowie, J.W., 1961. Contributions to the geology of North Greenland. *Medd. Grönl.*, 164(3): 47 pp.

Cowie, J.W. and Rozanov, A.Ye., 1973. Report of the International Working Group of the Symposium of the Cambrian—Precambrian Boundary. *Izv. Acad. Sci. USSR, Geol. Ser.*, 1973(12): 72—82.

Crawford, A.R., 1969. Reconnaissance Rb—Sr dating of southern Peninsular India. *J. Geol. Soc. India*, 10(2): 117—166.

Crawford, A.R. and Compston, W., 1970. The age of the Vindhyan System of Peninsular India. *Q. J. Geol. Soc. London*, 125(499): 3.

Crawford, A.R. and Compston, W., 1973. The age of the Cuddapah and Kurnool Systems, Southern India. *J. Geol. Soc. Aust.*, 19(4): 453—464.

Crittenden, M.D., Sharp, R.J. and Calkins, F.C., 1952. Geology of the Wasatch Mountains east of Salt Lake City, Parleys Canyon to the Traverse Range. *Geol. Surv. Utah, Guideb., No. 8*: 45 pp.

Cumming, G.L., Wilson, J.T., Farquhar, R.M. and Russell, R.D., 1955. Some dates and subdivisions of the Canadian Shield. *Geol. Assoc. Can. Proc.*, 7(2): 27.

Dana, J., 1872. Notice of the address of Prof. T. Sterry-Hunt. *Am. J. Sci., 3rd Ser.*, 3: 338—340.

Davidson, A., 1972(1971). Granite studies in the Slave province. *Geol. Surv. Can. Pap.*, 72-1: 109—115.
Davis, G.L. and Tilton, G.R., 1965. *Geochronology and Isotope Geochemistry*. Year Book Carnegie Institute, Annual Report, Director Geophysics Laboratory, pp. 165—177.
Davison, W.L., 1966. Caribon River map-area, Manitoba (54M). *Geol. Surv. Can. Pap.*, 65-25: 6 pp.
Dawes, P.R., 1970. The plutonic history of the Tasiussaq area, South Greenland. *Grönl. Geol. Unders. Bull.*, 88: 125 pp.
Dawes, P.R., Rex, D.C. and Jepsen, H.F., 1973(1972). K—Ar whole rock ages of dolerites from the Thule district, Western North Greenland. *Grönl. Geol. Unders. Rep. Act.*, 55: 61—66.
Dawson, J., 1868. *Acadian Geology*. London, 2nd ed.
Dawson, J., 1897. Note on *Cryptozoan* and other fossils. *Can. Rec. Sci.*, 7: 203—219.
Dawson, K.R., 1966. A comprehensive study of the Preissac—Lacorne batholith, Abitibi county, Quebec. *Geol. Surv. Can. Bull.*, 142: 76 pp.
Dearnley, R. and Dunning, F.W., 1968. Metamorphosed and deformed pegmatites and basic dykes in the Lewisian complex of the Outer Hebrides and their geological significance. *Q. J. Geol. Soc. London*, 123(492): 353—378.
Derry, D.R., 1930. Geology of the area from Minaki to Sydney Lake, District of Kenora. *Ont. Dep. Mines Annu. Rep.*, 39(3): 24 pp.
De Waard, D. and Walton, M., 1967. Precambrian geology of the Adirondack Highlands — a reinterpretation. *Geol. Rundsch.*, 56(2): 596—610.
Dibrov, V.E., 1964. *Geology of the Central Part of Eastern Sayan*. Nedra, Moscow, 334 pp.
Dimroth, E., 1970. Evolution of the Labrador geosyncline. *Geol. Soc. Am. Bull.*, 81(9): 2717—2742.
Dimroth, E., Baragar, W.R., Bergeron, R. and Jackson, G.D., 1970. The filling of the Circum-Ungava geosyncline. *Geol. Surv. Can. Pap.*, 70-40: 45—142.
Dobrokhotov, M.N., 1967. Stratigraphical chart on the Precambrian of the Ukrainian Shield. *Sov. Geol.*, 1967(6): 17—25.
Dobrokhotov, M.N., 1969. Certain problems of Precambrian geology in the Krivoy Rog—Kremenchug structural-facial zone. *Izv. Acad. Sci. USSR, Geol. Ser.*, 1969(4): 16—34.
Dodin, A.L., Konikov, A.Z. and Man'kovsky, V.K., 1968. *The Precambrian Stratigraphy of Eastern Sayan*. Nedra, Moscow, 280 pp.
Dol'nik, T.A., 1969. *Stratigraphy and Stromatolites of the Riphean, Vendian, and Lower Cambrian in North-Baikal and Patom Highlands*. Thesis, Irkutsk Univ., Irkutsk, 24 pp.
Dominikovsky, G.G. and Medushevskaya, I.A., 1973. Geologic evolution of crystalline basement in the western area of the East-European platform. *Dokl. Acad. Sci. USSR*, 17(2): 167—169.
Donaldson, J.A., 1963. Stromatolites in the Denault formation, Marion Lake, Coast of Labrador, Newfoundland. *Geol. Surv. Can. Bull.*, 102: 33 pp.
Donaldson, J.A., 1965. The Dubawnt Group, Districts of Keewatin and Mackenzie. *Geol. Surv. Can. Pap.*, 64-20: 11 pp.
Donaldson, J.A., 1967. Two Proterozoic clastic sequences: a sedimentological comparison. *Geol. Assoc. Can. Proc.*, 18: 33—54.
Donaldson, J.A., 1968(1967). Proterozoic sedimentary rocks of Northern Saskatchewan (74N). *Geol. Surv. Can. Pap.*, 68-1: 131.
Donaldson, J.A., 1968. Stratigraphy and sedimentology of the Hornby Bay Group, district of Mackenzie (pt. 86, I, K, L, M, N, O). *Geol. Surv. Can. Pap.*, 69-1: 155—157.
Donn, W.L., Donn, B.D. and Valentine, Ch.G., 1965. On the early history of the Earth. *Geol. Soc. Am. Bull.*, 76(3): 287—306.

Douglas, R.J. (Editor), 1970. *Geology and Economic Minerals of Canada*. Geol. Surv. Can., Ottawa, 838 pp.
Dragunov, V.I., 1967. Vendian, Lower- and Middle Cambrian strata of the right bank of the Lower Yenisei River. In: *Stratigrafiya Dokembriya i Kembriya Sredney Sibiri*. Krasnoyarsk. Knizhn. Izdatel'stvo, Krasnoyarsk, pp. 107—123.
Dragunov, V.I. and Smirnova, E.B., 1964. Certain characteristics of the tectonics of the north-western margin of the Median-Siberian plateau (with reference to the characters of the Lower Yenisei aulacogen). *Tr. VSEGEI, N.Ser.*, 97: 41—57.
Drannik, A.S., 1972. *The Precambrian Stratigraphy of the Ovruch Ridge*. Thesis, IGN Acad. Sci. Ukr. SSR, Kiev, 22 pp.
Drannik, A.S. and Bogatskaya, I.V., 1967. New data on composition, structure, and stratigraphic position of the Precambrian Ovruch effusive—sedimentary group. *Probl. Osad. Geol. Dokembr.*, *2*: 169—176.
Drevin, A.Ya., 1967. The results of Precambrian studies in the Middle Near-Bug area by the lithologic-structural method. *Probl. Osad. Geol. Dokembr.*, *2*: 88—96.
Drugova, G.M., 1971. Conglomerates of Monchegorsk area in Kola Peninsula. In: *Problemy Litologii Dokembriya*. Nauka, Leningrad, pp. 108—109.
Drugova, G.M. and Glebovitsky, V.A., 1971. *Granulite Facies of Metamorphism*. Nauka, Leningrad, 256 pp.
Eade, K.E., 1966a. Fort George River and Kaniapiskau River (west half) map-areas, New Quebec. *Geol. Surv. Can. Mem.*, 339: 84 pp.
Eade, K.E., 1966b. Kognak River (west half) district of Keewatin. *Geol. Surv. Can. Pap.*, 65-8: 12 pp.
Eade, K.E., 1970. Ennadai Lake and Nueltin Lake map-areas, district of Keewatin (65c and 65b west half). *Geol. Surv. Can. Pap.*, 70-1: 137—139 (abstr).
Eade, K.E., 1971. Geology of Ennadai Lake map-area, district of Keewatin. *Geol. Surv. Can. Pap.*, 70-45: 19 pp.
Eckelmann, W.R., 1956. Uranium—lead method of age determination, 1. Lake Athabasca problem. *Geol. Soc. Am. Bull.*, 67(1): 35—54.
Eckelmann, W.R. and Kulp, J.L., 1956. Uranium—lead method of age determination, 2. North-American localities. *Geol. Soc. Am. Bull.*, 68(9): 1117—1140.
Egiazarov, B.Kh., 1959. Geologic structure of North Land Archipelago. *Tr. NIIGA*, 94: 139 pp.
Egorova, L.Z., 1964. Structure and composition of crystalline basement and of Bavlin strata of Kuibyshev and Orenburg regions. *Tr. Kuibyshev NIINP*, 24: 3—207.
Eliseeva, G.D., Schcherbak, N.P. and Kazantseva, A.I., 1973. The Pb-isochron age of carbonate rocks of Near-Bug area. In: *Opred. Abs. Vozrasta Rudn. Mestor. i Molod. Magmat. Protsessov* (summaries). Acad. Sci. USSR, Moscow, pp. 25—26.
Elizar'ev, Yu.Z., 1964. On polyfacial regional metamorphism during the Archean in the south-west part of the Pri-Baikal area. *Izv. Acad. Sci. USSR, Geol. Ser.*, 1964(9): 21—29.
Emmons, E., 1888. *Letter to Persifal Franzer, 25 May 1887*. Int. Geol. Congr. Am., Comm. Rep.
Ermanovics, I.F., 1970a. Berens River—Deer Lake map-area, Manitoba and Ontario. *Geol. Surv. Can. Pap.*, 70-29: 139—141 (abstr.).
Ermanovics, I.F., 1970b. Precambrian geology of Hecla—Carroll Lake map-area, Manitoba—Ontario. *Geol. Surv. Can. Pap.*, 69-42: 26—27.
Ermanovics, I.F., 1973. Precambrian geology of the Berens River map-area (west half), Manitoba. *Geol. Surv. Can. Pap.*, 73-20: 17 pp.
Ershov, V.M., Markov, S.N. and Khairitdinov, R.K., 1969. The absolute age of the Zigalga formation rocks in the Urals. *Geokhimiya*, 1969(5): 623—627.
Esipchuk, K.E., 1968. *Stratigraphy and Petrogenesis of a Gneiss—Migmatite Complex in the Western Pri-Azov Region*. Thesis, Kiev Univ., Kiev, 21 pp.

Eskin, A.S., Obukhov, S.P. and Khrenov, P.M., 1971. Rapakivi-granite from the Western Pri-Baikal area. *Dokl. Acad. Sci. USSR*, 200(4): 921—924.

Eskola, P., 1965. The Precambrian of Finland. In: K. Rankama (Editor), *The Precambrian, 1*. Interscience, New York, N.Y., pp. 145—264.

Espendhade, G.H., 1970. Geology of the northern part of the Blue Ridge anticlinorium. In: *Studies of Appalachian Geology, Central and Southern*. Interscience, New York, N.Y., pp. 199—212.

Evernden, J.F., Curtis, C.M. and Kistler, R.W., 1960. Argon diffusion in glauconite, microcline, sanidine, leucite, and phlogopite. *Am. J. Sci.*, 258(8): 583—604.

Fahrig, W.F., 1961. The geology of the Athabasca Formation. *Geol. Surv. Can. Bull.*, 68: 31—32.

Fairbairn, H.W., Faure, G., Pinson, W.H. and Hurley, P.M., 1968. Rb—Sr whole-rock age of the Sudbury lopolith and basin sediments. *Can. J. Earth Sci.*, 5(3): 707—714.

Faure, G., 1964. The age of the Duluth gabbro complex and the Endion sill by the whole rock Rb—Sr method. *Mass. Inst. Technol., 12th Annu. Progr. Rep.*, 1384-12: 255—257.

Faure, G., Fairbairn, H.W., Hurley, P.M. and Pinson, W.H., 1964. Whole-rock Rb—Sr age of norite and micropegmatite at Sudbury, Ontario. *J. Geol.*, 72(6): 848—854.

Fedorovsky, V.S. and Leytes, A.M., 1968. On the geosynclinal troughs in the Early Proterozoic in Olekma—Vitim Mountain Land. *Geotektonika*, 4: 114—127.

Fiala, F., 1964. Eokambrische Tillite der Zelezné hory Ostböhmen. *Geol. Rundsch.*, 54(1): 102—115.

Firsov, L.V., 1965. The absolute age of granitoid in Taigonos Peninsula. *Dokl. Acad. Sci. USSR*, 162(2): 414—417.

Folinsbee, R., 1972. The Precambrian metallogenic epochs — are they atmospheric or centrospheric? In: *Ocherki Sovr. Geokh. i Analit. Khimii*. Nauka, Moscow, pp. 253—262.

Ford, T.D., 1958. The Pre-Cambrian fossils of Charnwood Forest. *Yorksh. Geol. Soc. Proc.*, 31(8): 211—217.

Ford, T.D. and Breed, W.Y., 1972. The Chuar Group of the Proterozoic, Grand Canyon, Arizona. *Proc. Int. Geol. Congr., 24 Sess., Montreal 1972, Sect. 1, Precambrian Geology*, pp. 3—18.

Fraser, J.A., 1965. Study of the Epworth Group. *Geol. Surv. Can. Pap.*, 61-1: 29 pp.

Fraser, J.A. and Tremblay, L.P., 1969. Correlation of Proterozoic strata in the northwestern Canadian Shield. *Can. J. Earth Sci.*, 6(1): 1—9.

Fraser, J.A., Donaldson, J.A., Fahrig, W.F. and Tremblay, L.P., 1970. Helikian basins and geosynclines of the Northwestern Canadian Shield. *Geol. Surv. Can. Pap.*, 70-40: 213—238.

French, B.M., 1967. Sudbury structure, Ontario; some petrographic evidence for origin by meteorite impact. *Science*, 156(3778): 1094—1098.

Frith, R.A. and Doig, R., 1975. Pre-Kenoran tonalitic gneisses in the Grenville Province. *Can. J. Earth Sci.*, 12: 844—849.

Frith, R.A., Frith, R., Helmstaedt, H., Hill, J. and Leatherbarrow, R., 1974. Geology of the Indian Lake area (86B), distr. of Mackenzie. *Geol. Surv. Can. Pap.*, 74-1A: 165—171.

Fryer, B.J., 1972. Age determination in the Circum-Ungava geosyncline and the evolution of the Precambrian banded iron-formation. *Can. J. Earth Sci.*, 9(6): 652—663.

Frumkin, I.M., 1968. The Olekma group in Middle Olekma basin and the stratigraphic position of the Archean Kurulta group in the Aldan shield. *Mater. Geol. Polezn. Iskop. Yakutsk. ASSR*, 1968(18): 129—139.

Furduy, R.S., 1968. Tillites in the Late Precambrian of Pri-Kolyma area. *Dokl. Acad. Sci. USSR*, 180(4): 948—951.

Furduy, R.S., 1969. On the Riphean strata of the Omolon Massif. *Dokl. Acad. Sci. USSR*, 188(1): 191—193.

Gabrielse, H., Blusson, S.L. and Roddick, J.A., 1973. Geology of Flat River, Glacier Lake and Wrigley Lake map-area, district of Mackenzie and Yukon Territory. *Geol. Surv. Can. Mem.*, 366: 153 pp.

Galimov, E.M., Kuzhetzova, N.G. and Prokhorov, V.S., 1968. On the composition of the older atmosphere with reference to the analysis of carbon from Precambrian carbonates. *Geokhimiya*, 1968(11): 1376—1381.

Gamaleya, Yu.N., 1968. Absolute age of granitoids from the Ulkan pluton. *Izv. Acad. Sci. USSR, Geol. Ser.*, 1968(2): 35—40.

Gamaleya, Yu.N., Losev, A.G. and Popov, M.Ya., 1969. The oldest cover formations in the southeastern part of the Siberian platform. *Sov. Geol.*, 1969(4): 137—144.

Garan', M.I., 1946. *The Age and Environment of the Older Formations in the Western Slopes of the Southern Urals.* Gosgeoltekhizdat, Moscow, 49 pp.

Garbar, D.J. and Mil'shtein, V.E., 1970. Stratigraphy and new forms of microphytolites in Jotnian strata of the southwestern part of the Pri-Onega area. *Dokl. Acad. Sci. USSR*, 195(1): 159—162.

Garifulin, L.L., 1971. Conglomerates of the Kolmozero—Voronya Group. In: *Stratigrafiya, Raschleneniye i Korrelyatsiya Dokembriya Sev.-Vost. Chasti Balt. Shchita.* Nauka, Leningrad, pp. 42—52.

Gavrilenko, V.A., Notkin, A.D. and Shipitsin, V.A., 1971. Age of the Chingasan group of Yenisei Ridge. *Dokl. Acad. Sci. USSR*, 197(6): 1387—1389.

Geier, P., 1963. Precambrian of Sweden. In: K. Rankama (Editor), *The Precambrian, 1.* Interscience, New York, N.Y., pp. 81—143.

Geisler, A.N., 1966. Geochronological correlation and paleogeography of Late Proterozoic strata in northern and central parts of the Russian platform. *Tr. VSEGEI, N. Ser.*, 114: 32—57.

Gerling, E.K. and Lobach-Zhuchenko, S.B., 1967. The radiological methods: their status and application in Precambrian mapping with reference to Karelia. In: *Problemy Izucheniya Geologii Dokembriya.* Nauka, pp. 36—47.

Gerling, E.K., Pushkarev, Yu.D. and Kotov, M.V., 1965a. Certain aspects of mineral behaviour under heat and high argon pressure. *Izv. Acad. Sci. USSR, Geol. Ser.*, 1965 (11): 3—13.

Gerling, E.K., Glebova-Kul'bakh, G.O. and Lobach-Zhuchenko, S.B., 1965b. New data on geochronology in Karelia. In: *Absolyutny Vozrast Dokembriyskikh Porod SSSR.* Nauka, Moscow—Leningrad, pp. 35—73.

Gerling, E.K., Lobach-Zhuchenko, S.B. and Borisenko, N.F., 1966. New data on Jotnian absolute age determinations in the Baltic shield. *Dokl. Acad. Sci. USSR*, 166(3): 674—677.

Gershoig, Yu.T., Kokhan, V.G. and Malakhov, Yu.T., 1974. Sandstone boulders in the rocks of the first slate horizon of the iron-ore formation of Krivoy Rog. *Geol. Zh.*, 7(6): 136—139.

Gibbins, W.A., Adams, C.Y. and McNutt, R.H., 1972. Rb—Sr isotopic studies of the Murray granite. *Geol. Assoc. Can. Spec. Pap.*, 10: 61—66.

Giletti, B.J. and Gast, P.W., 1961. Absolute age of Precambrian rocks in Wyoming and Montana. *Ann. N.Y. Acad. Sci.*, 91(2): 454—458.

Gilluly, J., 1949. Distribution of mountain building in geologic time. *Geol. Soc. Am. Bull.*, 60(4): 561—590.

Gintov, O.B., 1973. Precambrian ring structures in the Ukraine. *Geotektonika*, 5: 65—74.

Gladkovsky, A.K. and Khramtsov, V.N., 1967. Bauxites in the area of the Kursk Magnetic Anomaly. *Probl. Osad. Geol. Dokembr.*, 2: 133—162.

Glikson, A.Y., 1970. Geosynclinal evolution and geochemical affinities of Early Precambrian systems. *Tectonophysics*, 9(5): 397—433.

Glikson, A.Y., 1972. Early Precambrian evidence of a primitive ocean crust and island nuclei of sodic granite. *Geol. Soc. Am. Bull.*, 83(11): 3323—3344.
Goldich, S.S., 1968. Geochronology in the Lake Superior region. *Can. J. Earth Sci.*, 5(3): 715—724.
Goldich, S.S., Nier, A.O. and Baadsgaard, H., 1961. The Precambrian geology and geochronology of Minnesota. *Minn. Geol. Surv. Bull.*, 41: 18—34.
Goldich, S.S., Hedge, C.E. and Stern, T.W., 1970. Age of the Morton and Montevideo gneisses and related rocks, Southwestern Minnesota. *Geol. Soc. Am. Bull.*, 81(12): 3671—3696.
Golivkin, N.I., 1967a. Precambrian stratigraphy of Starooskol and Novooskol iron-ore deposits in the KMA area. In: *Geologiya i Polezn. Iscop. KMA*. Nedra, Moscow, pp. 60—75.
Golivkin, N.I., 1967b. Tectonic pattern of the Precambrian basement of the Starooskol and Novooskol areas of KMA. *Izv. Vuzov Geol. Razved.*, 1967(9): 29—30.
Golovanov, N.P. and Zlobin, M.N., 1966. On subdivision of the Riphean strata in East Taymyr. *Uch. Zap. NIIGA, Paleontol. Biostratigr.*, 13: 67—89.
Golovenok, V.K., 1957. On the stratigraphy of the northern part of Patom Highland. *Vestn. LGU, Ser. Geol. Geogr.*, 4: 54—64.
Golovenok, V.K., 1971. Relationships of gneiss and schist strata of the Keyvy group in Kola Peninsula. In: *Problemy Geologii Dokembriya Baltiyskogo Shchita i Pokrova Russkoy Platformy. Tr. VSEGEI*, 175: 206—220.
Goodwin, A.M., 1966. The relationship of mineralization to stratigraphy in the Michipicoten area, Ontario. *Geol. Assoc. Can. Spec. Pap.*, 3: 57—73.
Goodwin, A.M., 1968. Evolution of the Canadian Shield. *Geol. Assoc. Can. Proc.*, 19: 1—14.
Goodwin, A.M. and Schklanka, R., 1967. Archean volcano-tectonic basins: form and pattern. *Can. J. Earth Sci.*, 4(4): 777—795.
Gorokhov, I.M., 1964. Age determinations of the Korosten granite and Dnieper migmatite and metabasite in the Ukraine by Rb—Sr whole rock analysis. *Geokhimiya*, 1964(8): 744—753.
Gorokhov, I.M. and Gerling, E.K., 1971. The results of Rb—Sr analysis in geochronological studies of the eastern part of the Baltic shield. In: *Nov. Dannye po Geokhronol. Shkale v Absol. Letoischislenii*. Nauka, Moscow, pp. 67—69.
Gorokhov, I.M., Kutyavin, E.P. and Varshavskaya, E.S., 1973. The Rb—Sr rock age of Ovruch Ridge. In: *Opredeleniye Abs. Vozrasta Rudnykh Mestorozhd. i Molod. Magm. Protsessov*. Acad. Sci. USSR, Moscow, pp. 22—23.
Gorokhov, S.S., 1968. *Riphean of Ural-Tau Ridge*. Nauka, Moscow, 138 pp.
Goryunova, S.V., Rzhanova, G.N. and Orleansky, V.K., 1969. *Blue-Green Algae*. Nauka, Moscow, 228 pp.
Graindor, M. and Wasserburg, G., 1962. Détermination d'âges absolus dans le Nord du Massif armoricain. *C.R. Acad. Sci. Fr.*, 254(22): 3875—3877.
Graindor, M.J., 1957. Cayexidae nov. fam. organismes à squelette du Briovérien. *C.R. Acad. Sci. Fr.*, 224(15): 2075—2077.
Grechishnikov, N.P., 1971. Certain aspects of the Precambrian stratigraphy of the central part of the Ukrainian shield. *Geol. Zh.*, 31(2): 147—151.
Green, D.C., Baadsgaard, H. and Cumming, G.L., 1968. Geochronology of the Yellowknife area, Northwest Territories, Canada. *Can. J. Earth Sci.*, 5(3): 725—735.
Greene, B. and Williams, H., 1974. New fossil localities and the base of the Cambrian in Southeastern Newfoundland. *Can. J. Earth Sci.*, 11(2): 319—323.
Greer, L., 1930. Geology of the Shoal Lake (West) area, District of Kenora. *Ont. Dep. Mines Annu. Rep.*, 39(3): 42—56.
Griffin, W.L. and Heier, K.S., 1969. Paragenesis of garnet in granulite facies rocks, Lofoten—Vesteraalen, Norway. *Contrib. Mineral. Petrol.*, 28(2): 89—116.

Grigor'ev, V.N. and Semikhatov, M.A., 1958. On the age and origin of the so-called "tillites", in the northern part of Yenisei Ridge. *Izv. Acad. Sci. USSR, Geol. Ser.*, 1958 (11): 44—58.
Grinberg, G.A., 1968. *Precambrian of Okhotsk Massif*. Nauka, Moscow, 187 pp.
Gross, W.H. and Ferguson, S.A., 1965. The anatomy of an Archean greenstone belt. *Can. Min. Metall. Bull.*, 58(641): 940—946.
Grout, F.F., Gruner, J.W., Schwartz, G.M. and Thiel, G.A., 1951. Precambrian stratigraphy of Minnesota. *Geol. Soc. Am. Bull.*, 62(107): 1017—1078.
Gruner, J.W., 1923. Algae, believed to be Archean. *J. Geol.*, 31: 146—148.
Gruner, J.W., 1941. Structural geology of the Knife Lake of northeastern Minnesota. *Geol. Soc. Am. Bull.*, 52(10): 1577—1642.
Gunin, V.A., 1968. The older strata of the Yaroga and Subgan grabens (western part of the Aldan shield). *Mater. Geol. Polezn. Iskop. Yakutsk. ASSR*, 1968(18): 87—92.
Gunning, H.C. and Ambrose, J.W., 1940. Malarctic area, Quebec. *Geol. Surv. Can. Mem.*, 222: 129 pp.
Ham, W., Denison, R. and Merritt, C., 1964. Basement rocks and structural evolution of Southern Oklahoma. *Okla. Geol. Surv. Bull.*, 95: 302 pp.
Hamblin, W.K., 1961. Paleogeographic evolution of the Lake Superior region from Late Keweenawan to Late Cambrian time. *Geol. Soc. Am. Bull.*, 72(1): 1—18.
Hamilton, W.B., 1956. Precambrian rocks of Wichita and Arbuckle Mountains, Oklahoma. *Geol. Soc. Am. Bull.*, 67(10): 1319—1330.
Hanson, G.N. and Himmelberg, G.R., 1967. Ages of mafic dykes near Granite Falls, Minnesota. *Geol. Soc. Am. Bull.*, 78(11): 1429—1432.
Hanson, G.N., Goldich, S.S., Arth, J.G. and Yardley, D.H., 1971. Age of the Early Precambrian rocks of the Saganaga Lake—Northern Light Lake area, Ontario—Minnesota. *Can. J. Earth Sci.*, 9: 1100—1124.
Harland, W.B., 1961. An outline of the structural history of Spitsbergen. In: G.O. Raasch (Editor), *Geology of the Arctic*. Toronto, pp. 68—132.
Hart, S.R., Davis, G.L., Steiger, R.G. and Tilton, G.R., 1968. A comparison of the isotopic mineral ages variation and petrologic changes induced by contact metamorphism. In: *Radiometric Dating for Geologists*. Interscience, New York, N.Y., pp. 73—110.
Hedge, C.E., Peterman, Z.E. and Bradbock, W.A., 1967. Age of the major Precambrian metamorphism in the Northern Front Range, Colorado. *Geol. Soc. Am. Bull.*, 78(4): 551—557.
Heimlich, R.A. and Banks, P.O., 1968. Radiometric age determinations, Bighorn Mountains, Wyoming. *Am. J. Sci.*, 266(3): 180—192.
Henderson, G. and Pulvertaft, T.C., 1967. The stratigraphy and structure of the Precambrian rocks of the Umanak area, West Greenland. *Grönl. Geol. Unders. Misc. Pap.*, 52: 20 pp.
Henderson, J.B., 1970. Stratigraphy of the Archean Yellowknife Supergroup, Yellowknife Bay—Prosperous Lake area, district of Mackenzie. *Geol. Surv. Can. Pap.*, 70-26: 12 pp.
Henderson, J.B., 1972. Sedimentology of Archean turbidites at Yellowknife, Northwest Territories. *Can. J. Earth Sci.*, 9(7): 889—902.
Henderson, J.B., Cecile, M.P. and Kamineni, D.C., 1972(1971). Yellowknife and Hearne Lake map-areas, district of Mackenzie with notes on the Yellowknife Supergroup (Archean). *Geol. Surv. Can. Pap.*, 72-1: 117—118.
Heron, A., 1953. The geology of Central Rajputana. *Mem. Geol. Surv. India*, 79: 389 pp.
Heywood, W.W., 1968. Geological notes of Northeastern district of Keewatin and Southern Melville Peninsula, district of Franklin, Northwest Territories (pt. 46, 47, 56, 57). *Geol. Surv. Can. Pap.*, 66-40: 20 pp.
Hietanen, A., 1967. Scapolite in the Belt series in the St. Joe—Clearwater region, Idaho. *US Geol. Surv. Spec. Pap.*, 86: 56 pp.

Higgins, A.K., 1969. The Tartoq group on Nunaqaqertog and in the Iterdlak, SW Greenland. *Geol. Surv. Greenl. Rep.*, 17: 17 pp.
Higgins, A.K., 1970. The stratigraphy and structure of the Ketilidian rocks of Midternaes, South-West Greenland. *Grönl. Unders. Bull.*, 87: 96 pp.
Higgins, A.K. and Bondesen, E., 1966. Supracrustals of the Pre-Ketilidian age (the Tartoq group) and their relationships with Ketilidian supracrustals in the Ivigtut region, SW Greenland. *Geol. Surv. Greenl. Rep.*, 8: 21 pp.
Hills, F.A., Gast, P.W. and Houston, R.S., 1968. Precambrian geochronology of the Medicine Bow Mountain, Southeastern Wyoming. *Geol. Soc. Am. Bull.*, 79(12): 1757—1780.
Himmelberg, G.R., 1968. Geology of Precambrian rocks of Granite Falls—Montevideo area, Southwestern Minnesota. *Minn. Geol. Surv. Spec. Pap.*, 5: 33 pp.
Hjelmqvist, S., 1958. *Excursion guide, 7 Nordiska Geologmötet*. Stockholm, Excursion B.1, 28 pp.
Hoffman, P.F., 1968a. Precambrian stratigraphy, sedimentology, paleocurrents, and paleoecology in the East Arm of Great Slave Lake, district of Mackenzie (75L). *Geol. Surv. Can. Pap.*, 68-1: 140—142.
Hoffman, P.F., 1968b. Stratigraphy of the Lower Proterozoic (Aphebian) Great Slave supergroup, East Arm of Great Slave Lake, district of Mackenzie. *Geol. Surv. Can. Pap.*, 68-42: 93 pp.
Hoffman, P.F., 1969. Proterozoic paleocurrent and depositional history of the East Arm Fold Belt, Great Slave Lake, Northwest Territories. *Can. J. Earth Sci.*, 6(3): 441—462.
Hoffman, P.F., 1973. Aphebian supracrustal rocks of the Athapuscow aulacogen, East arm of Great Slave Lake, district of Mackenzie. *Geol. Surv. Can. Pap.*, 1A: 151—156.
Hoffman, P.F., Fraser, J.A. and McGlynn, J., 1970. The Coronation geosyncline of Aphebian age, district of Mackenzie. *Geol. Surv. Can. Pap.*, 70-40: 201—212.
Hofmann, H.J., 1969. Stromatolites from the Proterozoic Animikie and Sibley Groups, Ontario. *Geol. Surv. Can. Pap.*, 68-69: 77 pp.
Hofmann, H.J., 1971. Precambrian fossils, pseudofossils and problematica in Canada. *Geol. Surv. Can. Bull.*, 189: 146 pp.
Hofmann, H.J., 1974. The stromatolite *Archaeozoon acadiense* from the Proterozoic Green Head Group of Saint John, New Brunswick. *Can. J. Earth Sci.*, 11(8): 1095—1115.
Holtedahl, O., 1953. *Norges Geologi. Nor. Geol. Unders.*, 164: 450 pp. (Russian translation: *Geology of Norway*. Inostrannaya Literatura, Moscow, 1957, 424 pp.
Holubek, J., 1966. Stratigraphy of the Upper Proterozoic in a core of the Bohemian Massif. *Rozp. Cesk. Akad. Ved Rada Mat. Prir. Ved.*, 76(4): 62 pp.
Holubek, J., 1968. Structure and basement relationships of Archean metasediments in the Noranda—Malartic area of Quebec. *Geol. Surv. Can. Pap.*, 68-1: 143—144.
Horwood, H.C., 1935. A pre-Keewatin (?) tonalite. *Rep. Soc. Can. Trans., 3rd Ser.*, 29(4): 139—147.
Houston, R.S., 1968. A regional study of rocks of Precambrian age in the part of the Medicine Bow Mountains in southwestern Wyoming. *Geol. Surv. Wyo. Mem.*, 1: 167 pp.
Howell, B.F., 1956. Evidence from fossils of the age of the Vindhyan System. *J. Palaeontol. Soc. India*, 1(1): 108—112.
Howell, D.G., 1971. A stromatolite from the Proterozoic Pahrump group, eastern California. *J. Paleontol.*, 45(1): 48—51.
Hunt, C.B. and Mabey, D.R., 1966. Stratigraphy and structure of Death Valley, California. *US Geol. Surv. Prof. Pap.*, 484-A: 162 pp.
Hurst, R.W., 1973. The early Archean of Coastal Labrador. In: S.A. Morse (Editor), *The Nain Anorthosite Project, Labrador, Field Report*. Contrib. No. 13, Geol. Dep., Univ. Mass.

Hutchinson, R.D., 1953. Geology of Harbour Grace map-area, Newfoundland. *Geol. Surv. Can. Mem.*, 275: 43 pp.

Igolkina, N.S., 1961. The Precambrian strata of the sedimentary cover in the northern part of the Russian platform. *Inf. Sb. VSEGEI*, 43: 3—10.

Irving, R.D., 1887. Is there a Huronian group? *Am. J. Sci., 3rd Ser.*, 34.

Itskov, A.I., 1970. *Structure and Environment of the Late Precambrian Orogenic Complex of the Yenisei Ridge*. Thesis, MGU, Moscow, 24 pp.

Ivantishin, M.N. and Orsa, V.I., 1965. The gneiss—migmatite formations and granites in Zaporozh'ye—Mishurin Rog region. In: *Geokhronologia Dokembriya Ukrainy*. Naukova Dumka, Kiev, pp. 26—38.

Ivliev, A.I., 1971. Stratigraphy of the supracrustal complex in Salny Tundra area. In: *Stratigriya, Raschleniye i Korrelyatsiya dokembriya sev.-vost. chasti Baltiysk. shchita*. Nauka, Leningrad, pp. 52—60.

Jackson, G.D. and Taylor, F.C., 1972. Correlation of major Aphebian rock units in the Northern Canadian Shield. *Can. J. Earth Sci.*, 9(12): 1650—1669.

Jacobson, K.E., 1962. The Pre-Ordovician strata correlation in Volyn and Podolia. *Dokl. Acad. Sci. USSR*, 142(3): 663—666.

Jacobson, K.E., 1966. The problem of the Proterozoic—Paleozoic boundary in the western part of the Russian platform. *Izv. Acad. Sci. USSR, Geol. Ser.*, 1966(7): 88—106.

Jacobson, K.E., 1968. The major structural features of Riphean sedimentary strata in the Pri-Urals area. *Dokl. Acad. Sci. USSR*, 179(1): 175—178.

Jacobson, K.E., 1971a. Problems of correlation of Pre-Vendian non-metamorphosed strata of the Russian plate. In: *Problemy Geologii Dokembriya Baltiyskogo Shchita i Pokrova Russkoy Platformy, Izv. Acad. Sci. USSR, Geol. Ser.*, 1971(7): 88—106.

Jacobson, K.E., 1971b. The relationships of the Volyn group with other Upper Precambrian units in the western part of the Russian platform. *Sov. Geol.*, 1971(2): 66—74.

James, H.L., 1955. Zones of regional metamorphism in the Precambrian of northern Michigan. *Geol. Soc. Am. Bull.*, 66(12): 1455—1487.

James, H.L., 1958. Stratigraphy of Pre-Keweenawan rocks in parts of northern Michigan. *US Geol. Surv. Prof. Pap.*, 314-C: 27—44.

Jeffreys, H., 1959. *The Earth, its Origin, History, and Physical Constitution*. Cambridge Univ. Press, New York, N.Y., 4th ed., 420 pp.

Jenĉek, V. and Vajner, V., 1968. Stratigraphy and relations of the groups in the Bohemian part of the Moldanubicum. *Kristalinicum*, 6: 105—124.

Jepsen, H.F., 1971. The Precambrian, Eocambrian, and Early Paleozoic stratigraphy of the Jorgen Bronlund Fjord area, Peary Land, North Greenland. *Medd. Grönl.*, 192(2): 42 pp.

Johnson, M.R., 1969. Dalradian of Scotland. *Am. Assoc. Pet. Geol. Mem.*, 12: 151—158.

Jolliffe, A.W., 1966. Stratigraphy of the Steeprock Group, Steeprock Lake, Ontario. *Geol. Assoc. Can. Spec. Pap.*, 3: 75—98 (Precambrian Symposium).

Kairyak, A.I. and Khazov, R.A., 1967. Jotnian formations in the northeastern part of the Pri-Ladoga area. *Vestn. LGU, Ser. Geol. Geogr.*, 12(2): 62—72.

Kalyaev, G.I., 1965. *The Precambrian Tectonics of the Ukrainian Iron-Ore Formation Province*. Naukova Dumka, Kiev, 190 pp.

Kalyaev, G.I. and Komarov, A.N., 1969. Folded structures of Kirovograd block (central part of the Ukrainian shield). *Geol. Zh.*, 29(6): 29—39.

Kanasewich, E.R. and Farquhar, R.M., 1965. Lead isotope ratios from the Cobalt—Noranda area, Canada. *Can. J. Earth Sci.*, 2(4): 361—362.

Kao Chen-hsi, 1962. Preliminary studies of the Sinian stratigraphy in Northern China. In: *The Oldest Rocks in China*. Inostrannaya Literatura, Moscow, pp. 39—69 (translated from the Chinese).

Kargat'yev, V.A., 1970. Anhydrite in diopside rocks of Central Aldan area. In: *Miner. Syr'e*, 22: 65—74.

Katz, H.R., 1954. Einige Bemerkungen zur Lithologie und Stratigraphie der Tillitprofile im Gebiet der Kejser Josephs Fjord, Ostgrönland. *Medd. Grönl.*, 72(4): 63 pp.
Katz, H.R., 1961. Late Precambrian to Cambrian stratigraphy in East Greenland. In: G.O. Raasch (Editor), *Geology of the Arctic, 1*. Toronto, pp. 299—328.
Kay, M., 1955. Sediments and subsidence through time. *Geol. Soc. Am. Spec. Pap.*, 62: 665—684.
Kazakov, G.A., Knorre, K.G. and Strizhov, V.P., 1967. New age data on the lower formations of Nizhnebavlinsk group in Volga—Urals area. *Geokhimiya*, 1967(4): 482—485.
Kazansky, Yu.P., Katayeva, V.N. and Shugurova, N.A., 1973. Composition of the older atmospheres according to studies of gas inclusions in quartz rocks. In: *Geokhimiya Dokembriyskikh i Paleozoyskikh Otlozheniy Sibiri*. Novosibirsk, pp. 5—12.
Keller, B.M., 1964. The Riphean era. In: *Geologiya Dokembriya. Mezhdunar. Geol. Kongr., 22 Sess., Dokl. Sov. Geol. Nedra Probl.*, 10: 151—160.
Keller, B.M., 1968. The Upper Proterozoic of the Russian platform (Riphean and Vendian). *Ocherki Reg. Geol. SSSR.*, 2: 101 pp.
Keller, B.M. and Semikhatov, M.A., 1963. General problems of Upper Precambrian strata. In: *Stratigrafiya SSSR. Verkhniy Dokembriy*. Gosgeoltekhizdat, Moscow, pp. 578—586.
Keller, B.M. and Semikhatov, A.M., 1968. *Riphean Reference Sections on Continents. Stratigraphy and Paleontology*. VINITI, Moscow, pp. 5—108.
Keller, B.M., Kazakov, G.A. and Krylov, I.M., 1960. New data on the stratigraphy of the Riphean era (Upper Proterozoic). *Izv. Acad. Sci. USSR, Geol. Ser.*, 1960(12): 26—41.
Keller, B.M. Semikhatov, M.A. and Chumakov, H.M., 1967. The Upper Proterozoic of the Siberian platform and surrounding fold belts. In: *Stratigrafiya Dokembriya i Kembriya Sredney Sibiri*. Krasnoyarsk. Khizhn. Izdatel'stvo, Krasnoyarsk, pp. 247—291.
Keller, B.M., Aksenov, E.M., Korolev, B.G., Krylov, I.N., Rosanov, A.Yu., Semikhatov, M.A. and Chumakov, N.M., 1974. *Vendomian (Terminal Riphean) and its Regional Subdivisions*. VINITI, Moscow, 126 pp.
Kerr, J.W., 1967. Stratigraphy of central and eastern Ellesmere Island, Arctic Canada, I. Proterozoic and Cambrian. *Geol. Surv. Can. Pap.*, 67-27: 63 pp.
Khain, V.E., 1958. Large-scale cycles in Earth evolution. *Nauchn. Dokl. Vyssh. Shk., Geol.—Geogr. Nauki*, 1958(1): 25—33.
Kharitonov, L.Ya., 1941. Stratigraphy and tectonics of the Precambrian Karelian formation. *Tr. Leningr. Gor. Upr.*, 23: 46 pp.
Kharitonov, L.Ya., 1957. An attempt at tectonic subdivision of the eastern part of Baltic shield. *Uch. Zap. LGU, Ser. Geol. Nauk*, 9: 34—70.
Kharitonov, L.Ya., 1966. *Structure and Stratigraphy of the Karelides in the Eastern Part of the Baltic Shield*. Nedra, Moscow, 358 pp.
Kharitonov, L.Ya., Bogdanov, Yu.B. and Voinov, A.S., 1964. On the stratigraphy of iron-ore formations in Western Karelia. *Vestn. LGU*, 24: 35—43.
Khatuntseva, A.Ya., 1972. The Precambrian stratigraphy of the north-western (Volyn) part of the Ukrainian shield. *Geol. Zh.*, 32(1): 140—150.
Khomentovsky, V.V., 1974. Criteria for subdivision of the Vendian as a Paleozoic system. In: *Etudy po Stratigrafii*. Nauka, Moscow, pp. 33—70.
Khomentovsky, V.V., Shenfil', V.Yu. and Yakshin, M.S., 1968. Correlation of Late Precambrian strata in the outer belt of the Baikal—Patom folded area. *Geol. Geofiz.*, 1968(1): 3—12.
Kidd, D.F., 1933(1932). Great Bear Lake arc, N.W.T. *Geol. Surv. Can. Summ. Rep.*, B II: 76—150.
Kirichenko, G.I., 1967. Precambrian stratigraphy of the western margin of the Siberian platform and surrounding fold belts. *Tr. VSEGEI*, 112: 3—48.
Kirichenko, G.I., 1968. The Yenisei folded system. *Geol. Stroyeniye SSSR*, 1: 111—116.

Kirilyuk, V.P., 1966. *Geology and Environment of Precambrian Complexes on the South-western Margin of the Aldan—Vitim Shield*. Thesis, Lvov Univ., Lvov, 23 pp.
Kirsanov, V.V., 1968. Stratigraphy and correlation of strata of the Vendian complex in the eastern margin of the Russian platform. *Izv. Acad. Sci. USSR, Geol. Ser.*, 1968(6): 86—103.
Kirsanov, V.V., 1970. Vendian strata of central areas of the Russian platform. *Izv. Acad. Sci. USSR, Geol. Ser.*, 1970(12): 55—66.
Kir'yanov, V.V., 1969. Stratigraphical chart of Cambrian strata in Volyn. *Geol. Zh.*, 29(5): 48—62.
Kiselev, V.V. and Korolev, V.G., 1964. New data on Precambrian and Paleozoic stratigraphy in the western part of Kirgiz Ridge. In: *Materialy po Geologii. Tien Shan.* Acad. Sci. Kirgiz SSR, Frunze, pp. 3—44.
Klevtsova, A.A., 1963. Late Precambrian of the Pachelma trough and other parts of the Russian platform. *Dokl. Acad. Sci. USSR*, 150(3): 623—626.
Klevtsova, A.A., 1970. Structure of the Upper Precambrian sedimentary cover of the Russian platform. In: *Voprosy Tektoniki Dokembriya Kontinentov*. Nauka, Moscow, pp. 71—77.
Klevtsova, A.A., 1971. Major features of the Riphean evolution of the Russian platform. *Izv. Vuzov Geol. Razved.*, 1971(7): 3—13.
Knight, I., 1973. The Ramah Group between Nachvak Fjord and Bears Gut, Labrador. *Geol. Surv. Can. Pap.*, 73-1: 156—161.
Koeppel, V., 1968. Age and history of the uranium mineralization of the Beaverlodge area, Saskatschewan. *Geol. Surv. Can. Pap.*, 67-31: 111 pp.
Komar, V.A., 1964. Riphean columnar stromatolites in the north of the Siberian platform. *Uch. Zap. NIIGA, Paleontol. Biostratigr.*, 6: 84—105.
Komar, V.A., 1966. Upper Precambrian stromatolites in the north part of the Siberian platform and their stratigraphical importance. *Tr. GIN Acad. Sci. USSR*, 154: 122 pp.
Komar, V.A., Raaben, M.E. and Semikhatov, M.A., 1965. Conophytons in the Riphean of the USSR and their stratigraphical value. *Tr. GIN Acad. Sci. USSR*, 131: 73 pp.
Kondrat'eva, M.G., 1962. The Pre-Middle Devonian stratigraphy of Saratov and Volgograd Povolzh'ye. In: *Stratigr. Skhemy Paleozoiskikh Otlozheniy. Dodevon.* Gostoptekhizdat, Moscow, pp. 76—82.
Konikov, A.Z., 1974. New data on the stratigraphy of the Precambrian formations in Urik—Iya graben in the Pri-Sayan area. *Tr. VSEGEI, N. Ser.*, 199: 173—182.
Korolev, V.G., 1971. The Upper Precambrian stratigraphy of Tien-Shan and Karatau (abstracts). In: *Stratigriya Dokembriya Kazakhstana i Tien Shanya*. MGU, Moscow, pp. 117—118.
Korolyuk, I.K., 1963. Upper Precambrian stromatolites. In: *Stratigrafiya SSSR. Verkhniy Dokembriy*. Gosgeoltekhizdat, Moscow, pp. 479—497.
Korolyuk, I.K., 1966. The Riphean and Lower Cambrian problematic microorganisms in the Pri-Baikal area and Angara—Lena trough. *Vopr. Micropaleontol.*, 10: 174—198.
Korolyuk, I.K., Medvedeva, A.M. and Sidorov, A.L., 1961. On the age of the older formations of China — Alamat watershed in Vitim Highland. *Mater. Geol. Polezn. Iskop. Buryatsk.* ASSR, 1961(7): 170—172.
Korzhinsky, D.S., 1937. Crystalline rocks in South-western Pri-Baikal area. *Int. Geol. Congr., 17th Sess., Moscow 1937*, pp. 12—25.
Koster, F. and Baadsgaard, H., 1970. On the geology and geochronology of north-western Saskatchewan, 1. Tazin Lake region. *Can. J. Earth Sci.*, 7(3): 919—930.
Kostrikina, A.N., 1968. Stratigraphy of Proterozoic strata in Nognyazhek—Davangro—Khugda graben. *Mater. Geol. Polezn. Iskop. Yakutsk. ASSR*, 1968(18): 76—86.
Kouvo, O. and Tilton, G.R., 1966. Mineral ages from the Finnish Precambrian. *J. Geol.*, 74(4): 421—442.

Kozlov, M.T. and Latyshev, L.N., 1974. New data on the geology of the Pri-Imandra part of the Monchegorsk area. In: *Regional Geology, Metallogeny and Geophysics*. Apatity, pp. 64—74.

Krasil'shchikov, A.A., 1973. *Stratigraphy and Paleotectonics in Precambrian—Early Paleozoic in Spitsbergen*. Nedra, Leningrad, 120 pp.

Krasil'shchikov, A.A. and Vinogradov, V.A., 1960. New data on Precambrian stratigraphy and tectonics in the central part of Olenek Uplift. *Inf. Byull. NIIGA*, 22: 13—20.

Krasnobaev, A.A., 1967. α-lead age and some special characteristics of the structure of the Taratash augen-gneiss and Zilmerdak sandstone. In: *Mineraly Izverzhennykh Porod i Rud Urala*. Nauka, Leningrad, pp. 3—7.

Kratz, K.O., 1963. *Geology of the Karelides in Karelia*. Acad. Sci. USSR, Moscow—Leningrad, 210 pp.

Krishnan, M.S., 1960a. *Geology of India and Burma*. Higginbotham, Madras, 604 pp.

Krishnan, M.S., 1960b. Pre-Cambrian stratigraphy of India. *Int. Geol. Congr., Copenhagen 1960, Rep. No. 21*: 95—107.

Krishnaswamy, S., Lascar, B. and Murty, N., 1964. Geology of Madras—Mysore—Ootacamund area. *Int. Geol. Congr., New Delhi 1964, Rep. No. 22*: 24 pp. (Guide to excursions N A-24 and C-20).

Krogh, T.E. and Brooks, C., 1968—1969. The Grenville Front in the Chibougamau—Surprise Lake area, Quebec. Carnegie Institute, Annual Report, Director Geophysics Laboratory, pp. 313—315.

Krogh, T.E. and Davis, I.J., 1967. Indications by Rb—Sr isotopic studies of successive regional metamorphism in the Northwest Grenville area of Ontario. *Am. Geophys. Union Trans.*, 48(1): 242 (abstr.).

Krogh, T.E. and Davis, G.L., 1972. The lowering of zircon U—Pb ages and whole rock Rb—Sr ages during regional metamorphism in the Grenville Province, Ontario. *EOS*, 53(4): 542.

Krogh, T.E. and Hurley, P.M., 1968. Strontium isotope variation and whole-rock isochron studies, Grenville Province of Ontario. *J. Geophys. Res.*, 73(22): 7107—7125.

Krylov, I.N., 1963. Riphean columnar branching stromatolites of the Southern Urals and their stratigraphical value in the Upper Precambrian. *Tr. GIN Acad. Sci. USSR*, 69: 133 pp.

Krylov, I.N., 1966. On the columnar stromatolites of Karelia. In: *Ostatki Organizmov i Problematiki v Proterozoiskikh Obrazovaniyakh Karelii*. Karel'sk. Knizhn. Izdatel'stvo, Petrozavodsk, pp. 97—100.

Krylov, I.N., 1969. The Precambrian organic kingdom. In: *Tommotsky Yarus i Problema Nizhney Granitsy Kembriya*. Nauka, Moscow, pp. 250—263.

Krylov, I.N., 1971. Stromatolite implications in the Upper Precambrian stratigraphy of Kazakhstan and Central Asia. In: *Stratigriya Dokembriya Kazakhstana i Tien-Shanya*. MGU, Moscow, pp. 24—31.

Krylov, I.N., Shapovalova, E.G. and Fedonkin, P.N., 1971. Riphean strata of the Lower Lena River. *Sov. Geol.*, 1971(7): 85—95.

Krylova, M.D. and Neyelov, A.N., 1960. Conglomerate-like rocks in the Archean complex of Aldan. *Tr. Lab. Geol. Dokembr. Acad. Sci. USSR*, 9: 386—397.

Kukhareva, N.I., 1972. New data on the amphibolite—granite contact in the area of Saksagan belt. *Geol. Zh.*, 32(1): 56—64.

Kulish, E.A., 1971. Conglomerates in the lowermost Iyengra group of the Archean in Aldan area. *Dokl. Acad. Sci. USSR*, 198(4): 1216—1248.

Laberge, G.K., 1964. Development of magnetite in iron-formation of the Lake Superior region. *Econ. Geol.*, 59(7): 1313—1342.

Ladieva, V.D., 1965. The sedimentary—volcanogenic formations in Konsk—Belozero zone. In: *Geokhronologiya Dokembriya Ukrainy*. Naukova Dumka, Kiev, pp. 16—25.

Laitakari, I., 1969. On the set of olivine diabase dykes in Häme, Finland. *Bull. Comm. Géol. Finl.*, 241: 65 pp.
Lakshmanan, S., 1968. On the stratigraphic position of the Red shale (Iungel) series in the Son Valley. *Natl. Inst. Sci. India Proc.*, 34(1): 67—74.
Lambert, R.St.J. and Holland, J.G., 1972. A geochronological study of the Lewisian from Loch Laxford to Durness, Sutherland, NW Scotland. *J. Geol. Soc. London*, 128(1): 3—19.
Lambert, R.St.J. and Poole, A.B., 1964. The relationship of the Moine schists and Lewisian gneisses near Mallaigmore, Inverness-shire *Geol. Assoc. Proc.*, 75(1): 1—14.
Lanphere, M.A., 1968. Geochronology of the Yavapai Series of central Arizona. *Can. J. Earth Sci.*, 5(3): 757—762.
Latulippe, M., 1966. The relationship of mineralization to Precambrian stratigraphy in the Matagami Lake and Val d'Or districts of Quebec. *Geol. Assoc. Can. Spec. Pap.*, 3: 21—42.
Lauren, L., 1970. An interpretation of the negative gravity anomalies associated with the rapakivi granites and Jotnian sandstone in Southern Finland. *Geol. Fören. Stockh. Förh.*, 92(540): 21—34.
Lawson, A.C., 1913. A standard scale for the Pre-Cambrian rocks of North America. *Proc. Int. Geol. Congr., 12 Sess.*, 23 pp.
Laz'ko, E.M., 1956. *Geologic Structure of the Western Part of the Aldan Crystalline Massif*. Lvov Univ., Lvov, 195 pp.
Laz'ko, E.M., Kirilyuk, V.P., Sivoronov, A.A. and Yatstenko, G.M., 1970. Archean supracrustal formations of the southwestern part of the Ukrainian shield. *Dokl. Acad. Sci. USSR*, 194(6): 1168—1171.
Laz'ko, E.M., Krilyuk, V.P., Sivoronov, A.A. and Yatsenko, G.M., 1970. Geological Precambrian complexes in the southwestern part of the Ukrainian shield and the criteria for their subdivision. *Sov. Geol.*, 1970(6): 29—43.
Lee, J.S., 1934. *The Geology of China*. Murby, London, 528 pp.
Leech, A.P., 1966. Potassium—argon dates of basic intrusive rocks of the district of Mackenzie, N.W.T. *Can. J. Earth Sci.*, 3(3): 389—412.
Leith, C.K., Lund, R.J. and Leith, A., 1935. Precambrian rocks of the Lake Superior region. *US Geol. Surv. Prof. Pap.*, 184: 34 pp.
Lendzion, K., Mikhnyak, R. and Rozanov, A., 1965. Lithostratigraphic correlation of Late Precambrian and Lower Cambrian in Holy Cross Mts. and in the north-western part of the Russian platform. *Izv. Acad. Sci. USSR, Geol. Ser.*, 1965(8): 85—96.
Lepp, H., 1968. The distribution of manganese in the Animikian iron formation of Minnesota. *Econ. Geol.*, 63(1): 61—75.
Lepp, H. and Goldich, S.S., 1964. Origin of Precambrian iron formation. *Econ. Geol.*, 59(6): 1025—1068.
Lindsey, D.A., 1969. Glacial sedimentology of the Precambrian Gowganda formation, Ontario, Canada. *Geol. Soc. Am. Bull.*, 80(9): 1685—1701.
Li P'u, 1965. The results of K—Ar absolute age determinations on rocks of China. *Sci. Sin.*, 4(11): 1663—1672 (translated from the Chinese).
Livingston, D.E. and Damon, P.E., 1968. The ages of stratified Precambrian rock sequences in Central Arizona and northern Sonora. *Can. J. Earth Sci.*, 5(3): 763—772.
Logan, B., 1961. *Cryptozoon* and associate stromatolites from the recent Shark Bay, Western Australia. *J. Geol.*, 69(5): 517—533.
Long, L.E., 1961. Isotopic ages from Northern New Jersey and Southeastern New York. *Ann. N.Y. Acad. Sci.*, 91(2): 400—407.
Lü Hung-yün and Sha Tzu-an, 1965. Boundaries, classification, and paleogeography of Sinian strata in Southern China. *Dichzhi Kesyue*, 4: 12—60 (translated from the Chinese).

Lyubimova, E.A., 1960. The Earth heating. In: *Geologich. Resultaty Prikladn. Geokhimii i Geophisiki*. Gosgeoltekhizdat, Moscow, pp. 14—19.
McCartney, W.D., 1967. Whitbourne map-area, Newfoundland. *Geol. Surv. Can. Mem.*, 341: 135 pp.
McGregor, V.R., 1968. The Early Precambrian gneisses of the Godthaab district, West Greenland. *Philos. Trans. R. Soc. London, Ser. A*, 273(1235): 343—358.
McGregor, V.R. and Bridgwater, D., 1973(1972). Field mapping of the Precambrian basement in the Godthäbsfjord district, Southern West Greenland. *Grønl. Geol. Unders. Rep. Act.*, 55: 29—32.
MacLaren, A.S., 1952. Preliminary map of Kinojevis, Timiskaming country, Quebec. *Geol. Surv. Can. Pap.*, 52-6: 20 pp.
MacMannis, W.S., 1964. La Hood formation — a coarse facies of the Belt series in Southwestern Montana. *Geol. Soc. Am. Bull.*, 74(4): 407—436.
Magnusson, N.H., 1964. Jotniska och Subjotniska bergarternas oldersstallning. *Geol. Fören. Stockh. Förh.*, 86(516): 28—32.
Magnusson, N.H., 1965. Pre-Cambrian history of Sweden. *Q.J. Geol. Soc. London*, 121: 1—30.
Ma Hsing-yuan, 1962. Major aspects of geologic structure of Wutaishan Mts. In: *Drevneishiye Porody Kitaya*. Inostrannaya Literatura, Moscow, pp. 154—221 (translated from the Chinese).
Makhnach, A.S. and Veretennikov, N.S., 1970. *The Upper Proterozoic Volcanogenic Formation in Byelorussia*. Nauka i Tekhnika, Minsk, 80 pp.
Makiyevsky, S.J. and Nikolayeva, K.A., 1971. Conglomerates and older crusts of weathering in the Precambrian metamorphic strata in the northwestern part of the Kola Peninsula. In: *Stratigrafiya, Raschleneniye i Korrelyatsiya Dokembriya Sev.-Vost. Chasti Balt. Shchita*. Nauka, Leningrad, pp. 28—42.
Maksumova, R.A., 1967. *Stratigraphy and Lithology of Upper Proterozoic Rocks in the Southeastern Part of the Talas—Kara-tau Zone*. Thesis, Frunze, 27 pp.
Mal'kov, B.A., 1969. The diabase age of the Riphean Bystrukhino formation in the Middle Timan River. *Dokl. Acad. Sci. USSR*, 189(4): 810—814.
Manuylova, M.M. (Editor), 1968. *Geochronology of Precambrian Strata in the Siberian Platform and Surrounding Fold Belts*. Nauka, Leningrad, 331 pp.
Mardla, A.K., Mens, K.A. and Kala, E.A., 1968. *On the Stratigraphy of Cambrian Strata in Estonia*. Mintis, Vilnyus, 25 pp.
Masaytis, V.L. (Editor), 1964. *Geology of Korea*. Nedra, Moscow, 264 pp. (translated from the Korean).
Maslenikov, V.A., 1968. Precambrian absolute geochronology in the eastern part of the Baltic shield. In: *Geologiya i Glub. Stroyeniye Vost. Chasti Balt. Shchita*. Nauka, Moscow—Leningrad, pp. 60—77.
Mathur, S.M., 1960. A note on the Bijawar Series in the eastern part of type area, Chhatarpur district, N.P. *Geol. Surv. India Rec.*, 86(3): 539—544.
Mats, V.D. and Taskin, A.P., 1971. Proterozoic of the Pri-Sayan area and in marginal zones in the Baikal mountain area. *Dokl. Acad. Sci. USSR*, 200(2): 422—425.
Mattew, G.F., 1890. *Eozoon* and other low organisms in Laurentian rocks of St. John. *Nat. Hist. Soc. N.B. Bull.*, 9: 36—41.
Mattews, P.S. and Scharrer, R.H., 1968. A graded unconformity at the base of the Early Pre-Cambrian Pongola System. *Geol. Soc. S.Afr. Trans.*, 71(3): 257—272.
Mawdsley, J.B. and Norman, G.W., 1935. Chibougamau Lake map-area, Quebec. *Geol. Surv. Can. Mem.*, 185: 95 pp.
Menner, V.V., 1963. Paleontological basis of Upper Precambrian stratigraphy. In: *Stratigrafiya SSSR. Verkhniy Dokembriy*. Gosgeoltekhizdat. Moscow, pp. 476—507.
Miller, J.A. and Brown, P.E., 1965. Potassium—argon age studies in Scotland. *Geol. Mag.*, 102(2): 106—134.

Miller, W.G. and Knight, C.W., 1908. The Grenville—Hastings unconformity. *Ont. Dep. Mines Annu. Rep.*, 16(1).
Miller, W.G. and Knight, C.W., 1914. The Precambrian geology of southeastern Ontario. *Ont. Dep. Mines Annu. Rep.*, 22(2).
Mironyuk, E.P., 1959. The Eopaleozoic (Sinian) strata of the left bank of the Middle Olekma River. *Inf. Sb. VSEGEI*, 17: 5—10.
Mironyuk, E.P., Lyubimov, B.K. and Mangushevsky, E.L., 1971. *Geology of the Western Part of Aldan Shield*. Nedra, Moscow, 236 pp.
Mirskaya, D.D., 1971. New data on rocks of the Lebyazhye formation. In: *Materialy po Geologii i Metallogenii Kol'skogo Poluostrova*. Kol'sk. Acad. Sci. USSR, Apatity, pp. 25—34.
Misra, S.B., 1971. Stratigraphy and depositional history of Late Precambrian coelenterate-bearing rocks, Southeastern Newfoundland. *Geol. Soc. Am. Bull.*, 82(4): 979—987.
Mitrofanov, F.P. and Kol'tsova, T.V., 1965. The age of certain Post-Precambrian intrusive rocks of Eastern Sayan. In: *Absolyutny Vozrast Dokembriyskikh Porod SSSR*. Nauka, Moscow—Leningrad, pp. 142—148.
Mladshikh, S.V. and Ablizin, B.D., 1967. Stratigraphy of the Upper Precambrian in the western slopes of the Middle Urals. *Izv. Acad. Sci. USSR, Geol. Ser.*, 1967(2): 67—80.
Money, P.L., 1968. The Wollaston Lake fold-belt system, Saskatchewan—Manitoba. *Can. J. Earth. Sci.*, 5(6): 1489—1505.
Moorbath, S., O'Nions, R.K. and Pankhurst, R.J., 1972. Further rubidium—strontium age determinations on the very Early Precambrian rocks of the Godthaab district, West Greenland. *Nature Phys. Sci.*, 240(100): 78—82.
Moorbath, S., O'Nions, R.K. and Pankhurst, R.J., 1973. Early Archean age for the Isua Iron Formation, West Greenland. *Nature*, 245(5421): 138—139.
Moore, E.S., 1914(1912). Region east of the south end of Lake Winnipeg. *Geol. Surv. Can. Summ. Rep.*, 262.
Moore, J.M. and Thompson, P.H., 1972. The Flinton group. Grenville Province, Eastern Ontario, Canada. *Proc. Int. Geol. Congr., 24 Sess., Montreal 1972, Sect. 1, Precambrian Geology*, pp. 221—229.
Moralev, V.M. and Perfil'ev, Yu.S., 1972. The Precambrian geology of Southern India. *Sov. Geol.*, 1972(6): 98—107.
Morey, G.B., Green, J.G., Ojakangas, R.W. and Sims, P.K., 1970. Stratigraphy of the Lower Precambrian rocks in Vermillion district, northeastern Minnesota. *Minn. Geol. Surv. Rep.*, 14: 33 pp.
Morozov, S.G. and Revenko, E.A., 1969. On the age of carbonate strata of Bavlin rocks in Bashkiria. *Dokl. Acad. Sci. USSR*, 184(4): 1012—1015.
Morse, S.A., 1964. Age of Labrador anorthosites. *Nature*, 203(4944): 509—510.
Morse, S.A., 1969. The Kiglapait layered intrusion, Labrador. *Geol. Soc. Am. Mem.*, 112: 146 pp.
Muir, M.D. and Sutton, J., 1970. Some fossiliferous Pre-Cambrian chert pebbles within the Torridonian of Britain. *Nature*, 226(5244): 443—445.
Murchison, R., Verneuil, E. and Keyserling, A., 1845. *The Geology of Europe and Ural Mountains*. London — Paris.
Murray, R.C., 1955. Late Keweenawan or Early Cambrian glaciation in Upper Michigan. *Geol. Soc. Am. Bull.*, 66(3): 341—344.
Nalivkin, A.B., 1962. Stratigraphy and tectonics of metamorphic strata in Timan. In: *Stratigr. Skhemy Paleozoisk. Otlozheny. Dodevon*. Gostoptekhizdat, Moscow, pp. 122—123.
Narozhnykh, L.I. and Postnikova, I.E., 1971. Comparison of microphytolites of the Polessian and Serdobsk groups. *Dokl. Acad. Sci. USSR*, 198(6): 1411—1414.
Nath, M., Raja Rao, C.S. and Srikantan, B., 1964. Geology of Jaipur—Ajmer—Udaipur area, Rajasthan. Guide to excursion. *Proc. Int. Geol. Congr.*, 22 Sess., New Delhi 1964, 22 pp.

Nautiyal, S.P., 1965. Precambrian of Mysore Plateau. *Proc. Indian Sci. Congr., 53 Sess., Calcutta 1965, Pt. 2, Sect. Geol., Geogr.*, pp. 15—20.

Negrutza, V.Z., 1971. Stratigraphy of the hyperborean strata in Rybachy, Sredny Peninsulas, and Kildin Island. *Tr. VSEGEI*, 175: 153—186.

Negrutza, V.Z. and Negrutza, T.F., 1968. The Jatulian geology problem. *Tr. VSEGEI, N. Ser.*, 143: 81—96.

Nikitina, A.P., 1971. Bauxites in Kursk Magnetic Anomaly. In: *Platformenye Boksity SSSR*. Nauka, Moscow, pp. 120—123.

Norin, E., 1937. Geology of Western Qurug Tagh, Eastern Tien-Shan. *Rep. Sci. Exped. N.W. Prov. China under lead Dr. Sven Hedin. III. Geology 1*. Stockholm, 195 pp.

Nuzhnov, S.V., 1967. *The Riphean Strata in the Southeastern Part of the Siberian Platform*. Nauka, Moscow, 160 pp.

Nuzhnov, S.V., 1968. Regional stratigraphical chart of Proterozoic strata in the Aldan shield. *Mater. Geol. Polezn. Iskop. Yakutsk. ASSR*, 1968(18): 19—37.

Nuzhnov, S.V. and Mikhailov, V.A., 1968. Lower Proterozoic stratigraphy of the southern margin of the Aldan shield. *Mater. Geol. Polezn. Iskop. Yakutsk. ASSR*, 1968(18): 70—75.

Nuzhnov, S.V. and Yarmolyuk, V.A., 1968. New data on Precambrian stratigraphy with reference to the Aldan shield. *Sov. Geol.*, 1968(5): 3—20.

Nuzhnov, S.V., Kudryavtsev, V.P. and Akhmetov, R.N., 1968. Subdivision of the Sakhabor (Late Archean) strata in the Aldan shield. *Dokl. Acad. Sci. USSR*, 182(1): 164—166.

Obradovich, J.D. and Peterman, Z.E., 1962. Geochronology of the Belt Series, Montana. *Can. J. Earth. Sci.*, 5/3(2): 737—747.

Obruchev, S.V., 1958. Criteria for Precambrian correlation in the Siberian platform and its surrounding fold belts. In: *Trudy Mezhduvedom. Soveshch. po Razrabot. Unifits. Strat. Skhem Sibiri*. Acad. Sci. USSR, Leningrad, pp. 129—138.

Obruchev, S.V., 1963. Proterozoic correlation in the fold belts surrounding the Siberian platform. In: *Russkaya i Sibirskaya Platformy i Ikh Obrambleniye. Tr. Geol. Mus. Acad. Sci. USSR*, 14: 44—58.

Obruchev, S.V., Neyelov, A.N. and Nikitina, L.P., 1967. Stratigraphy and correlation of the Lower Precambrian in Eastern Siberia. In: *Stratigrafiya Dokembriya i Kembriya Sredney Sibiri*. Krasnoyarsk. Khizhn. Izdatel'stvo, Krasnoyarsk, pp. 5—28.

Obruchev, V.A., 1935—1938. *Geology of Siberia*. Acad. Sci. USSR, Moscow—Leningrad, 1358 pp.

Ödman, O.H., 1972. Översikt av konglomeratförekonster i Norrbottens urberg och den stratigrafiska betydelsen av dessa. *Sver. Geol. Unders.*, 66(8): 12 pp.

Osborne, F.P. (Editor), 1964. Geochronology in Canada. *R. Soc. Can. Spec. Publ.*, 8: 156 pp.

Ovchinnikov, L.N., Dunayev, V.A., Krasnobayev, A.A. and Stepanov, A.J., 1968. Age zonation in the Urals by radiologic data. *Dokl. Acad. Sci. USSR*, 180(7): 1230—1233.

Pankhurst, R.J., Moorbath, S., Rex, D.C. and Turner, G., 1973. Mineral age pattern in ca. 3700 m.y. old rocks from West Greenland. *Earth Planet. Sci. Lett.*, 20(2): 155—170.

Pap, A.M., 1967. Certain peculiarities of Precambrian studies of crystalline basement in Byelorussia. *Probl. Osad. Geol. Dokembr.*, 2: 248—256.

Pap, A.M., 1971. Precambrian stratigraphy of Byelorussia. *Dokl. Acad. Sci. USSR*, 201(4): 923—926.

Parfenov, Yu.I., 1963. Tectonics of the southern part of the Yenisei Ridge. *Tektonika Sib.*, 2: 94—96.

Pasteels, P. and Silver, L.T., 1966. Geochronologic investigation in the crystalline rocks of the Grand Canyon, Arizona, *Geol. Soc. Am. Spec. Pap.*, 87: 56 (abstr.).

Pavlovsky, E.V., 1958. An outline of the Precambrian and Lower Paleozoic of Highlands. *Izv. Acad. Sci. USSR, Geol. Ser.*, 1958(6): 23—47.

Pavlovsky, E.V., 1962. A peculiar pattern of tectonic evolution in the Earth's crust during the Early Precambrian. *Tr. Vost. Sib. Geol. Inst., Geol. Ser.*, 5: 77—108.

Peacock, J.D., 1956. The geology of Dronning Louise Land, NE Greenland. *Medd. Grönl.*, 137(7): 37 pp.

Pedersen, K.P., 1970(1969). Late Precambrian microfossils from Peary Land. *Grönl. Geol. Unders. Rep. Act.*, 2: 16—17.

Pedersen, K.R. and Lam, J., 1968. Precambrian organic compounds from the Ketilidian of South West Greenland. *Grönl. Geol. Unters. Bull.*, 74: 16 pp.

Perry, E., Monster, J. and Reimer, T., 1971. Sulfur isotopes in Swaziland System barites and the evolution of the Earth's atmosphere. *Science*, 171(3975): 1015—1016.

Perry W.J. and Roberts, H.G., 1968. Late Precambrian glaciated pavements in the Kimberley region, Western Australia. *J. Geol. Soc. Aust.*, 15(1): 51—57.

Petrov, V.G., 1972. On the structural relationships of the Riphean troughs in central areas of East European platform. *Byull. MOIP, Otd. Geol.*, 2: 5—16.

Pettijohn, F.J., 1943. Archean sedimentation. *Geol. Soc. Am. Bull.*, 54: 925—972.

Pichamuthu, C.S., 1967. The Precambrian of India. In: K. Rankama (Editor), *The Precambrian, 2*. Interscience, New York, N.Y., pp. 1—96.

Pichamuthu, C.S., 1971. Precambrian geochronology of Peninsular India. *J. Geol. Soc. India*, 12(3): 262—273.

Pidgeon, R.T. and Compston, W., 1965. The age and origin of the Cooma granite and associated metamorphic zones, New South Wales. *J. Petrol.*, 6(2): 193—222.

Polishchuk, V.D. (Editor), 1970. *Geology, Hydrogeology, and Iron-Ores in Kursk Magnetic Anomaly Basin, 1. Geology, 1. Precambrian.* Nedra, Moscow, 439 pp.

Polkanov, A.A. and Gerling, E.K., 1961. Geochronology and geologic evolution of the Baltic shield and surrounding fold belts: problems of geochronology and geology. *Tr. LAGED, Acad. Sci. USSR*, 12: 7—102.

Polovinkina, Yu.I., 1960. Stratigraphic subdivision of the older gneissic strata in the Ukraine. *Dokl. Acad. Sci. USSR*, 134(4): 909—912.

Polunovsky, R.M., 1969. Characters of gneissic rocks in Central Pri-Azov area and the problems of bedding of this group. *Dokl. Acad. Sci. USSR*, 187(6): 1360—1363.

Postnikov, V.G. and Postnikova, I.E., 1968. Stratigraphy and correlation of Upper Riphean and Vendian strata in the south of the Siberian platform and surrounding fold belts. *Izv. Acad. Sci. USSR, Geol. Ser.*, 1968(7): 85—93.

Potapenko, Yu.Ya., 1969. On the Pre-Devonian formations and evolutionary stages in Northern Pri-Elbrus area. *Dokl. Acad. Sci. USSR*, 187(1): 153—155.

Potapenko, Yu.Ya. and Momot, S.P., 1966. Lithology and age of the Urlesh formation in Northern Pri-Elbrus area. *Sov. Geol.*, 1966(4): 133—137.

Poulsen, C., 1956. The Cambrian of the East Greenland geosyncline. *Proc. Int. Geol. Congr., 20 Sess.*, 1: 59—69.

Poulsen, C., 1967. Fossils from the Lower Cambrian of Bornholm. *Mat. Fys. Medd. Dan. Vid. Selsk.*, 36(2): 48 pp.

Poulsen, V., 1964. The sandstones of the Precambrian Eriksfjord formation in South Greenland. *Geol. Surv. Greenl. Rep.*, 2: 16 pp.

Prasada Rao, G.H., Murty, Y.G. and Dekkshitulu, M., 1964. Stratigraphic relations and associated sedimentary sequences in part of Keionjhar, Cuttack, and Sundargarh districts, Orissa. *Proc. Int. Geol. Congr., 22 Sess., New Delhi 1964, Pt. 10*: 72—88.

Prest, V.K., 1952. Geology of the Carr Township area. *Ont. Dep. Mines Annu. Rep.*, 60(4): 15—32.

Price, R.A., 1964. The Precambrian Purcell system in the Rocky Mountains of southern Alberta and British Columbia. *Can. Pet. Geol. Bull.*, 12: 399—426.

Pringle, J.R., 1973. Rb—Sr age determinations on shales associated with the Varanger Ice Age. *Geol. Mag.*, 109(6): 465—472.

Pronin, A.A., 1965. *Major Aspects of Tectonic Evolution in the Urals. The Variscan Cycle*. Nauka, Moscow, 160 pp.

Pulvertaft, T.C., 1973. Recumbent folding and flat-lying structure in the Precambrian of northern West Greenland. *Philos. Trans. R. Soc. London, Ser. A.*, 273(1235): 535—545.

Pyke, D.R., Waldrett, A.J. and Eckstrand, O.R., 1973. Archean ultramafic flows in Munro township, Ontario. *Bull. Geol. Surv. Am.*, 84(3): 955—978.

Raaben, M.E., 1966. Sedimentation rates in the Riphean. *Izv. Acad. Sci. USSR, Geol. Ser.*, 1966(9): 117—129.

Rabkin, M.I., 1960. Precambrian in the Anabar shield area. In: *Dokl. Sov. Geol. Stratigr. i. Korrelyats. Dokembriya (Int. Geol. Congr. 21 Sess.)*. Acad. Sci. USSR, Moscow, pp. 69—76.

Rabkin, M.I. and Lopatin, B.G., 1966. Metamorphic and magmatic formations of the Anabara shield. In: *Magmat. i Metamorf. Obrazovan. Sibiri*. Nedra, Moscow, pp. 156—168.

Rabotnov, V.T., 1964. The Late Precambrian stratigraphy of the watershed of the Olekma and Tokko Rivers. *Dokl. Acad. Sci. USSR*, 156(6): 1351—1354.

Rabotnov, V.T., Komar, Vl.A., Narozhnykh, L.I. and Gorbachev, V.F., 1970. Upper Precambrian stratigraphy in the Middle Kolyma River. *Dokl. Acad. Sci. USSR*, 194(5): 1157—1160.

Radhakrishna, B.R., 1967. Reconsideration of some problems in the Archean complex of Mysore. *J. Geol. Soc. India*, 8: 102—109.

Raja, R.C. and Iqbaludhin, M.R., 1968. Algae structures from Aravalli beds near Dakan Kotru, Udaipur district, Rajasthan. *Curr. Sci.*, 37(19): 560—561.

Rama Rao, B., 1940. The Archean complex of Mysore. *Mysore Dep. Mines Geol. Bull.*, 17: 80 pp.

Rama Rao, B., 1962. *Handbook of the Geology of Mysore State, Southern India*. Bangalore, 264 pp.

Ramdohr, P., 1961. The Witwatersrand controversy: a final comment on the review by Professor C.F. Davidson. *Min. Mag.*, 105(1): 18—21.

Rankama, K., 1970. Global Precambrian stratigraphy: background and principles. *Scientia*, 105(699-700): 382—421.

Rankin, D.W., 1969. Late Precambrian glaciation in the Blue Ridge Province of the Southern Appalachian Mountains. *Geol. Soc. Am. Spec. Pap.*, 121: 246 (abstr.).

Rankin, D.W., 1970. Stratigraphy and structure of Precambrian rocks in North-Western North Carolina. In: *Studies of Appalachian Geology, Central and Southern*. Interscience, New York, N.Y., pp. 227—246.

Rankin, D.W., Stern, T.W., Reed, I.C. and Newell, M.F., 1969. Zircon age of felsic volcanic rocks in the Upper Precambrian of the Blue Ridge, Appalachian Mountains. *Science*, 166(3906): 741—743.

Ravich, M.G., 1963. Stratigraphy of Taymyr and Northern Land. In: *Stratigrafiya SSSR. Nizhny Dokembry. Aziatskaya Chast' USSR*. Gosgeoltekhizdat, Moscow, pp. 152—159.

Reesor, J.E., 1957. The Proterozoic of the Cordillera in Southern British Columbia and Southwestern Alberta. *R. Soc. Can. Spec. Publ.*, 2: 150—177.

Regional Stratigraphy of China, 1963(1960). Inostrannaya Literatura, Moscow, 659 pp. (translated from the Chinese).

Reitlinger, E.A., 1959. Atlas of microscopic organic remains in the older strata of Siberia. *Tr. GIN Acad. Sci. USSR.*, 25: 62 pp.

Resolution of a General Plenary Session of the three Constant Commissions of the Interdepartmental Stratigraphic Committee on the Lower Precambrian, Upper Precambrian, and on Absolute Age Determinations, 1965. Moscow, 7: 15 pp.

Roach, R., Adams, C. and Brown, M., 1972. The Precambrian stratigraphy of the Armorican Massif, NW France. *Proc. Int. Geol. Congr., 24 Sess., Montreal 1972, Sect. 1, Precambrian geology*, pp. 246—252.

Roberts, J.D., 1971. Late Precambrian glaciation: an anti-greenhouse effect. *Nature*, 234(5326): 216—217.

Roscoe, S.M., 1957. Stratigraphy of Quirke Lake—Elliot Lake, Blind River area, Ontario. *Trans. R. Soc. Can. Spec. Publ.*, 2: 54—58.

Roscoe, S.M., 1969. Huronian rocks and uraniferous conglomerates in the Canadian Shield. *Geol. Surv. Can. Pap.*, 68-40: 205 pp.

Ross, C.P., 1963. The Belt series in Montana. *US Geol. Surv. Prof. Pap.*, 346: 122 pp.

Ross, C.P., 1970. Precambrian of the USA: Northwestern United States, the Belt series. In: K. Rankama (Editor), *The Precambrian, 4*. Interscience, New York, N.Y., pp. 145—252.

Rotar', A.F., 1974. Mashak formation (Riphean) in the Southern Urals. *Sov. Geol.*, 1974 (4): 116—122.

Rozanov, Yu.A., Missarzhevsky, V.V. and Volkova, N.A., 1969. The Tommotian stage and the problem of the Cambrian lower boundary. *Tr. GIN Acad. Sci. USSR*, 206: 380 pp.

Rubinshtein, M.M., 1967. *Ar-method Applied to Certain Problems of Regional Geology*. Metsnnereba, Tbilisi, 239 pp.

Rudnik, V.A. and Sobotovich, E.V., 1968. On the age of the Archean Timpton and Dzheltula groups of the Aldan shield. *Dokl. Acad. Sci. USSR*, 189(3): 607—616.

Rukhin, L.B., 1959. *General Principles of Paleogeography*. Gostoptekhizdat, Moscow, 557 pp.

Ryan, B.D. and Blenkinsop, J., 1971. Geology and geochronology of the Hellroaring Creek Stock, British Columbia. *Can. J. Earth Sci.*, 8(1): 85—95.

Ryka, W., 1968. The charnockites of the crystalline basement in the Polish Lowland. *Inst. Geol. Warszawa Bull.*, 237: 57—63.

Ryka, W., 1970. Development of the crystalline basement of North-Eastern Poland. *Mater. Prace*, 34: 23—35.

Sakko, M., 1971. Varhais—Karjalaisten metadiabaasien radiometrisia zirconikiä. *Geology*, 23(9/10): 117—119.

Salmo, British Columbia. Map 1145 A, 1: 63, 360, 1965. Geol. Surv. Canada, Dep. Mines, Techn. Surv.

Salop, L.J., 1958a. Precambrian of the USSR. On the correlation of the Precambrian strata. In: *Geol. Stroyeniye SSSR, 1. Stratigrafiya*. Gosgeoltekhizdat, Moscow, pp. 133—138.

Salop, L.J., 1958b. Major aspects of tectonic evolution of the USSR territory. Archean and Proterozoic eras. In: *Geol. Stroyeniye SSSR, 3. Tektonika*. Gosgeoltekhizdat, Moscow, pp. 237—261.

Salop, L.J., 1960. Major aspects of geologic evolution of the USSR during the Precambrian. In: *Dokl. Sov. Geologov. Stratigrafiya i Korrelyatsiya Dokembriya*. Acad. Sci. USSR, Moscow—Leningrad, pp. 186—187.

Salop, L.J., 1963. Geologic interpretation of Ar absolute age data for rocks. *Geol. Geofiz.*, 1963(1): 3—21.

Salop, L.J., 1964. Pre-Cambrian geochronology and some features of the early stage of the geological history of the Earth. *Proc. Int. Geol. Congr., 22 Sess., New Delhi 1964, Pt. 10, Archean and Precambrian Geology*, pp. 131—149.

Salop, L.J., 1964—1967. *Geology of Baikal Mountain Area*. Nedra, Moscow, Vol. 1 (1964) 515 pp., Vol. 2 (1967) 699 pp.

Salop, L.J., 1966. Contribution to the stratigraphy of the Lower Precambrian in Southern India. In: *Problemy Geologii na 22 Sess. IGC*. Nauka, Moscow, pp. 59—70.

Salop, L.J., 1968a. Archean and Proterozoic of the USSR. In: *Geol. Stroyeniye SSSR, 1. Stratigrafiya*. Nedra, Moscow, pp. 76—229.

Salop, L.J., 1968b. Pre-Cambrian of the USSR. *Proc. Int. Geol. Congr., 23 Sess., Prague 1968, Geology of the Pre-Cambrian*, pp. 61—73.

Salop, L.J., 1969. Certain criteria for creation of a unified stratigraphical scale of Precambrian. In: *Geol. Stroyeniye SSSR, 5. Osnovnye Problemy Geologii*. Nedra, Moscow, pp. 63—81.

Salop, L.J., 1970a. General criteria for stratigraphic and geochronologic subdivision of the Precambrian. In: *Geokhronologiya Dokembriya*. Nauka, Moscow, pp. 112—130.

Salop, L.J., 1970b. Revision of the geochronological scale of the Precambrian. *Byull. MOIP*, 45(4): 115—135; 45(5): 5—26.

Salop, L.J., 1971a. Basic features of the stratigraphy and tectonics of the Precambrian in the Baltic shield. In: *Problemy Geologii Dokembriya. Baltiyskogo Shchita i Pokrova Russkoy Platformy. Tr. VSEGEI*, 175: 6—87.

Salop, L.J., 1971b. Two types of Precambrian structures: gneiss folded ovals and domes. *Byull. MOIP*, 46(4): 5—30 (English transl. in: *Int. Geol. Rev., Am. Geol. Inst.*, 1972, 1: 1209—1228).

Salop, L.J., 1972a. A unified stratigraphic scale of the Precambrian. *Proc. Int. Geol. Congr., 24 Sess., Montreal 1972. Precambrian Geology*, pp. 253—259.

Salop, L.J., 1972b. The problem of gold-uraniferous conglomerates: its geological aspects. *Tr. VSEGEI*, 178: 150—174.

Salop, L.J., 1974a. On certain debatable problems of geology of the Baikal folded area. *Geol. Geofiz.*, 1974(1): 11—24.

Salop, L.J., 1974b. On the stratigraphy and tectonics of the Mama—Chuya mica-bearing area in the Precambrian. *Tr. VSEGEI, N.Ser.*, 199: 83—143.

Salop, L.J. and Murina, G.A., 1970. An age of the Berdyaush rapakivi pluton and the problem of geochronological boundaries of the Lower Riphean. *Sov. Geol.*, 1970(6): 15—27.

Salop, L.J. and Scheinmann, Yu.M., 1969. Tectonic history and structures of platforms and shields. *Tectonophysics*, 7(5/6): 565—597.

Salop, L.J. and Travin, L.V., 1971. The Archean stratigraphy of the central part of the Aldan shield. *Sov. Geol.*, 1971(3): 5—19.

Salop, L.J. and Travin, L.V., 1974. New data on stratigraphy and tectonics of the central part of the Aldan shield. *Tr. VSEGEI, N. Ser.*, 199: 5—82.

Salop, L.J., Travin, L.V. and Shalek, E.A., 1974. Stratigraphy and tectonics of the southern part of Baikal Ridge in the Precambrian. *Tr. VSEGEI, N.Ser.*, 144:172.

Sarkar, S.N., 1972. Present status of Precambrian geochronology of Peninsular India. *Proc. Int. Geol. Congr., 24 Sess., Montreal 1972, Sect. 1, Precambrian Geology*, pp. 260—272.

Sarkar, S.N. Saha, A.K. and Miller, J.A., 1967. Potassium—argon ages from oldest metamorphic belt in India. *Nature*, 215(5104): 946—948.

Savitsky, V.E., 1962. The relationships of Cambrian and Upper Precambrian in the Anabar shield. In: *Soveshch. po Stratigr. Otlozh. Pozdnego Dokembr. Sibiri i Dal'nego Vostoka*. Tezisy Dokl., Novosibirsk, pp. 53—54.

Sayyah, T.A., 1965. *Geochronological Studies of the Kinsley Stock, Nevada, and the Raft River Range, Utah*. Thesis Utah Univ., Salt Lake City, 112 pp. (unpublished).

Schnitzer, W.A., 1969a. Die jung-algonkischen Sedimentationsträume Peninsula-Indiens. *Neues Jahrb. Geol. Paläontol. Abh.*, 133(2): 191—198.

Schnitzer, W.A., 1969b. Zur Stratigraphie und Lithologie des nördlichen Chhattisgarh-Beckens (zentral Indien) unter besonderer Berücksichtigung von Algenriff-Komplexen. *Z. Dtsch. Geol. Ges.*, 118: 290—295.

Schopf, J.W. and Barghoorn, E.S., 1967. Alga-like fossils from the Early Precambrian of South Africa. *Science*, 156(3774): 508—511.

Schopf, J.W., Ford, T.D. and Breed, W.J., 1973. Microorganisms from the Late Precambrian of the Grand Canyon, Arizona. *Science*, 179(4080): 1319—1321.
Sederholm, J., 1899. Über eine archäische Sediment-Formation im südwestlichen Finnland und ihre Bedeutung für die Erklärung der Entstehungsweise des Grundgebirges. *Bull. Comm. Géol. Finl.*, 6: 254 pp.
Sederholm, J., 1930. *Pre-Quaternary Rocks of Finland*. Explanatory notes to accompany a general geological map of Finland, No. 91 : 47 pp.
Sedgwick, A., 1838. A synopsis of the English series of stratified rocks interior to the old red sandstone; with an attempt to determine the successive natural groups and formations. *Proc. Geol. Soc. London*, 21(58): 684.
Semenenko, N.P., 1965. The Precambrian stratigraphical scale for the Ukrainian shield. In: *Geokhronologiya Dokembriya Ukrainy*. Naukova Dumka, Kiev, pp. 174—181.
Semenenko, N.P., 1970. Intercontinental correlation of the Precambrian. In: *Geokhronologiya Dokembriya*. Nauka, Moscow, pp. 5—22.
Semenenko, N.P., Rodionov, S.P., Usenko, I.S., Lichak, I.L. and Tsarovsky, J.D., 1960. The Precambrian stratigraphy of the Ukrainian crystalline shield. In: *Dokl. Sov. Geol., Stratigrafiya i Korrelyatsiya Dokembriya*. Acad. Sci. USSR, Moscow, pp. 36—45.
Semikhatov, M.A., 1962. The Riphean and Lower Cambrian of Yenisei Ridge. *Tr. GIN Acad. Sci. USSR*, 68: 210 pp.
Semikhatov, M.A., 1964. On Proterozoic problems. *Izv. Acad. Sci. USSR, Geol. Ser.*, 1964(2): 66—84.
Semikhatov, M.A., 1974. *Stratigraphy and Geochronology of the Proterozoic*. Nauka, Moscow, 300 pp.
Semikhatov, M.A., Komar, V.A. and Serebryakov, S.N., 1970. *The Judomian Complex of the Stratotype Locality*. Nauka, Moscow, 208 pp.
Sen, S., 1970. Some problems of Precambrian geology of the Central and Southern Aravalli Range, Rajasthan. *J. Geol. Soc. India*, 11(3): 217—231.
Sethuraman, K. and Moore Jr., J.M., 1973. Petrology of metavolcanic rocks in the Bishop Corners—Donaldson area, Grenville Province, Ontario. *Can. J. Earth Sci.*, 10(5): 589—614.
Shafeyev, A.A., 1970. *The Precambrian of Southwestern Pri-Baikal Area and of Khamar-Daban*. Nauka, Moscow, 179 pp.
Shatalov, E.T. (Editor), 1968. *Geological Structure of the USSR, 1*. Nedra, Moscow, 711 pp.
Shatsky, N.S., 1945. Outlines of tectonics in the Volga—Urals oil-bearing area and in the adjacent western slope of the Southern Urals. *Mater. Poznaniyu Geol. Stroyeniya USSR, N.Ser.*, 2(6): 129 pp.
Shcherbak, N.P., Polovko, N.I. and Levkovskaya, N.Yu., 1969. Isotopic composition of accessory minerals from the lower formation of Krivoy Rog group. *Geol. Zh.*, 29(3): 23—32.
Shemyakin, V.M., 1972. The Early Precambrian strata of intrusive charnockite in the Baltic shield. *Izv. Acad. Sci. USSR, Geol. Ser.*, 1972(9): 40—45.
Shotsky, I.I., 1967. New data on stratigraphy and lithology of the Teterev—Bug group. In: *Problemy Glubinnoy Geologii Dokembriya*. Nedra, Moscow, pp. 118—125.
Shride, A.F., 1967. Younger Precambrian geology in Southern Arizona. *US Geol. Surv. Prof. Pap.*, 556: 89 pp.
Shul'ga, P.L., 1952. Paleozoic stratigraphical chart for the south-western margin of the Russian platform. *Geol. Zh.*, 12(4): 22—40.
Shurkin, K.A., Gorlov, N.V., Salye, M.E., Duk, A.A. and Nikitin, Yu.V., 1962. The White Sea complex in Northern Karelia and in SW Kola Peninsula. *Tr. Lab. Dokembr. Geol. Acad. Sci. USSR*, 14: 306 pp.
Siedlecka, A., 1973. The Late Precambrian Ost-Finnmark supergroup — a new lithostratigraphic unit of high rank. *Nor. Geol. Unders.*, 289: 55—60.

Siedlecka, A. and Roberts, D., 1972. A late Precambrian tilloid from Varangerhalvoya — evidence of both glaciation and subaqueous mass movement. *Nor. Geol. Tidsskr.*, 52(2): 135—141.

Siedlecka, A. and Siedlecki, S., 1967. Some new aspects of the geology of Varanger Peninsula (Northern Norway). *Nor. Geol. Unders.*, 247: 288—306.

Sidorenko, A.V. and Sidorenko, Sv.A., 1971. Organic matter in Precambrian sedimentary—metamorphic rocks and certain geological problems. *Sov. Geol.*, 1971(5): 3—20.

Sidorenko, Sv.A., 1971. *Organic Matter in Sedimentary—Metamorphic Pre-Cambrian Rocks.* Thesis, GIN Acad. Sci. USSR, Moscow, 22 pp.

Silver, L.T., 1969. Precambrian batholiths of Arizona. *Geol. Soc. Am. Spec. Pap.*, 121: 558—559 (abstr.).

Silver, L.T., MacKinney, C.R. and Wright, M.C., 1962. Some Precambrian ages in the Panamit Range, Death Valley, California. *Geol. Soc. Am. Spec. Pap.*, 68(55): 102—104.

Simonen, A., 1960. Pre-Quaternary rocks in Finland. *Bull. Comm. Géol. Finl.*, 191: 49 pp.

Sinitsyn, V.M., 1967. *Introduction to Paleoclimatology.* Nedra, Leningrad, 232 pp.

Sklyarov, R.Ya., 1968. *Proterozoic and Paleozoic of the Irkineyev—Chadobets Zone.* Thesis, Kazan. Univ., Kazan', 24 pp.

Skvor, U., 1968. The main principles of the pre-platform geological development of the Bohemian Massif. *Kristallinicum*, 6: 79—92.

Slawson, W.E., Kanasewich, E.R. and Ostic, R.G., 1963. Age of the North American crust. *Nature*, 4905: 413.

Smirnov, Yu.D., 1964. The evolution of the Urals fold belt during the Precambrian. In: *Dokl. Sov. Geol. Probl., 10. Geologiya Dokembriya.* Nedra, Moscow, pp. 195—207.

Smith, A.G. and Barnes, W.C., 1966. Correlation and facies changes in the carbonaceous, calcareous, and dolomite formation of the Precambrian Belt—Purcell supergroup. *Geol. Soc. Am. Bull.*, 77(12): 1399—1426.

Smyslov, A.A., 1969. Heat regime of the Earth's crust and of the subcrustal masses. In: *Geologicheskoye Stroyeniey SSSR*, 5. Nedra, Moscow, pp. 261—278.

Sobolevskaya, R.F. and Mil'shtein, V.E., 1961. Stratigraphy of the Sinian strata in Central Taymyr. *Tr. NIIGA*, 125: 20—30.

Sobotovich, E.V., Grashchenko, S.M. and Lovtsyus, A.V., 1963. The rock age in the Tarom quarry of Pri-Dniester area by the lead isochron method. In: *Trudy XI Sess. Kom. po Opredeleniyu Abs. Vozr. Geol. Formatsy*, Moscow, pp. 38—41.

Sobotovich, E.V., Grashchenko, S.M. and Lovtsyus, A.V., 1965. The age of the Sharyzhalgay group rocks (the Baikal block). *Izv. Acad. Sci. USSR, Geol. Ser.*, 1965 (9): 38—41.

Sobotovich, E.V., Iskanderova, A.D. and Kamenev, E.N., 1973. New data on the Azoic age of the oldest strata of the Earth (Enderby Land and Okhotsk Massif). In: *Opred. Abs. Vozrasta Rudnykh Mestor. i Molodykh Magmat. Protsessov.* (summaries). Acad. Sci. USSR, Moscow, pp. 87—89.

Sokolov, B.S., 1952a. On the age of the oldest sedimentary cover of the Russian platform. *Izv. Acad. Sci. USSR*, Geol. Ser., 1952(5): 12—20.

Sokolov, B.S., 1952b. Stratigraphical chart of the Lower Paleozoic (Pre-Devonian) strata in the northwestern part of the Russian platform. In: *Devon Russkoy Platformy.* Gostoptekhizdat, Leningrad, pp. 16—38.

Sokolov, B.S., 1971. Vendian in the north of Eurasia. *Geol. Geofiz.*, 1971(6): 3—23.

Sokolov, B.S., 1972. A working meeting on Vendian stratigraphy of the Siberian platform. *Geol. Geofiz.*, 1972(3): 141—147.

Solontsov, L.F., 1960. Results of Riphean studies in the eastern part of the Russian platform and present concepts on its stratigraphical subdivision within the Tartarian area. *Izv. Kazan. Fil. Acad. Sci. USSR, Ser. Geol. Nauk*, 9: 209—224.

Solontsov, L.F. and Aksenov, E.M., 1969. The Riphean of the East European platform. *Izv. Vuzov Geol. Razved.*, 1969(10): 3—14.

Solontsov, L.F. and Aksenov, E.M., 1970. Stratigraphy of the Valday group of the East-European platform. *Izv. Vuzov, Geol. Razved.*, 1970(6): 3—13.

Solontsov, L.F., Klevtsova, A.A. and Aksenov, E.M., 1966. New data on stratigraphy of Riphean strata in the eastern part of the Russian platform. *Sov. Geol.*, 1966(1): 70—77.

Springer, G.D., 1949. Geology of the Cat Lake, Winnipeg River area. *Manit. Mines Bur. Prelim. Rep.*, 4-7: 15 pp.

Srinivasan, R. and Sreenivan, B.L., 1972. Dharwar stratigraphy. *J. Geol. Soc. India*, 13(1): 75—85.

Stepanov, O.A. and Shkol'nik, E.L., 1972. On the stratigraphical value of microphytilites in the older strata of the Shantar Islands. *Izv. Acad. Sci. USSR, Geol. Ser.*, 1972(5): 146—149.

Stern, T.W., Newell, M.F. and Hunt, C.B., 1966. Uranium—lead and potassium—argon ages of part of the Amargosa thrust complex, Death Valley, California. *US Geol. Surv. Prof. Pap.*, 550-13: 24—65.

Stevens, R.D., 1965. K—Ar age of Cambrian glauconites from Alberta. *Geol. Surv. Can. Pap., Rep. Act.*, 65-1: 32—34.

Stevenson, J. (Editor), 1962. The tectonics of the Canadian Shield. *R. Soc. Can. Spec. Publ.*, 8: 180 pp.

Stewart, J.H., 1970. Upper Precambrian and Lower Cambrian strata in the Southern Great Basin, California and Nevada. *US Geol. Surv. Prof. Pap.*, 620: 206 pp.

Stille, H., 1948. Die assyntische Ära und vor-, mit- und nachassyntische Magmatismus. *Z. Dtsch. Geol. Ges.*, 98: 152—165.

Stille, H., 1949. Das Leitmotiv der geotektonischen Erdentwicklung. *Dtsch. Akad. Wiss., Vortr. Schr.*, 32: 27 pp.

Stockwell, C.H., 1938. Rice Lake—Gold Lake area, Southeastern Manitoba. *Geol. Surv. Can. Mem.*, 210: 79 pp.

Stockwell, C.H., 1961. Structural provinces, orogenesis, and time-classification of rocks of the Canadian Precambrian Shield. *Geol. Surv. Can. Pap.*, 61-17: 29—77.

Stockwell, C.H., 1964. Fourth report on structural provinces, orogenesis, and classification of rocks of the Canadian Precambrian Shield. *Geol. Surv. Can. Pap.*, 64-17: 1—21.

Stockwell, C.H., 1968. Geochronology of stratified rocks of the Canadian Shield. *Can. J. Earth Sci.*, 5(3): 693—698.

Stockwell, C.H., 1973. Revised Precambrian time scale for the Canadian Shield. *Geol. Surv. Can. Pap.*, 72-52: 4 pp.

Stockwell, C.H., McGlynn, J.C., Emslie, R.F., Sanford, B.V., Norris, A.W., Donaldson, J.A., Fahrig, W.F. and Currie, K.L., 1970. Geology of the Canadian Shield. In: R.J. Douglas (Editor), *Geology and Economic Minerals of Canada*. Geol. Surv. Can., Ottawa, 838 pp.

Strakhov, N.M., 1963. *Lithogenesis, its Types and Evolution During the Earth's History*. Gosgeoltekhizdat. Moscow, 535 pp.

Sudovikov, N.G., Glebovitsky, V.A., Drugova, G.M. and Krylova, M.D., 1965. *Geology and Petrology of the Southern Margin of the Aldan Shield*. Nauka, Leningrad, 290 pp.

Sutton, J. and Watson, J., 1951. The Pre-Torridonian metamorphic history of the Torridon and Scourie areas in the North-West Highlands. *Q. J. Geol. Soc. London*, 106 (241): 241—307.

Sveshnikov, K.I., 1969. Description of a normal section in the lowermost Udokan group. *Zapiski Zabaikal. Fil. Geogr. Obshchestva. SSSR*, 35: 22—28.

Svetov, A.P., 1972. Jatulian paleovolcanology in central Karelia. *Tr. Inst. Geol. Karelsk. Fil. Acad. Sci. USSR*, 11: 120 pp.

Taskin, A.P., 1971. *The Upper Precambrian of Eastern Pri-Sayan Area*. Thesis, Irkyt. Univ., Irkutsk, 22 pp.
Tectonic Map of Canada, 1968. Geol. Surv. Can., Scale 1 : 5,000,000.
Thompson, G.R. and Hower, J., 1973. An explanation for low radiometric age for glauconite. *Geochim. Cosmochim. Acta*, 37(6): 1473—1491.
Thorpe, R.L., 1971. Comments on rock ages in the Yellowknife area, district of Mackenzie. *Geol. Surv. Can. Pap.*, 1: 76—79.
Thorsteinsson, R. and Tozer, E.T., 1962. Banks, Victoria and Stefansson Islands. Arctic Archipelago. *Geol. Surv. Can. Mem.*, 330: 85 pp.
Tikhomirov, C.N. and Yanovsky, A.S., 1970. New data on the Precambrian of the South-East Pri-Ladoga area. *Dokl. Acad. Sci. USSR*, 194(3): 660—663.
Tikhonov, V.L. and Anosov, V.S., 1966. On the Bodaybo and Kadalikan complexes of Baikal—Patom Highland. In: *Geologiya i Polezn. Iskop. Baikalo—Patomskogo Nagor'ya Vost*. Sib. Knizhn. Izdatel'stvo, Irkutsk, pp. 14—23.
Tilton, G.R., 1968. Sphene: uranium—lead ages. *Science*, 159(3822): 1458—1461.
Timergazin, K.R., 1959. *The Pre-Devonian Formations of Western Bashkiria and Their Possible Oil and Gas Potential*. Acad. Sci. USSR, Bashkirsky filial, Ufa, 312 pp.
Tkachenko, B.V. (Editor), 1970. *Reference Section for Upper Precambrian Strata of the Western Slopes of the Anabar Uplift. Collected reports.* NIIGA, Leningrad, 146 pp.
Tremblay, L.P., 1971. Geology of Beechey Lake map-area, district of Mackenzie (76G). *Geol. Surv. Can. Mem.*, 365: 56 pp.
Tugarinov, A.I. and Voytkevich, G.V., 1970. *Geochronology of the Precambrian of the Continents*. Nedra, Moscow, 2nd ed. suppl., 431 pp.
Tugarinov, A.I., Shanin, L.L., Kazakov, G.A. and Arakel'yants, M.M., 1965a. On glauconite age of the Vindhyan system rocks (India). *Geokhimiya*, 165(6): 652—660.
Tugarinov, A.I., Stupnikova, N.I. and Zykov, S.I., 1965b. Geochronology in the southern part of the Siberian platform. *Izv. Acad. Sci. USSR, Geol. Ser.*, 1965(1): 21—36.
Tugarinov, A.I., Bibikova, E.V. and Gorashchenko, G.L., 1968. Granulite ages in the Baltic shield. *Geokhimiya*, 1968(9): 1052—1060.
Tugarinov, A.I., Bibikova, E.V. and Krasnobayev, A.A., 1970. Precambrian geochronology in the Urals. *Geokhimiya*, 1970(4): 501—509.
Usenko, I.S., Esipchuk, K.E. and Tsukanov, V.A., 1971. Stratigraphy of the Precambrian gneiss—migmatite complex in the Pri-Azov area. *Geol. Zh.*, 31(2): 134—146.
Valdiya, K.S., 1969. Stromatolites of the Lesser Himalayan carbonate formations and the Vindhyans. *J. Geol. Soc. India*, 10: 1—25.
Van Breemen, O. and Dodson, M.H., 1972. Metamorphic chronology of the Limpopo Belt, Southern Africa. *Geol. Soc. Am. Bull.*, 83(7): 2005—2018.
Van Breemen, O. and Upton, B.G.J., 1972. Age of some Gardar intrusive complexes, South Greenland. *Geol. Soc. Am. Bull.*, 83(11): 3381—3390.
Van Breemen, O., Dodson, M.H. and Vail, J.R., 1966. Isotopic age measurements on the Limpopo orogenic belt, Southern Africa. *Earth Planet. Sci. Lett.*, 1(6): 401—406.
Van Breemen, O., Allaart, J.A. and Aftalion, M., 1971—1972. Rb—Sr whole rock and U—Pb zircon age studies on granites of the Early Proterozoic mobile belt of South Greenland. *Grønl. Geol. Unders. Rep. Act.*, 45: 45—48.
Van Gundy, C.E., 1951. Nankoweap Group of the Grand Canyon, Algonkian of Arizona. *Geol. Soc. Am. Bull.*, 62(8): 953—960.
Van Hise, C.R., 1908. The problem of the Pre-Cambrian. *Geol. Soc. Am. Bull.*, 119.
Van Hise, C.R. and Leith, C.K., 1911. The geology of the Lake Superior region. *US Geol. Surv. Mem.*, 52.
Van Niekerk, C.B. and Burger, A.J., 1969. A note on the minimum age of the acid lava of the Onverwacht Series of the Swaziland System. *Geol. Soc. S.Afr. Trans.*, 72(1): 9—21.

Van Schmus, K., 1965. The geochronology of the Blind River—Bruce Mines area, Ontario, Canada. *J. Geol.*, 73(5): 755—780.

Väyrynen, H., 1954. *Suomen Kalliopera*. Helsinki, 300 pp.

Verhoogen, J., Turner, F.J., Weiss, L.E., Wahrhoftig, C. and Fyfe, W.S., 1970. *The Earth. An Introduction to Physical Geology*. Holt—Rinehart—Winston, New York, N.Y., 748 pp.

Verma, K.K. and Prasadk, N., 1968. On the occurrence of some trace fossils in Bhander Limestone (Upper Vindhyan) of Rewa district. Madhya-Pradesh. *Curr. Sci.*, 37(19): 72—86.

Viljoen, M.J. and Viljoen, R.P., 1969. The geology and geochemistry of the lower ultramafic units of the Onverwacht Group and a proposed new class of igneous rocks. *Geol. Soc. S.Afr. Spec. Publ.*, 2: 55—85.

Vinogradov, A.P., 1957. Age of the Precambrian rocks in the Ukraine. *Geokhimiya.*, 1957(7): 10—17.

Vinogradov, A.P., 1964. Gas regime of the Earth. In: *Khimiya Zemnoy Kory*, 2. Nauka, Moscow, pp. 5—21.

Vinogradov, A.P., 1967. *Geochemistry of the Ocean. Introduction*. Nauka, Moscow, 214 pp.

Vinogradov, A.P. and Tugarinov, A.I., 1964. Problems of the Precambrian geochronology in Eastern Asia. In: *Absolut. Vozrast Geol. Formatsyi*. Nauka, Moscow, pp. 177—184.

Vinogradov, A.P., Tarasov, L.S. and Zykov, S.I., 1959. Isotopic ratio of the lead ore in the Baltic shield. *Geokhimiya*, 1959(7): 18—25.

Vinogradov, A.P., Tugarinov, A.I. and Zykov, S.I., 1960a. The age of pegmatite from the Stanovoy complex. *Geokhimiya*, 1960(5): 383—391.

Vinogradov, A.P., Tugarinov, A.I., Zykov, S.I. and Stupnikova, N.I., 1960b. On the rock age of the Aldan shield area. *Geokhimiya*, 1960(7): 563—569.

Vinogradov, V.A. and Sobolevskaya, R.F., 1958. Sinian strata in the northern part of Kharaulakh Mts. *Tr. NIIGA*, 85: 64—66.

Vinogradov, V.I., Ivanov, I.B., Litsarev, M.A., Pertsev, N.N. and Shanin, L.I., 1969. On the age of the oxygen-rich atmosphere of the Earth. *Dokl. Acad. Sci. USSR*, 185(5): 1144—1147.

Volborth, A., 1962. Rapakivi-type granites in the Precambrian complex of Gold Butte, Clark county, Nevada. *Geol. Soc. Am. Bull.*, 73(7): 813—831.

Volobuev, M.I., Zykov, S.I. and Stupnikova, N.I., 1964. Geochronology of the Yenisei Ridge. In: *Absol. Vozrast Geol. Formatsyi*. Nauka, Moscow, pp. 108—127.

Volobuev, M.I., Zykov, S.I. and Stupnikova, N.I., 1970. Geochronological studies of the Yenisei—East-Sayan folded area. In: *Trudy XV Sess. Kom po Opred. Absol. Vozrasta Geol. Formatsyi*. Nauka, Moscow, pp. 85—106.

Volobuev, M.I., Zykov, S.I. and Stupnikova, N.I., 1973. Geochronology of Grenvillides basement and geosynclinal formations of the Yenisei Ridge. In: *Opred. Abs. Vozrasta Rudnykh Mestorozhd. i Molod. Magmat. Protsessov*. Acad. Sci. USSR, Moscow, pp. 74—75.

Volochayev, F.Ya., Kukushkin, A.I. and L'vov, K.A., 1967. Stratigraphy of the older strata of Timan. *Dokl. Acad. Sci. USSR*, 173(6): 1389—1392.

Vologdin, A.G., 1965. The occurrence of Upper Sinian algae in the rocks of Udokan Ridge of Chita area. *Dokl. Acad. Sci. USSR*, 160(2): 446—449.

Vorona, I.D., Dukin, V.A. and Ivensen, Yu.V., 1968. Gold-bearing older conglomerate of the Aldan shield. In: *Mater. Geol. Polezn. Iskop. Yakutsk. ASSR*, 1968(18): 191—200.

Walcott, Ch., 1889. The fauna of the Lower Cambrian. *US Geol. Surv. Annu. Rep.*, 10: 549.

Walcott, Ch.D., 1912. Notes on fossils from limestone of Steeprock series, Ontario, Canada. *Geol. Surv. Can. Mem.*, 28: 10 pp.

Wanless, R.K., Stevens, R.D. and Lachance, G.R., 1968. Age determination and geological studies. K–Ar isotopic ages. *Geol. Surv. Can. Pap.*, 67-2 (Pt. A, Rep. 8): 101 pp.
Wasserburg, G.J. and Lanphere, M.A., 1965. Age determination in the Precambrian of Arizona and Nevada. *Geol. Soc. Am. Bull.*, 76(7): 735—758.
Watts, W.W., 1947. *Geology of the Ancient Rocks of Charnwood Forest*. Leicester, 150 pp.
Weeks, L.J., 1957. The Proterozoic of Eastern Canadian Appalachia. *R. Soc. Can. Spec. Publ.*, 2: 141—149.
Welin, E., 1972. The Svecofennian folded zone in Northern Sweden. *Geotektonika*, 5: 53—60.
Welin, E. and Blomqvist, G., 1964. Age measurements on radioactive minerals from Sweden. *Geol. Fören. Stockh. Förh.*, 86/1(516): 33—50.
Welin, E. and Lundqvist, Th., 1970. New Rb—Sr age data for the Sub-Jotnian volcanics (Dala porphyries) in the Los-Hamra region, Central Sweden. *Geol. Fören. Stockh. Förh.*, 92/1(540): 35—39.
Welin, E., Christiansson, K. and Nilsson, Ö., 1971. Rb—Sr radiometric ages of extrusive and intrusive rocks in Northern Sweden, I. *Sver. Geol. Unders.*, 65(12): 38 pp.
Wheeler, H.E., 1965. Ozark Precambrian—Paleozoic relations. *Am. Assoc. Pet. Geol. Bull.*, 49(10): 1647—1665.
Wheeler, H.E., 1966. Ozark Precambrian—Paleozoic relations: discussions and replies. *Am. Assoc. Petr. Geol. Bull.*, 50(5): 1033—1061.
Williams, N.C., 1953. Late Precambrian and Early Paleozoic geology of western Uinta Mountains. *Am. Assoc. Pet. Geol. Bull.*, 37(12): 2734—2742.
Wilson, H.D., Andrew, P., Moxham, R.L. and Ramlal, K., 1965. Archean volcanism in the Canadian shield. *Can. J. Earth Sci.*, 2: 161—175.
Wilson, M.E., 1958. Precambrian classification and correlation in the Canadian Shield. *Geol. Soc. Am. Bull.*, 69(6): 757—773.
Wilson, M.E., 1965. Precambrian of Canada (Canadian Shield). In: K. Rankama (Editor), *The Precambrian, 2*. Interscience, New York, N.Y., pp. 263—415.
Windley, B.F., 1969a. Anorthosites of Southern West Greenland. In: North Atlantic — geology and continental drift. *Soc. Econ. Paleontol. Miner. Mem.*, 12: 899—915.
Windley, B.F., 1969b. Evolution of the Early Precambrian basement complex of Southern West Greenland. *Geol. Assoc. Can. Spec. Pap.*, 5: 155—162.
Winsnes, T.C., 1965. Precambrian of Spitsbergen and Bären-Insel. In: K. Rankama (Editor), *The Precambrian, 2*. Interscience, New York, N.Y., pp. 1—24.
Woodward, L.A., 1967. Stratigraphy and correlation of Late Precambrian rocks of Pilot Range, Elko county, Nevada, and Box Elder county, Utah. *Am. Assoc. Pet. Geol. Bull.*, 51(2): 235—243.
Wright, L.A. and Troxel, B.W., 1966. Strata of Late Precambrian—Cambrian age, Death Valley region, California—Nevada. *Am. Assoc. Pet. Geol. Bull.*, 50(5): 846—857.
Wynne-Edwards, H.R., 1967. Westport map-area, Ontario, with special emphasis on the Precambrian rocks. *Geol. Surv. Can. Mem.*, 346: 142 pp.
Wynne-Edwards, H.R., Gregory, A.F., Hay, P.W., Giovanella, G.A. and Reinhardt, E.W., 1966. Mont Laurier and Kempt Lake map-areas, Quebec. A preliminary report on the Grenville project. *Geol. Surv. Can. Pap.*, 66-32: 32 pp.
Yarosh, A.Ya., 1970. Structural relationships in the eastern margin of the Russian platform and the Urals geosyncline. In: *Voprosy Tektoniki Dokembriya Kontinentov*. Nauka, Moscow, pp. 23—89.
Yatsenko, G.M., Vernikovskaya, V.N. and Markovsky, V.M., 1969. New data on the structure of the crystalline basement of the Volyn'—Podolia margin of the Russian platform. *Sov. Geol.*, 1969(2): 128—132.
Yatskevich, S.V., 1970. The Riphean stratigraphy of Povolzhye in the Saratov region. *Dokl. Acad. Sci. USSR*, 195(5): 1183—1186.

Yorath, C.I., Balkwill, H.R. and Klassen, R.W., 1969. Geology of the eastern part of the northern interior and Arctic coastal plains, Northwest Territories. *Geol. Surv. Can. Pap.*, 68-27: 29 pp.

Young, G.M., 1966. Huronian stratigraphy of the MacGregor Bay area, Ontario: relevance to the paleogeography of the Lake Superior region. *Can. J. Earth Sci.*, 3(2): 203–210.

Young, G.M., 1968. Sedimentary structures in Huronian rocks of Ontario. *Palaeogeogr., Palaeoclimat., Palaeoecol.*, 4(2): 125—153.

Young. G.M., 1973. Tillites and aluminous quartzites as possible time markers for Middle Precambrian (Aphebian) rocks of North America. *Geol. Assoc. Can. Spec. Pap.*, 12: 97—127.

Young. G.M., 1974. Stratigraphy, paleocurrents and stromatolites of Hadrynian (Upper Precambrian) rocks of Victoria Island, Arctic Archipelago, Canada. *Precambr. Res.*, 1: 13—41.

Young, G.M. and Church, W.R., 1966. The Huronian system in the Sudbury district and adjoining areas of Ontario: A review. *Geol. Assoc. Can. Proc.*, 17: 65—82.

Zabirov, Yu.A., 1966. New data on the geologic structure of the Chadobets Uplift. In: *Chetvertaya Krasnoyarsk. Krayevaya Geol. Konfer.* Krasnoyarsk, pp. 18—22 (abstracts).

Zabiyaka, A.I., 1974. The Precambrian stratigraphy of Northwestern Taymyr. *Tr. VSEGI, N; Ser.*, 199: 183—198.

Zabrodin, V.E., 1966. The early stages in the evolution of sedimentary cover in the South-east of the Siberian platform. *Izv. Acad. Sci. USSR, Geol. Ser.*, 1966 (8): 121—125.

Zagorodny, V.G., 1962. On the stratigraphy of the Pechenga Formation. In: *Geology of Kola Peninsula.* Acad. Sci. USSR, Moscow—Leningrad, pp. 84—89.

Zaytsev, Yu.A. and Filatova, L.N., 1971. New data on the Precambrian structures in Ulu-Tau. In: *Voprosy Geol. Centr. Kazakhstana, 10.* MGU, Moscow, pp. 21—92.

Zaytsev, Yu.A., Zykov, S.I. and Krasnobayev, A.A., 1972. Geochronological results of Precambrian studies in Central Kazakhstan. *Izv. Acad. Sci. USSR, Geol. Ser.*, 1972(8): 3—19.

Zhadnova, T.P., 1961. Stratigraphy in the northeast part of the Patom Highland. *Tr. CNIGRI*, 38: 49—85.

Zhuravleva, Z.A., 1964. The Riphean and Lower Cambrian oncolites and catagraphs in Siberia and their stratigraphic importance. *Trudy Geol. Inst. Acad. Sci. USSR*, 114: 75 pp.

Zhuravleva, Z.A., 1968. The diagnostic features of oncolites and catagraphs and their distribution in the Riphean section of the Southern Urals. *Tr. GIN Acad. Sci. USSR*, 188: 18—25.

Zhuravleva, Z.A., Komar, V.A. and Chumakov, N.M., 1959. Stratigraphical relationships of the Patom complex and sedimentary strata of the western and northern slopes of the Aldan shield. *Dokl. Acad. Sci. USSR*, 128(5): 1026—1029.

Ziegler, P.A., 1959. Frühpaleozoische tillite im ostlichen Yukon Territorium (Canada). *Eclogae Geol. Helv.*, 52(2): 435—741.

Zoricheva, A.I., 1956. The Paleozoic stratigraphy of the northern part of the Russian plate. *Mater. VSEGEI, N.Ser.*, 14: 153—168.

Zoubek, V., 1965. Moldanubicum und seine Stellung in geologischen Bau Europas. *Freiberg. Forschungsh.*, 190: 110 pp.

Zubtsov, E.I., 1972. Precambrian tillites of Tien-Shan and their stratigraphical value. *Byull. MOIP Otd. Geol.*, 1: 42—56.

Zwart, H.J., 1968. The Paleozoic crystalline rocks of the Pyrenees in their structural setting. *Kristalinicum*, 6: 125—140.

INDEX OF LOCAL STRATIGRAPHIC NAMES*

* The tables (correlation charts) in which the stratigraphic units occur, are referred to by Roman numerals, followed by the numbers of the columns in parentheses.